Kevin P. Knudson
Algebraic Topology

Also of Interest

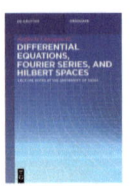
Differential Equations, Fourier Series, and Hilbert Spaces
Lecture Notes at the University of Siena
Raffaele Chiappinelli, 2023
ISBN 978-3-11-129485-8, e-ISBN (PDF) 978-3-11-130252-2

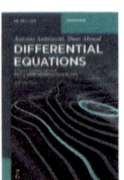
Differential Equations
A first course on ODE and a brief introduction to PDE
Shair Ahmad, Antonio Ambrosetti, 2023
ISBN 978-3-11-118524-8, e-ISBN (PDF) 978-3-11-118567-5

Dynamics of Discrete Group Action
The Legacy of David R. Adams
Boris N. Apanasov, 2024
ISBN 978-3-11-078403-9, e-ISBN (PDF) 978-3-11-078410-7
in: Advances in Analysis and Geometry
ISSN 2511-0438, e-ISSN 2511-0543

Topics in Complex Analysis
Dan Romik, 2023
ISBN 978-3-11-079678-0, e-ISBN (PDF) 978-3-11-079681-0

Advanced Mathematics
An Invitation in Preparation for Graduate School
Patrick Guidotti, 2022
ISBN 978-3-11-078085-7, e-ISBN (PDF) 978-3-11-078092-5

Kevin P. Knudson

Algebraic Topology

A Toolkit

DE GRUYTER

Mathematics Subject Classification 2020
Primary: 55-01; Secondary: 55N10, 55N31, 55P10, 55R20

Author
Prof. Kevin P. Knudson
University of Florida
Department of Mathematics
PO Box 118105
Gainesville FL 32611
USA
kknudson@ufl.edu

ISBN 978-3-11-101481-4
e-ISBN (PDF) 978-3-11-101485-2
e-ISBN (EPUB) 978-3-11-101486-9

Library of Congress Control Number: 2024938448

Bibliographic information published by the Deutsche Nationalbibliothek
The Deutsche Nationalbibliothek lists this publication in the Deutsche Nationalbibliografie;
detailed bibliographic data are available on the Internet at http://dnb.dnb.de.

© 2024 Walter de Gruyter GmbH, Berlin/Boston
Cover image: Kevin P. Knudson
Typesetting: VTeX UAB, Lithuania
Printing and binding: CPI books GmbH, Leck

www.degruyter.com

This one goes out to the one I love...

Preface

When I entered my PhD program in the fall of 1991, I was pretty sure I would study differential geometry, even though I had spent a lot of time as an undergraduate learning general topology and a lot of algebra. My first semester did not clarify things for me–the standard first course in measure theory, more algebra, a topics course in algebraic geometry that was a bit advanced for me at the time. My mathematical trajectory changed forever that spring, however, when I was introduced to algebraic topology. Finally, here was a branch of mathematics that appealed to my topological interests and my affinity for algebraic calculation. I was hooked.

Our professor, Richard Hain (who would eventually become my PhD advisor), told very good stories. His geometric intuition was (and still is) otherworldly, but he brought it down to earth. I still have the course notes I took that semester, and they formed the basis for the notes I have used when I teach algebraic topology. They have expanded to include more material and topics that are especially interesting to me, and this book is the LaTeX'ed result.

My goal here is to present a streamlined introduction to what I view as the essential elements of algebraic topology, a toolkit as it were. This text is not intended to serve as an encyclopedic reference work, but to be a solid foundation and an invitation to further study. An ambitious instructor could, with judicious omissions, cover it in one semester. I prefer a more leisurely two-semester amble, but that is partially a function of my department's course requirements.

Readers familiar with standard texts in algebraic topology will notice some influences. That first course for me was based on the book of Greenberg & Harper [1]. It has much to recommend it, but it is quite terse (perhaps too much so). The current standard text is Hatcher's excellent book [2], and I have used it successfully in the past. Indeed, his direct calculation of the cohomology ring of projective space (rather than appealing to Poincaré duality) appears here with little modification; the extra technical work required to do it is actually quite illuminating. Hatcher's book is rather encyclopedic, however, and as mentioned above, I am aiming for a more pared-down version. Much of the material on cellular homology and spectral sequences is modeled on the place where I learned it, the book of Griffiths & Morgan [3]. Spectral sequences are an important tool that every topologist needs in their kit, and that is why they are included here. I will say that this presentation is analogous to what we show undergraduates in the introductory calculus sequence: a focus on *how to use* these techniques over the details of how they work. Other sources I consulted include Spanier's classic text [4], Munkres's algebraic topology book [5], Steenrod's treatise on fiber bundles [6], and Weibel's book on homological algebra [7].

The book contains four chapters. Chapter 1 focuses on preliminary material: continuous maps and homotopy, lifting properties and fibrations, basic category theory, and the fundamental group. Chapter 2 is about homology in all its forms–simplicial, singular,

https://doi.org/10.1515/9783111014852-201

cellular–and the properties thereof. Manifolds and their homology are covered in detail. This chapter also includes a section about persistent homology and topological data analysis. This area has gained prominence over the last couple of decades, bringing what was once a very theoretical branch of mathematics into a world of applications. Chapter 3 is about cohomology and duality, including the cup product and cohomology ring. Poincaré duality appears here, of course. Finally, Chapter 4 introduces homotopy groups and the basics of the homotopy theory of cell complexes. The Leray–Serre spectral sequence is introduced and used to compute many interesting examples of cohomology rings and homotopy groups.

Here is one topic that I have omitted that may be controversial: a detailed study of covering spaces. While this theory is beautiful and interesting, every time I teach the course I find myself regretting spending so much time on it. It is true that covering spaces are needed from time to time later in the text, but the limited discussion here is sufficient. I have included an exploration of covering spaces as a project at the end of the first chapter, however, if an instructor wishes to cover it.

Each chapter concludes with one or two projects for students to explore topics in greater depth. Exercises are included throughout as well.

I hope the reader finds this book to be a clear and compelling introduction to a subject that continues to find new applications. If mathematicians, both emerging and established, find it useful, then I have been successful.

Gainesville, Florida, July 2024   K.P.K.

Acknowledgments

This is my third book, but it never seems to get easier. As anyone who has done this knows, there are periods of intense enthusiasm interspersed with nagging feelings of doubt. But here it is, so enthusiasm must win out in the end.

I am indebted to many excellent mathematicians from whom I have learned so much over the years. These include Peter Fletcher, Richard Hain, John Harer, Herbert Edelsbrunner, Henry King, Neža Mramor, Ulrich Bauer, Robert Ghrist, Peter Bubenik, Chuck Weibel, Gunnar Carlsson, Andrei Suslin, Eric Friedlander, Mark Walker, Nick Scoville, Vidit Nanda, and on and on. My apologies to anyone I may have omitted.

Many thanks are due to the fine folks at De Gruyter, especially to Steve Elliot for recruiting me to do this, and to Ms. Ute Skambraks for her excellent editorial assistance. I am also grateful to a pair of anonymous referees for their comments on a draft of this text. Any errors or omissions are entirely my responsibility, not theirs.

Finally, I am always grateful for the love and support of my wife Ellen and our son Gus, the best people I know.

https://doi.org/10.1515/9783111014852-202

Contents

1 Preliminaries

1.1 Topological spaces

We assume that the reader already has a familiarity with the definitions and basic properties of topological spaces, including the notions of continuity, connectivity, compactness, separation axioms, etc. For completeness, we give a brief summary here.

1.1.1 Some notions from general topology

Definition 1.1.1. A *topological space* is a set X equipped with a collection $\mathcal{T}$ of subsets called *open sets* such that the collection $\mathcal{T}$ is closed under arbitrary unions and finite intersections. The sets X and $\emptyset$ are required to belong to $\mathcal{T}$.

Definition 1.1.2. A function $f : X \to Y$ between topological spaces is *continuous* if for each open set $V \subset Y$ the inverse image $f^{-1}(V)$ is open in X.

Example 1.1.3. Consider the set $\mathbb{R}$ of real numbers with its standard topology having basis the usual open intervals. The definition of continuity for a function $f : \mathbb{R} \to \mathbb{R}$ that one learns in a first analysis course reads as follows. A function $f : \mathbb{R} \to \mathbb{R}$ is continuous at a if for every $\varepsilon > 0$ there exists a $\delta > 0$ such that if $|x - a| < \delta$ then $|f(x) - f(a)| < \varepsilon$. Note that this is equivalent to Definition 1.1.2 above.

Definition 1.1.4. A *metric* on a set X is a function $d : X \times X \to \mathbb{R}$ satisfying the following.
- $d(x,y) \geq 0$ for all $x, y \in X$, with equality if and only if $x = y$;
- $d(x,y) = d(y,x)$ for all $x, y \in X$;
- (triangle inequality) for all $x, y, z \in X$, we have $d(x,z) \leq d(x,y) + d(y,z)$.

Given a metric d on X, $x \in X$, and $\varepsilon > 0$, define the *ε-neighborhood* of x to be the set $B_\varepsilon(x) = \{y \in X : d(x,y) < \varepsilon\}$. The sets $B_\varepsilon(x)$, $x \in X$, form the basis of a topology on X and we call (X, d) a *metric space*.

Example 1.1.5. The euclidean space $\mathbb{R}^n$ is a metric space with $d(x,y)$ being the standard euclidean distance between points.

Definition 1.1.6. A topological space X is *Hausdorff* if for every pair of points $x \neq y$ in X there exist disjoint open sets U and V such that $x \in U$ and $y \in V$.

Example 1.1.7. The euclidean space $\mathbb{R}^n$ is Hausdorff: given points $x \neq y$, let $\varepsilon = \|x - y\|/2$ and take $U = B_\varepsilon(x)$, $V = B_\varepsilon(y)$.

Definition 1.1.8. A space X is *connected* if it is impossible to write $X = A \cup B$, where A and B are nonempty disjoint open sets.

https://doi.org/10.1515/9783111014852-001

Example 1.1.9. The real line $\mathbb{R}$ is connected, as is any interval in $\mathbb{R}$. In general, $\mathbb{R}^n$ is connected for all n.

By an *open cover* of a space X we mean a collection $\{U_\lambda\}_{\lambda \in \Lambda}$ of open subsets of X whose union is all of X. A *subcover* is a subcollection of the U_λ that also covers X.

Definition 1.1.10. A topological space X is *compact* if every open cover of X has a finite subcover.

Example 1.1.11. In $\mathbb{R}$ a closed interval $[c, d]$ is compact. In general, any closed and bounded set in a euclidean space $\mathbb{R}^n$ is compact (this is the Heine–Borel Theorem).

Quotient spaces

A common construction in topology is to begin with a space and make identifications among some elements in the space to create a new object. This is often done via some equivalence relation on the original space and since there is a canonical map to the set of equivalence classes, we want to ensure that the topology on the new space is compatible with this map.

Definition 1.1.12. A surjective function $p : X \rightarrow Y$ is a *quotient map* if a subset $V \subset Y$ is open if and only if $p^{-1}(V)$ is open in X.

Note that a quotient map is continuous by definition; it is the condition that V is open in Y only if its inverse image is open in X that is new here. This condition imposes a topology on Y. What one should really be thinking of here is that we have an equivalence relation $\sim$ on X defined by $x_1 \sim x_2 \Leftrightarrow p(x_1) = p(x_2)$ and then the quotient map p defines a topology on the set Y of equivalence classes.

The quotient topology is characterized by the following *universal mapping property*.

Proposition 1.1.13. *Suppose $p : X \rightarrow Y$ is a quotient map and suppose $f : X \rightarrow Z$ is a continuous map with $f(x_1) = f(x_2)$ whenever $p(x_1) = p(x_2)$. Then there exists a unique continuous map $\bar{f} : Y \rightarrow Z$ such that $f = \bar{f} \circ p$.*

Proof. If $y \in Y$, write $y = p(x)$ for some $x \in X$ and then set $\bar{f}(y) = f(x)$. The map $\bar{f}$ is well-defined, since if $p(x') = y$, then $f(p(x')) = f(p(x))$ and hence $\bar{f}(y)$ is independent of the choice of lift. The map $\bar{f}$ is unique: Suppose $g : Y \rightarrow Z$ satisfies $f = g \circ p$. Then if $y \in Y$, choose a lift x and then

$$g(y) = g(p(x)) = f(x) = \bar{f}(p(x)) = \bar{f}(y).$$

Finally, to see that the map $\bar{f}$ is continuous, suppose $U \subseteq Z$ is open. We must show $\bar{f}^{-1}(U)$ is open in Y. But $\bar{f}^{-1}(U)$ is open in Y if and only if $p^{-1}(\bar{f}^{-1}(U))$ is open in X. The latter set is simply $f^{-1}(U)$, and this is open in X since f is continuous. □

Examples

We first list some standard spaces. The *n-dimensional euclidean space* $\mathbb{R}^n$ consists of n-tuples of real numbers $(x_1, \ldots, x_n)$. The topology on $\mathbb{R}^n$ is the usual one, in which an ε-ball around a point x consist of all points y with $\|x - y\| < \varepsilon$, where $\|\cdot\|$ denotes the standard euclidean norm. The *n-disc*, D^n, is the set $\{x = (x_1, \ldots, x_n) \in \mathbb{R}^n \,|\, \|x\| \le 1\}$, with the subspace topology. The *n-sphere*, S^n is the boundary of the disc D^{n+1}; it is the set $\{x = (x_1, \ldots, x_{n+1}) \,|\, \|x\| = 1\}$. Note that S^0 consists of two points $\{\pm 1\}$.

(1) D^n is homeomorphic to the quotient space $S^{n-1} \times [0,1]/(S^{n-1} \times \{0\})$. Here, if A is a subspace of X, the notation X/A denotes the space obtained by "collapsing A to a point." Formally, this is the set of equivalence classes for the relation $\sim$, where $a \sim a'$ for all $a, a' \in A$, $x \sim x$ for $x \in X \setminus A$. The space $S^{n-1} \times [0,1]$ is a "cylinder" and the subspace $S^{n-1} \times \{0\}$ is the base of that cylinder. Collapsing that base to a point creates a cone over $S^{n-1} \times \{1\}$; viewing this from above one sees a disc. See Figure 1.1.

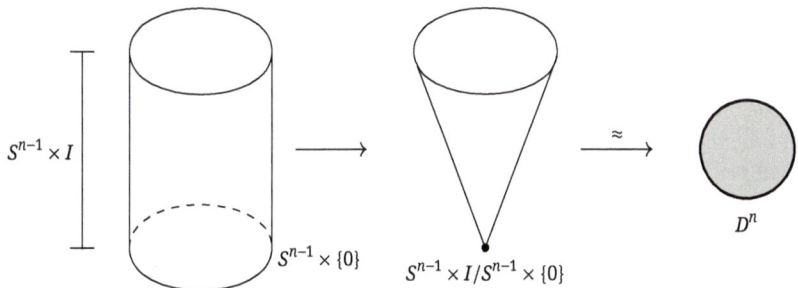

Figure 1.1: The disc as a quotient of $S^{n-1} \times [0,1]$.

That is the intuition anyway. To prove this formally, we proceed as follows. First put the quotient topology on $S^{n-1} \times [0,1]/S^{n-1} \times \{0\}$; call the quotient map p. Now define $\varphi : S^{n-1} \times [0,1] \to D^n$ by $\varphi(x, t) = tx$. Observe that this is well-defined since $S^{n-1} \subset \mathbb{R}^n$ and $\|tx\| = t \le 1$ for all $x \in S^{n-1}$. This function is clearly continuous and surjective. Moreover, $\varphi(x, t) = 0$ if and only if $(x, t) \in S^{n-1} \times \{0\}$. By the universal mapping property, we get a unique map $\overline{\varphi} : S^{n-1} \times [0,1]/S^{n-1} \times \{0\} \to D^n$ with $\varphi = \overline{\varphi} \circ p$. One checks easily that $\overline{\varphi}$ is a homeomorphism (see the exercises for this section).

(2) D^n/S^{n-1} is homeomorphic to S^n. This is easy to visualize when $n = 2$. In this case, we have a disc and we are collapsing the boundary circle to a single point. The result is a sphere: imagine a plastic trash bag with a drawstring; pulling the drawstring tight seals the bag. See Figure 1.2.

To prove this, we will use the previous example. Define $\psi : S^{n-1} \times [0,1] \to S^n$ by $\psi(x, t) = (\cos(\pi t), \sin(\pi t)x) \in \mathbb{R} \times \mathbb{R}^n = \mathbb{R}^{n+1}$. Note that since $\|x\| = 1$, $\|\psi(x, t)\| = 1$ and so the image of ψ lies in S^n. Note that $\psi(S^{n-1} \times \{0\}) = (1, 0, 0, \ldots, 0)$. So by the universal mapping property, we get a continuous map $\overline{\psi} : S^{n-1} \times [0,1]/S^{n-1} \times \{0\} = D^n \to S^n$. Moreover, $S^{n-1} \subset D^n$ is the image of $S^{n-1} \times \{1\}$ under $\overline{\psi}$. Also, $\overline{\psi}(S^{n-1} \times$

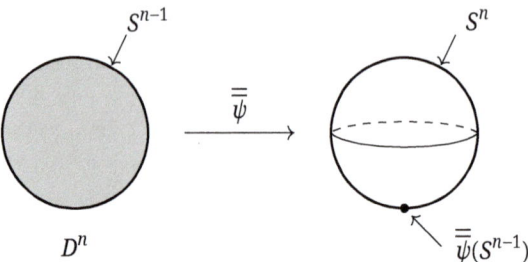

Figure 1.2: The sphere as a quotient of the disc.

$\{1\}) = (-1, 0, 0, \ldots, 0)$. Using the universal property again, $\overline{\psi}$ induces a continuous $\overline{\overline{\psi}} : D^n / S^{n-1} \to S^n$. The map $\overline{\overline{\psi}}$ is a homeomorphism.

(3) The torus T. This space is the surface of a donut and is defined as the product space $S^1 \times S^1$. However, it is more often thought of as a quotient of the unit square as follows. Given the square $[0, 1] \times [0, 1]$, identify points on the boundary in the following way: $(a, 0) \sim (a, 1)$ and $(0, b) \sim (1, b)$ (see Figure 1.3). There are two ways to see that the resulting space is $S^1 \times S^1$. First, products and quotients commute (sometimes): if $p_1 : X_1 \to Y_1$ and $p_2 : X_2 \to Y_2$ are quotient maps and Y_1 and X_2 are locally compact, then $p_1 \times p_2 : X_1 \times X_2 \to Y_1 \times Y_2$ is a quotient map. In this case, we have $p_j : [0, 1] \to S^1$ as above and then $p_1 \times p_2 : [0, 1] \times [0, 1] \to S^1 \times S^1$ is a quotient. Alternatively, we could note that the composition of quotient maps is a quotient; again see Figure 1.3.

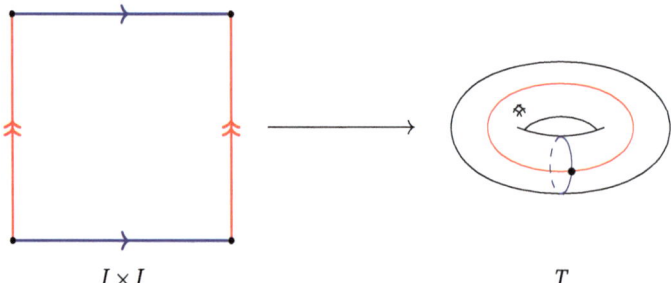

Figure 1.3: The torus as a quotient of the unit square.

(4) Adjunction spaces. Suppose X and Y are spaces and that $f : A \to Y$ is a continuous map, where A is a subspace of X. The *adjunction space* is the quotient

$$X \cup_f Y = X \amalg Y / \sim,$$

where $\sim$ is the relation $a \sim f(a)$ for $a \in A$. Intuitively, this is the space obtained by gluing X to Y along the image of the subspace A. In particular, if A consists of a single point, then this space is often denoted by $X \vee Y$ and called the *wedge* of X and Y. For example, $S^1 \vee S^1$ is two circles joined at a single point to form a figure eight. A common construction used in the sequel is to take $X = D^n$ with $A = S^{n-1}$, the boundary sphere. Then the space $D^n \cup_f Y$ is the space Y with an n-cell attached along the *attaching map* $f : S^{n-1} \to Y$.

Exercises

Exercise 1.1.1. Prove that a continuous bijection from a compact space to a Hausdorff space is a homeomorphism.

Exercise 1.1.2. Prove that the map $\overline{\varphi} : S^{n-1} \times [0,1]/S^{n-1} \times \{0\} \to D^n$ is a homeomorphism.

Exercise 1.1.3. Prove that the map $\overline{\overline{\psi}} : D^n/S^{n-1} \to S^n$ is a homeomorphism.

Exercise 1.1.4. Prove that if $f : X \to Y$ is a quotient map and Z is locally compact, then $f \times \mathrm{id}_Z : X \times Z \to Y \times Z$ is a quotient map. Deduce that if $p_1 : X_1 \to Y_1$ and $p_2 : X_2 \to Y_2$ are quotient maps with Y_1 and X_2 locally compact, then $p_1 \times p_2 : X_1 \times X_2 \to Y_1 \times Y_2$ is a quotient map.

Exercise 1.1.5. Prove that the composition of quotient maps is a quotient map.

Exercise 1.1.6. Prove that the sphere S^n is the quotient space obtained from two disjoint copies of D^n by gluing them along their boundary spheres S^{n-1}.

1.1.2 Simplicial complexes

Simplicial complexes are fundamental objects in algebraic topology. Most of the spaces we are interested in may be realized in this way, and simplicial complexes have nice combinatorial descriptions that make them easy to implement on a computer. We begin with a very concrete description of these objects.

Recall that a collection of points $v_0, \ldots, v_n$ in a euclidean space $\mathbb{R}^k$ is in *general position* if for a set of real scalars $t_0, \ldots, t_n$ the equations $\sum t_i = 0$ and $\sum t_i v_i = 0$ imply that $t_i = 0$ for all $i = 0, \ldots, n$. Geometrically this means that no three of the v_i lie on the same line, no four lie in the same plane, etc.

Definition 1.1.14. Let n be a nonnegative integer and let $v_0, \ldots, v_n$ be points in general position in some euclidean space $\mathbb{R}^k$. The *n-simplex* σ spanned by $v_0, \ldots, v_n$ is the convex hull of these points. This consists of the following set:

$$\sigma = \left\{ x \mid x = \sum_{i=0}^n t_i v_i, \sum_{i=0}^n t_i = 1, t_i \geq 0 \right\}.$$

The real numbers $t_0, \ldots, t_n$ are called the *barycentric coordinates* of the point x. The points $v_0, \ldots, v_n$ are called the *vertices* of the simplex. We write $\sigma = \langle v_0, \ldots, v_n \rangle$.

A 1-simplex is a line segment; a 2-simplex is a triangle; a 3-simplex is a tetrahedron; etc. In general, an n-simplex has $(n + 1)$ vertices and is an n-dimensional object.

Example 1.1.15. The *standard n-simplex,* Δ^n is the n-simplex in $\mathbb{R}^{n+1}$ with vertices the standard basis vectors $e_1, \ldots, e_{n+1}$. See Figure 1.4.

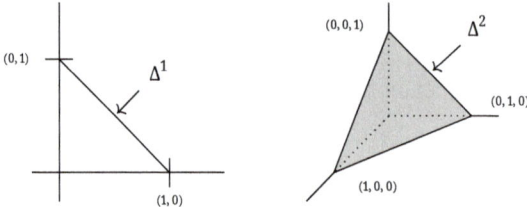

Figure 1.4: The standard n-simplex Δ^n for $n = 1, 2$.

Remark 1.1.16. Note that we can realize Δ^n in $\mathbb{R}^n$ as the convex hull of $\{0, e_1, \ldots, e_n\}$. This is sometimes more convenient. However, the resulting n-simplex does not have sides of equal length, whereas the standard $\Delta^n \subset \mathbb{R}^{n+1}$ does.

Definition 1.1.17. Let σ be an n-simplex with vertices $v_0, \ldots, v_n$. A *face* τ of σ is a k-simplex spanned by a subset of the vertices of σ; that is, $\tau = \langle v_{i_0}, v_{i_1}, \ldots, v_{i_k} \rangle$. We denote this relationship by $\tau < \sigma$.

Loosely speaking, a simplicial complex is obtained by gluing together simplices along common faces. Formally, we have the following.

Definition 1.1.18. A *simplicial complex* K in $\mathbb{R}^n$ is a collection of simplices such that
(1) Every face of a simplex of K is in K; and
(2) The intersection of any two simplices of K is in K.

A subset $L \subseteq K$ is a *subcomplex* if L is itself a simplicial complex.

Note that this definition prohibits oddities such as gluing a vertex of a triangle to a point on a side of another. See Figure 1.5 for some examples of simplicial complexes.

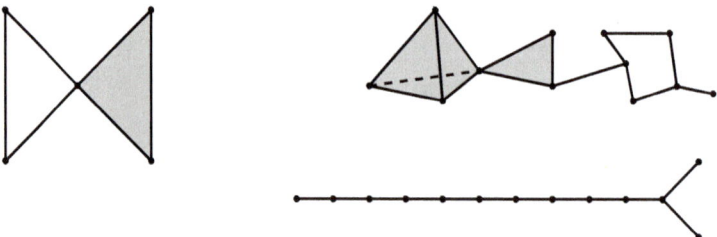

Figure 1.5: Some simplicial complexes.

Definition 1.1.19. Let K be a simplicial complex. The p-skeleton of K, denoted $K^{(p)}$, consists of all k-simplices of K with $k \leq p$. It is a subcomplex of K.

For example, $K^{(0)}$ consists of all vertices of K, $K^{(1)}$ is a graph whose vertex set is $K^{(0)}$, and so on.

Definition 1.1.20. Suppose $\sigma = \langle v_0, \ldots, v_n \rangle$ is a simplex and let v be a vertex disjoint from σ. The *join*, $v * \sigma$, is the $(n+1)$-simplex $v * \sigma = \langle v, v_0, \ldots, v_n \rangle$.

Example 1.1.21. Consider the standard n-simplex Δ^n. The $(n-1)$-skeleton consists of all faces of dimension $\leq n-1$. This is the set of $n+1$ codimension-one faces, each identifiable with Δ^{n-1} glued together along their common faces. We often denote this by $\partial\Delta^n$ and call it the *boundary* of Δ^n. See Figure 1.6.

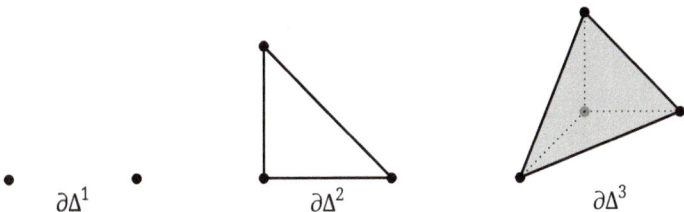

$\partial\Delta^1$ $\qquad\qquad\qquad$ $\partial\Delta^2$ $\qquad\qquad\qquad$ $\partial\Delta^3$

Figure 1.6: The boundary of Δ^n, $n = 1, 2, 3$, as realized in $\mathbb{R}^n$.

Definition 1.1.22. Let K be a simplicial complex in $\mathbb{R}^n$. Denote by $|K|$ the union of the simplices in K. The space $|K|$, equipped with the subspace topology inherited from $\mathbb{R}^n$, is called the *polytope* or *underlying space* of K.

Note that if K consists of finitely many simplices, $|K|$ is compact.

Simplicial maps
Because we want our constructions to respect whatever structures we have in place, we need to specify the types of maps we allow between objects.

Definition 1.1.23. Suppose K and L are simplicial complexes. A map $f : K \to L$ is *simplicial* if for every simplex σ in K, $f(\sigma)$ is a simplex in L.

Note that if f is a simplicial map, then $f(K^{(0)}) \subseteq L^{(0)}$; that is, f takes vertices to vertices. It then takes edges to edges, and so on. In particular, f is determined by what it does to vertices. More is true, however.

Theorem 1.1.24. *Let K and L be simplicial complexes and suppose $f : K^{(0)} \to L^{(0)}$ is a map such that if $v_0, \ldots, v_k$ span a simplex σ in K, then $f(v_0), \ldots, f(v_k)$ span a simplex in L. Then there is a continuous map $g : |K| \to |L|$ extending f such that if $x = \sum t_i v_i \in \sigma$, we have $g(x) = \sum t_i f(v_i)$.*

Proof. This is more or less clear. We extend the barycentric coordinates on each simplex of K to all of K by noting that if $x \in |K|$, then x is either in $K^{(0)}$ or it lies in the interior of exactly one simplex $\sigma = \langle w_0, \ldots, w_r \rangle$ of K. In that case, $x = \sum t_i w_i$ with $t_i > 0$ for each i and $\sum t_i = 1$. This allows us to give coordinates for x relative to all the vertices by setting

$t_v(x) = 0$ if $v \neq w_j$ and $t_v(x) = t_j$ if $v = w_j$. We then define g as in the statement of the theorem. The continuity of g is left as an exercise. □

The map g is called the *simplicial map induced by f*.

Barycentric subdivision

A useful construction when working with simplicial complexes is the *barycentric subdivision*.

Definition 1.1.25. Suppose $\sigma = \langle v_0, \ldots, v_k \rangle$ is a k-simplex. The *barycenter $\hat{\sigma}$* of σ is the point in the interior of σ with barycentric coordinates

$$\hat{\sigma} = \frac{1}{k+1} \sum_{i=0}^{k} v_i.$$

Geometrically, the barycenter is the centroid of the simplex. So the barycenter of an edge is its midpoint, of a 2-simplex is the intersection of the three edge bisectors, etc. The barycenter of a vertex is the vertex itself. We now define the *barycentric subdivision* of a simplex $\sigma = \langle v_0, \ldots, v_k \rangle$ inductively by assuming the $(k-1)$-skeleton has been subdivided. Let $\hat{\sigma}$ be the barycenter of σ and then for each $(k-1)$-simplex τ in the subdivided $(k-1)$-skeleton build the k-simplex $\hat{\sigma} * \tau$. See Figure 1.7.

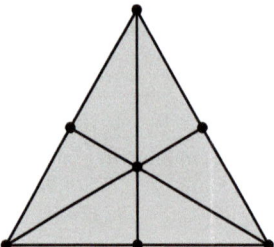

Figure 1.7: The barycentric subdivision of a 2-simplex.

Now, given a simplicial complex K, the *barycentric subdivision of K*, denoted sd(K), is obtained by taking the barycentric subdivision of each simplex in K. See Figure 1.8.

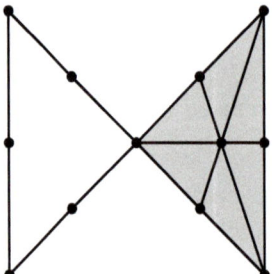

Figure 1.8: The barycentric subdivision of a simplicial complex.

Remark 1.1.26. Note that we can iterate this process to obtain the k-th barycentric subdivision $\mathrm{sd}^k(K)$ for all $k \geq 0$. As $k \to \infty$ observe that the simplices of $\mathrm{sd}^k(K)$ get smaller in the sense that their diameters become arbitrarily small. This is an important fact that we will need later.

Abstract simplicial complexes

It will often be useful to take a more abstract point of view in dealing with complexes. To that end let us make the following definition.

Definition 1.1.27. Let V be a set and let K be a collection of subsets of V. Then K is an *abstract simplicial complex* if it is closed under taking subsets; that is, if σ is an element of K, then any subset of σ is also in K (including the empty set!). An element σ of K is called a *simplex*, and a subset of σ is called a *face* of σ. The *dimension* of σ is one less than the number of elements of σ.

While nothing forbids allowing the set V to be infinite, we most often work with finite complexes. Note that a simplicial complex, as we have defined it above, is an abstract simplicial complex. Indeed, if the simplicial complex K has vertex set $K^{(0)}$, then each simplex is determined by its vertices and each face is determined by its vertices. That is, the set of simplices of K forms an abstract simplicial complex on the set $V = K^{(0)}$. Conversely, given an abstract simplicial complex K on a set V of $n + 1$ elements, we can realize it as a subcomplex of Δ^n by enumerating the elements of V as $v_1, v_2, \ldots, v_{n+1}$ and then identifying v_i with $e_i \in \mathbb{R}^{n+1}$. In particular, we note the following fact:

A simplicial complex with $n + 1$ vertices may be realized as a subcomplex of Δ^n.

Exercises

Exercise 1.1.7. Let K be a simplicial complex in $\mathbb{R}^n$. Prove that $|K|$ is a Hausdorff space.

Exercise 1.1.8. Prove the simplicial map g in Theorem 1.1.24 is continuous.

1.1.3 Cell complexes

Alas, as nice as simplicial complexes are, they have one fatal flaw: they can have an enormous number of simplices. For example, the internet may be modeled as a 1-dimensional simplicial complex (i. e., a graph) where each webpage is a vertex and there is an edge joining two vertices if one page links to the other. (Note: this is actually a *directed graph*, since the linking relationship may not be reciprocal, but we can ignore that for the sake of simplicity.) This is a very large complex, and working with it is quite cumbersome.

A more efficient representation of many spaces may be obtained by realizing them as a *cell complex*. Here, the basic building blocks are k-dimensional cells e^k, which are homeomorphic to a k-simplex or the closed unit disc D^k, but we allow more flexibility with how the cells are glued together.

Let us illustrate this with a familiar example. Recall that the torus T was realized as the quotient of the unit square $[0,1] \times [0,1]$ by identifying opposite edges. This allows us to construct a cell structure on T. First, observe that the four corners get identified together, yielding a single 0-cell v. Each pair of opposite edges get identified together to give two 1-cells a and b and they are attached together by gluing their ends to v. Finally, the interior of the square, which is homeomorphic to a 2-cell e, is attached to the edges a and b via the quotient map. As a result, we see that T is built of four cells: v, a, b, and e. See Figure 1.9.

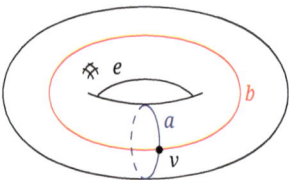

Figure 1.9: A cell decomposition of the torus.

Definition 1.1.28. A *cell complex* (or *CW complex*) is a space X constructed inductively as follows.
(1) Take a discrete set $X^{(0)}$, the points being regarded as 0-cells.
(2) Inductively, form the *n-skeleton* $X^{(n)}$ by attaching n-cells e_α^n via *attaching maps* $\varphi_\alpha :$ $S_\alpha^{n-1} = \partial e_\alpha^n \to X^{(n-1)}$. Formally, this means that we form the quotient space

$$X^{(n)} = X^{(n-1)} \amalg_\alpha D_\alpha^n / \sim$$

where $x \sim \varphi_\alpha(x)$ for all $x \in S_\alpha^{n-1} = \partial D_\alpha^n$.
(3) Stop at some n, setting $X = X^{(n)}$; or continue indefinitely, setting $X = \bigcup_n X^{(n)}$. In the latter case, X has the *weak topology*: $A \subseteq X$ is open (closed) if and only if $A \cap X^{(n)}$ is open (closed) in $X^{(n)}$ for each n.
(4) The complex X is *regular* if each attaching map φ_α is a homeomorphism onto its image.

Definition 1.1.29. If X and Y are cell complexes, then a continuous map $f : X \to Y$ is *cellular* if $f(X^{(i)}) \subset Y^{(i)}$ for all $i \geq 0$.

We describe several examples of cell complexes in Section 1.1.5 below.

Definition 1.1.30. A *subcomplex* of a cell complex X is a closed subspace $A \subseteq X$ that is a union of cells of X. Since A is closed, each attaching map φ_α for $e_\alpha \subset A$ has image in A and so A is itself a cell complex. A pair (X, A) with A a subcomplex is called a *CW-pair*.

Definition 1.1.31. Given two cell complexes X and Y, the *product complex* $X \times Y$ has n-cells $e^i \times e^j$, where e^i is an i-cell in X, e^j is a j-cell in Y, and $i + j = n$.

Examples

(1) The n-skeleton $X^{(n)}$ is a subcomplex of the cell complex X.
(2) The equatorial sphere S^{n-1} is a subcomplex of S^n, provided we build it inductively as follows. Begin with S^0, which is two disjoint points. Then attach two 1-cells to obtain S^1, two 2-cells to obtain S^2, and so on. We then have a filtration of subcomplexes $S^0 \subset S^1 \subset S^2 \subset \cdots \subset S^n$. This provides a *regular* cell complex structure on S^n. Note that the simpler cell decomposition $S^n = e^0 \cup e^n$ is not regular.
(3) The sphere S^{n-1} is a subcomplex of the disc D^n.
(4) The torus T is the product complex $S^1 \times S^1$. If we think of S^1 as $e^0 \cup e^1$, then T has one 0-cell, two 1-cells, and one 2-cell, obtained as the products of the cells in each copy of S^1. Note that this is not a regular cell decomposition, as the attaching map $\partial e^2 \to S^1 \vee S^1$ is not a homeomorphism.
(5) Any simplicial complex is a regular cell complex.

1.1.4 Manifolds

Loosely speaking, a *manifold* is a topological space in which every point has a neighborhood homeomorphic to an open set in some euclidean space $\mathbb{R}^n$. We are familiar with this concept as inhabitants of our planet–locally, we see a two-dimensional disc around us (at least when we're not on top of a mountain or the edge of a cliff). And while one possible space we could live on is a flat plane (spoiler: it is not), it is also possible to build more complicated structures in this way, such as a sphere or the surface of a donut. The actual definition is a bit more technical, of course.

Definition 1.1.32. A topological space M is an *n-dimensional manifold* if it is a paracompact Hausdorff space in which each point has an open neighborhood homeomorphic to an open ball in $\mathbb{R}^n$, called a *coordinate neighborhood*.

The simplest example of an n-manifold is $\mathbb{R}^n$ itself. More generally, an open subset of $\mathbb{R}^n$ is an n-manifold. Note that the dimension n of the manifold is part of the structure; that is, coordinate neighborhoods are all homeomorphic to $\mathbb{R}^n$ for the same n as one moves around the manifold. More examples of manifolds are given in Section 1.1.5.

It is possible to put additional structure on manifolds. To this end, let us be a bit more specific about coordinates. By an *atlas* for a manifold M we mean a family of injective mappings $\varphi_\alpha : U_\alpha \to M$ of open sets $U_\alpha \subset \mathbb{R}^n$ into M satisfying the following:

(1) $\bigcup_\alpha \varphi_\alpha(U_\alpha) = M$;

(2) for any pair α, β with $\varphi_\alpha(U_\alpha) \cap \varphi_\beta(U_\beta) = W \neq \emptyset$, the sets $\varphi_\alpha^{-1}(W)$ and $\varphi_\beta^{-1}(W)$ are open sets in $\mathbb{R}^n$ and the map $\varphi_\beta^{-1} \circ \varphi_\alpha$ is continuous;

(3) the family $\{(U_\alpha, \varphi_\alpha)\}$ is maximal relative to these conditions.

The pair $(U_\alpha, \varphi_\alpha)$ with $p \in \varphi_\alpha(U_\alpha)$ is called a *parametrization* or *system of coordinates* at p, and the set $\varphi_\alpha(U_\alpha)$ is then a *coordinate neighborhood*. Through a simple linear change of coordinates in $\mathbb{R}^n$, one can assume that $0 \in U_\alpha$ and $\varphi_\alpha(0) = p$.

Now, if the compositions $\varphi_\beta^{-1} \circ \varphi_\alpha$ are differentiable, then we call M a *differentiable manifold*, and if these maps are smooth (i. e., they have derivatives of all orders) then we call M a *smooth manifold*.

In this text we will mostly be concerned with topological manifolds, although many of our examples are in fact smooth.

Homogeneous spaces

A *Lie group* is a differentiable manifold G that is also a group. The group multiplication and inversion operations must be differentiable in the sense that if $g \in G$, the map $G \to G, x \mapsto gx$ is differentiable and the map $G \to G, x \mapsto x^{-1}$ is also differentiable.

Examples of Lie groups:

(1) $\mathbb{R}^n$ under addition.

(2) The circle S^1, viewed as the set of complex numbers of modulus 1, is a Lie group under complex multiplication.

(3) If G and H are Lie groups, so is the product $G \times H$.

(4) The n-torus $T^n = (S^1)^n$ (see Section 1.1.5) is then a Lie group.

(5) Let $F = \mathbb{R}$ or $\mathbb{C}$. Denote by $\mathrm{GL}_n(F)$ the set of $n \times n$ invertible matrices with entries in F. This is a Lie group under matrix multiplication. To see that this space is a manifold, note that it is the open set in the Euclidean space F^{n^2} (which is either $\mathbb{R}^{n^2}$ or $\mathbb{C}^{n^2} = \mathbb{R}^{4n^2}$) consisting of those matrices with nonzero determinant; that is, $\mathrm{GL}_n(F) = \det^{-1}(F \setminus \{0\})$, and since det is a continuous map, the pullback of the open set $F \setminus \{0\}$ is open. For the group operations, note that matrix multiplication is simply a polynomial in the entries of the matrices, and matrix inversion is a rational function of these entries via the cofactor expansion formula.

(6) The space $\mathrm{SL}_n(F)$ consisting of matrices of determinant 1 is a closed subgroup of $\mathrm{GL}_n(F)$. It has dimension $n^2 - 1$.

(7) The space $O(n)$ is the set of real $n \times n$ matrices X satisfying $XX^T = I$ and is called the *orthogonal group*. It is a closed subgroup of $\mathrm{GL}_n(\mathbb{R})$. The subgroup $SO(n)$, consisting of matrices of determinant 1, is called the *special orthogonal group*.

(8) The space $U(n)$ is the set of complex $n \times n$ matrices satisfying $XX^* = I$, where X^* is the conjugate transpose, and is called the *unitary group*. It is a closed subgroup of $\mathrm{GL}_n(\mathbb{C})$. The subgroup $SU(n)$, consisting of matrices of determinant 1, is called the *special unitary group*.

Recall that if G is a group, and X is a nonempty set, an *action* of G on X is a map $G{\times}X \to X$, denoted $(g, x) \mapsto gx$ such that (1) $ex = x$ for all $x \in X$ (e is the identity element of G), and (2) $(gh)x = g(hx)$ for all $g, h \in G$, $x \in X$. A G-action is *transitive* if for any $x, y \in X$ there exists a $g \in G$ such that $gx = y$. Note that if $gx = gy$, then $x = ex = (g^{-1}g)x = g^{-1}(gx) = g^{-1}(gy) = (g^{-1}g)y = ey = y$ so that each element $g \in G$ induces a permutation of the set X.

Definition 1.1.33. Let G be a group. A topological space X is a *homogeneous space* for G if G acts transitively on X. Note that each element of g acts as a homeomorphism of X.

Let X be a homogeneous space for G and suppose that x_0 is a marked point in X. The *stabilizer* of x_0 is the subgroup G_0 of G consisting of those elements that fix x_0. Then there is a bijective correspondence

$$X \leftrightarrow G/G_0$$

defined by $x \leftrightarrow gG_0$, where g is an element of G taking x_0 to x. Conversely, if H is a subgroup of G, then the coset space G/H is a homogeneous space for G with marked point eH.

Example 1.1.34. The sphere S^n is a homogeneous space for $G = O(n{+}1)$. To see this, note that $S^n \subset \mathbb{R}^{n+1}$, and that given any vector $v \in S^n$, there is an orthogonal transformation taking it to any other $w \in S^n$. In fact, one could also do this with a rotation (that is, with an element of $SO(n + 1)$) and so S^n is also a homogeneous space for $G = SO(n + 1)$. Given any $v \in S^n$, the stabilizer G_v consists of those (special) orthogonal matrices that leave the n-plane orthogonal to v invariant; that is, G_v is either $O(n)$ or $SO(n)$ and so we see that $S^n \cong O(n + 1)/O(n)$ and $S^n \cong SO(n + 1)/SO(n)$.

More generally, if G is a Lie group acting on X, then the stabilizer G_0 is a closed subgroup of G and is hence a Lie subgroup. It follows that G/G_0 is a smooth manifold and that X carries a smooth structure compatible with the action of G. Note also that if H is a Lie subgroup of G, then $\dim G/H = \dim G - \dim H$.

1.1.5 Dramatis personæ

In this section, we describe several spaces that we will see often in the sequel.

Spheres

The n-sphere, S^n is the space of unit vectors in the Euclidean space $\mathbb{R}^{n+1}$:

$$S^n = \left\{ (x_1, x_2, \ldots, x_{n+1}) \in \mathbb{R}^{n+1} \mid \sum_{i=1}^{n+1} x_i^2 = 1 \right\}.$$

It appears in many guises.

(1) S^n is homeomorphic to the boundary of the standard $(n + 1)$-simplex Δ^{n+1}. Viewed in this way, S^n may be given the structure of a simplicial complex of dimension n.

(2) S^n has a particularly efficient cellular structure consisting of one 0-cell x and one n-cell. The attaching map $\varphi : \partial e^n \to \{x\}$ collapses the boundary of e^n to a point. Of course, there are other cellular decompositions (e. g. any triangulation of S^n as a simplicial complex). One particularly nice one begins with two 0-cells, which we think of as S^0 and then inductively viewing S^i as the equator in S^{i+1}, attaching two i-cells as the northern and southern hemispheres at each stage. The end result is $S^n = (e^0 \cup e^0) \cup (e^1 \cup e^1) \cup \cdots \cup (e^n \cup e^n)$. This is a *regular* cell decomposition of S^n, since each attaching map $\varphi_i : \partial e^i \to (S^n)^{(i-1)}$ is a homeomorphism onto its image.

(3) S^n is a compact n-manifold. A particularly nice collection of charts is given via *stereographic projection*. Denote the north pole $(0, 0, \ldots, 0, 1)$ by N. The map $s_N : S^n - \{N\} \to \mathbb{R}^n$ obtained by setting $s_N(x)$ to be the unique point of intersection of the ray joining N to x with the coordinate plane $\mathbb{R}^n = \{(x_1, \ldots, x_n, 0)\} \subset \mathbb{R}^{n+1}$ is a homeomorphism. Similarly, denoting by S the south pole, the analogous map s_S is also a homeomorphism.

The *n*-torus

The n-dimensional torus, T^n, is defined to be the product space

$$T^n = \underbrace{S^1 \times \cdots \times S^1}_{n}.$$

As such it has a cellular structure with i-cells consisting of the $\binom{n}{i}$ i-fold products of the e^1 in each factor, viewing S^1 as the 1-sphere with one 0-cell and one 1-cell. This space is also the quotient of the unit cube in $\mathbb{R}^n$ obtained by identifying opposite faces.

The surface of genus *g*, M_g

The torus T^2 is the surface M_1. It is obtained as the quotient of the unit square by identifying opposite sides. As a cell complex, it has a decomposition with one 0-cell, two 1-cells, and one 2-cell. Generalizing this a bit, consider a regular $4g$-gon, $g \geq 1$, and make the identifications of the boundary edges as shown in Figure 1.10. The resulting space is called the *orientable surface of genus g* (we will explain the meaning of "orientable" later) and has a cell decomposition with one 0-cell, $2g$ 1-cells, and one 2-cell.

The Klein bottle

The torus T^2 is obtained from the square by identifying opposite sides with the same orientation. If instead we give one pair of sides a twist, we obtain a space called the *Klein bottle*. It cannot be embedded in $\mathbb{R}^3$ without self-intersection. It has a cell decomposition

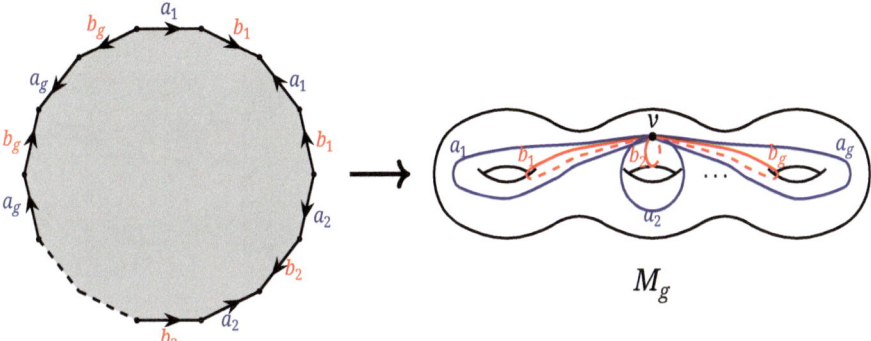

Figure 1.10: The identifications of the boundary of a regular $4g$-gon and the resulting surface.

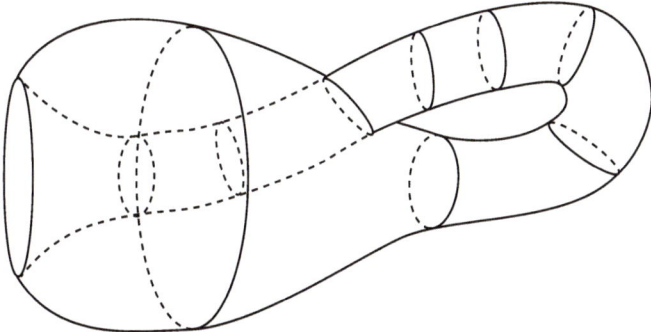

Figure 1.11: The Klein bottle.

similar to T^2: one 0-cell, 2 1-cells, and a single 2-cell, but the attaching map is different. See Figure 1.11.

Real projective space

As a set, real projective space, $\mathbb{R}P^n$ consists of all lines through the origin in $\mathbb{R}^{n+1}$. We topologize this set as follows. Each line through 0 in $\mathbb{R}^{n+1}$ intersects the sphere S^n in exactly two points, say x and its antipode $-x$. This yields a 2 : 1 surjective map $\pi : S^n \to \mathbb{R}P^n$. We declare this map to be a quotient map; that is, a set $U \subseteq \mathbb{R}P^n$ is open if and only if $\pi^{-1}(U)$ is open in S^n (see Figure 1.12). This puts a topology on $\mathbb{R}P^n$. Moreover, since the sphere S^n is compact, this implies that the space $\mathbb{R}P^n$ is compact as well. We note the following facts about $\mathbb{R}P^n$.

- $\mathbb{R}P^n$ may also be described as the following quotient. If ℓ is a line through 0 in $\mathbb{R}^{n+1}$, then ℓ is determined by a nonzero vector, unique up to scalar multiplication. It follows that $\mathbb{R}P^n = \mathbb{R}^{n+1} - \{0\}/ \sim$, where $v \sim \lambda v$ for all real numbers $\lambda \neq 0$. If we denote a point in $\mathbb{R}^{n+1}$ by $(x_0, \ldots, x_n)$, then the corresponding point in $\mathbb{R}P^n$ will be denoted by $[x_0 : \cdots : x_n]$. These are called *homogeneous coordinates*.

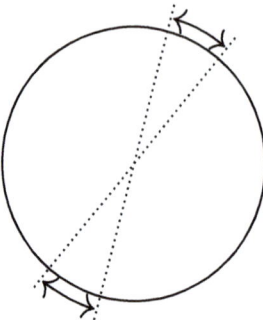

Figure 1.12: An open set in $\mathbb{R}P^1 = S^1$.

- $\mathbb{R}P^n$ may also be realized as a quotient of the n-dimensional ball D^n via the identification $x \sim -x$ for $x \in S^{n-1} = \partial D^n$. In the case $n = 2$, $\mathbb{R}P^2$ is the quotient of a square identifying each pair of sides with opposite orientation.
- $\mathbb{R}P^n$ has a particularly nice cellular structure. First observe that $\mathbb{R}P^1$ is homeomorphic to the circle S^1. It follows that $\mathbb{R}P^1 = e^0 \cup e^1$. Now assume inductively that $\mathbb{R}P^n = e^0 \cup e^1 \cup \cdots \cup e^n$. View S^n as the equator of S^{n+1} and observe that the identification $x \sim -x$ preserves this equator. Under the map $\pi : S^{n+1} \to \mathbb{R}P^{n+1}$, the northern and southern hemispheres get identified and the equatorial S^n maps to $\mathbb{R}P^n \subset \mathbb{R}P^{n+1}$. It follows that $\mathbb{R}P^{n+1} = \mathbb{R}P^n \cup_\pi e^{n+1} = e^0 \cup e^1 \cup \cdots \cup e^n \cup e^{n+1}$.
- $\mathbb{R}P^n$ is in fact a smooth manifold. First, note that if $x_i \neq 0$, then in $\mathbb{R}P^n$ we have

$$[x_0 : \cdots : x_n] = \left[\frac{x_1}{x_i} : \cdots : \frac{x_{i-1}}{x_i} : 1 : \frac{x_{i+1}}{x_i} : \cdots : \frac{x_n}{x_i} \right].$$

Define sets V_i, $i = 0, \ldots, n$ in $\mathbb{R}P^n$ by

$$V_i = \{ [x_0 : \cdots : x_n] \mid x_i \neq 0 \}.$$

Geometrically, the set V_i is the set of lines in $\mathbb{R}^{n+1}$ passing through 0 that do not belong to the hyperplane $x_i = 0$. Now define $\varphi_i : \mathbb{R}^n \to V_i$ by $\varphi_i(y_0, \ldots, y_{n-1}) = [y_0 : \cdots : y_{i-1} : 1 : y_i : \cdots : y_{n-1}]$. We leave it as an exercise to show that $\{(\mathbb{R}^n, \varphi_i)\}$ is a smooth structure on $\mathbb{R}P^n$.

Observe that we have an increasing sequence of subcomplexes $\mathbb{R}P^1 \subset \mathbb{R}P^2 \subset \cdots \subset \mathbb{R}P^n \subset \cdots$ and so we obtain the infinite real projective space $\mathbb{R}P^\infty = \bigcup_n \mathbb{R}P^n$. It has a cell decomposition with one cell in each dimension and is a $2 : 1$ quotient of $S^\infty = \bigcup_n S^n$.

Complex projective space

This is the complex version of $\mathbb{R}P^n$. As a set $\mathbb{C}P^n$ consists of complex lines through the origin in $\mathbb{C}^{n+1}$. As with $\mathbb{R}P^n$, we can visualize this in many ways.

– $\mathbb{C}P^n$ is the quotient of $\mathbb{C}^{n+1} - \{0\}$ obtained by identifying v with λv for all complex numbers $\lambda \in \mathbb{C}$.
– $\mathbb{C}P^n$ is also the quotient $S^{2n+1}/\sim$, where $v \sim \lambda v$ for complex numbers λ with $|\lambda| = 1$. This shows that $\mathbb{C}P^n$ is compact.
– $\mathbb{C}P^n$ is the quotient $D^{2n}/\sim$ where $v \sim \lambda v$ for $v \in \partial D^{2n} = S^{2n-1}$ and $|\lambda| = 1$. This point of view is convenient for describing a cell decomposition of $\mathbb{C}P^n$. Note that the vectors in $S^{2n+1} \subset \mathbb{C}^{n+1}$ with last coordinate real and nonnegative have the form $(w, \sqrt{1 - |w|^2}) \in \mathbb{C}^n \times \mathbb{C}$, with $|w| \leq 1$. This is the graph of the function $w \mapsto (w, \sqrt{1 - |w|^2})$, which is a disc D_+^{2n} bounded by $S^{2n-1} \subset S^{2n+1}$ consisting of vectors of the form $(w, 0) \in \mathbb{C}^n \times \mathbb{C}$, $|w| = 1$. Each vector in S^{2n+1} is equivalent to one in D_+^{2n} under $v \sim \lambda v$, uniquely if the last coordinate is nonzero. If the last coordinate is zero, then the identifications restrict to S^{2n-1}. It follows that $\mathbb{C}P^n = \mathbb{C}P^{n-1} \cup_\varphi e^{2n}$, where $\varphi : S^{2n-1} \to \mathbb{C}P^{n-1}$ is the quotient. Given that $\mathbb{C}P^1 = S^2$, we deduce that there is a cell decomposition $\mathbb{C}P^n = e^0 \cup e^2 \cup \cdots \cup e^{2n}$.

As in the real case, we have $\mathbb{C}P^1 \subset \mathbb{C}P^2 \subset \cdots \subset \mathbb{C}P^n \subset \cdots$, and if we take the union, we obtain the infinite complex projective space $\mathbb{C}P^\infty$. It has a cell decomposition with one cell in each even dimension.

Grassmann manifolds

By a *k-frame* in $\mathbb{R}^n$ we mean a k-tuple of linearly independent vectors. Denote the set of all such frames by $V_k(\mathbb{R}^n)$ and observe that $V_k(\mathbb{R}^n)$ is an open subset of the k-fold cartesian product $\mathbb{R}^n \times \cdots \times \mathbb{R}^n$. The *Grassmann manifold* $G(k, n)$ is the set of all k-planes through the origin in $\mathbb{R}^n$. It is topologized via the quotient map $V_k(\mathbb{R}^n) \to G(k, n)$, taking a k-frame to the plane which it spans. The case $k = 1$ recovers $G(1, n)$ as the projective space $\mathbb{R}P^{n-1}$.

– The spaces $G(k, n)$ and $G(n - k, n)$ are homeomorphic via the map $X \mapsto X^\perp$ taking a k-plane X to its orthogonal complement.
– The space $G(k, n)$ is a compact manifold of dimension $k(n - k)$. Perhaps the simplest way to see this is to note that $G(k, n)$ may also be realized as a quotient $O(n)/(O(k) \times O(n - k))$, where $O(n)$ is the *orthogonal group* consisting of $n \times n$ orthogonal matrices and $O(k) \times O(n - k)$ is the subgroup consisting of matrices having diagonal block form with the first block an orthogonal $k \times k$ matrix and the second block an orthogonal $(n - k) \times (n - k)$ matrix. Observe that $O(n)$ is a compact manifold since it is the set of $n \times n$ matrices X satisfying $XX^T = I$. This condition shows that $O(n)$ is the solution set to a system of polynomial equations and is therefore a closed subset of the set of $n \times n$ matrices (a space homeomorphic to $\mathbb{R}^{n^2}$). Moreover, the set $O(n)$ is bounded and so it is compact. There is an obvious action of $O(n)$ on $G(k, n)$ and the stabilizer of the k-frame $(e_1, \ldots, e_k)$ (where the e_i are the standard basis vectors in $\mathbb{R}^n$) is the group $O(k) \times O(n - k)$ described above. It follows that $G(k, n)$ is the orbit space and

$\dim G(k, n) = \dim O(n) - \dim(O(k) \times O(n - k))$. We leave the rest of the calculation as an exercise.

- The Grassmann manifold $G(k, n)$ has a cell decomposition in terms of *Schubert cells*. Consider the sequence of subspaces of $\mathbb{R}^n$: $\mathbb{R}^0 \subset \mathbb{R}^1 \subset \mathbb{R}^2 \subset \cdots \subset \mathbb{R}^n$, where $\mathbb{R}^i$ is the span of the first i standard basis vectors. A k-plane X then determines a sequence of integers

$$0 \leq \dim(X \cap \mathbb{R}^1) \leq \dim(X \cap \mathbb{R}^2) \leq \cdots \leq \dim(X \cap \mathbb{R}^n) = k.$$

Consecutive integers differ by at most 1. A *Schubert symbol* $\sigma = (\sigma_1, \ldots, \sigma_k)$ is a sequence of k integers satisfying $1 \leq \sigma_1 < \sigma_2 < \cdots < \sigma_k \leq n$. Given a Schubert symbol σ, let $e(\sigma) \subset G(k, n)$ denote the set of k-planes X such that

$$\dim(X \cap \mathbb{R}^{\sigma_i}) = i, \dim(X \cap \mathbb{R}^{\sigma_i - 1}) = i - 1.$$

Note that each X belongs to exactly one of the sets $e(\sigma)$.

In terms of matrices, $X \in e(\sigma)$ if and only if it can be described as the row space of a $k \times n$ matrix of the form

$$\begin{bmatrix} * & \cdots & * & 1 & 0 & \cdots & 0 & 0 & 0 & \cdots & 0 & 0 & 0 & \cdots & 0 \\ * & \cdots & * & * & * & \cdots & * & 1 & 0 & \cdots & 0 & 0 & 0 & \cdots & 0 \\ \vdots & & & & & & & & & & & & & & \vdots \\ * & \cdots & * & * & * & \cdots & * & * & * & \cdots & * & 1 & 0 & \cdots & 0 \end{bmatrix},$$

where the i-th row has σ_i-th entry positive (say equal to 1) and all subsequent entries zero. Equivalently, we could consider the column space of the transpose of this matrix. From this we deduce that the set $e(\sigma)$ is an open cell of dimension $d(\sigma) = (\sigma_1 - 1) + (\sigma_2 - 2) + \cdots + (\sigma_k - k)$. The $\binom{n}{k}$ sets $e(\sigma)$ form the cells of a CW-decomposition of $G(k, n)$ and it is easy to see that the number of r-cells in $G(k, n)$ is equal to the number of partitions of r into at most k integers, each of which is $\leq n - k$.

- There is also a complex version of this manifold, built from complex k-planes in $\mathbb{C}^n$. The discussion above works just as well over $\mathbb{C}$ to yield a cell decomposition, but the dimensions get doubled (since $\mathbb{C}$ is a 2-dimensional real vector space).

Exercises

Exercise 1.1.9. Prove that stereographic projection $s_N : S^n - \{N\} \to \mathbb{R}^n$ is a homeomorphism.

Exercise 1.1.10. Prove that $\mathbb{R}P^n$ is homeomorphic to the quotient of the n-dimensional ball D^n via the identification $x \sim -x$ for $x \in S^{n-1}$.

Exercise 1.1.11. Prove that $\mathbb{R}P^1$ is homeomorphic to S^1.

Exercise 1.1.12. Let $T_1S^2 = \{(x, v) \in S^2 \times S^2 : x \cdot v = 0\}$. Prove that T_1S^2 is homeomorphic to $\mathbb{R}P^3$.

Exercise 1.1.13. Prove that $\mathbb{C}P^1$ is homeomorphic to S^2.

Exercise 1.1.14. Prove that the atlas $\{(\mathbb{R}^n, \varphi_i)\}$ defined above on $\mathbb{R}P^n$ is a smooth structure.

Exercise 1.1.15. Find an atlas on $\mathbb{C}P^n$ analogous to the one on $\mathbb{R}P^n$ and show that the resulting maps $\varphi_j^{-1} \circ \varphi_i$ are analytic.

Exercise 1.1.16. Let $\mathbb{H}P^n$ be the set of quaternionic lines through the origin in *quaternionic space* $\mathbb{H}^{n+1}$. Show that $\mathbb{H}P^n$ may be realized as a quotient of S^{4n+3} and describe a cell decomposition for this space. Which familiar space is homeomorphic to $\mathbb{H}P^1$?

Exercise 1.1.17. Prove that $\dim O(n) = n(n-1)/2 = n^2 - n(n+1)/2$. Deduce that a $\dim G(k, n) = k(n-k)$.

Exercise 1.1.18. Prove that if the Lie group G acts on X, then the stabilizer G_0 of $x_0 \in X$ is a closed subgroup of G. Prove further that G/G_0 is homeomorphic to X, and if H is a Lie subgroup of G, then $\dim G/H = \dim G - \dim H$.

Exercise 1.1.19. Enumerate the cells in the Schubert cell decomposition of $G(2, 4)$.

Exercise 1.1.20. Prove that the Grassmann manifold $G(k, n)$ is homeomorphic to the space of $n \times n$ symmetric, idempotent matrices of trace k. (Recall that a matrix X is symmetric if $X = X^T$ and idempotent if $X^2 = X$.)

1.2 Continuous maps and homotopy

1.2.1 Homotopy of maps

Suppose that X is a topological space and that A is a subspace. Then we call (X, A) a *pair*. If we have a continuous map $f : X \to Y$ that maps the subspace $A \subset X$ into the subspace $B \subset Y$, then we will write $f : (X, A) \to (Y, B)$.

Definition 1.2.1. Suppose $f_0, f_1 : (X, A) \to (Y, B)$ are continuous maps that agree on A. A *homotopy from f_0 to f_1 relative to A* is a continuous map $F : X \times I \to Y$ ($I = [0, 1]$) such that

(1) $F(x, 0) = f_0(x)$ for all $x \in X$;
(2) $F(x, 1) = f_1(x)$ for all $x \in X$; and
(3) $F(a, t) = f_0(a)$ for all $a \in A$ and $t \in [0, 1]$.

In this case, we write $f_0 \simeq f_1$ rel A.

If the subspace A is empty, we simply omit it from the notation and the third condition above is then vacuous. Note that the particular choice of interval I is irrelevant. We often use $I = [0,1]$, but we could just as well (and often do!) use any closed interval $[a, b]$.

Homotopy represents continuous deformation. That is, if two maps are homotopic, then one should imagine a continuous morphing of the image of f_0 into the image of f_1. Or, in terms of graphs of real-valued functions on the real line, one should picture the graph of f_0 shifting continuously to the graph of f_1.

Example 1.2.2. Consider the map $F : \mathbb{R} \times [-1,1] \to \mathbb{R}$ given by $F(x,t) = x^3 + tx^2 + 1$. This is clearly continuous and is therefore a homotopy from $f_{-1}(x) = x^3 - x + 1$ and $f_1(x) = x^3 + x + 1$. The graphs of a few $f_t(x)$ are shown in Figure 1.13.

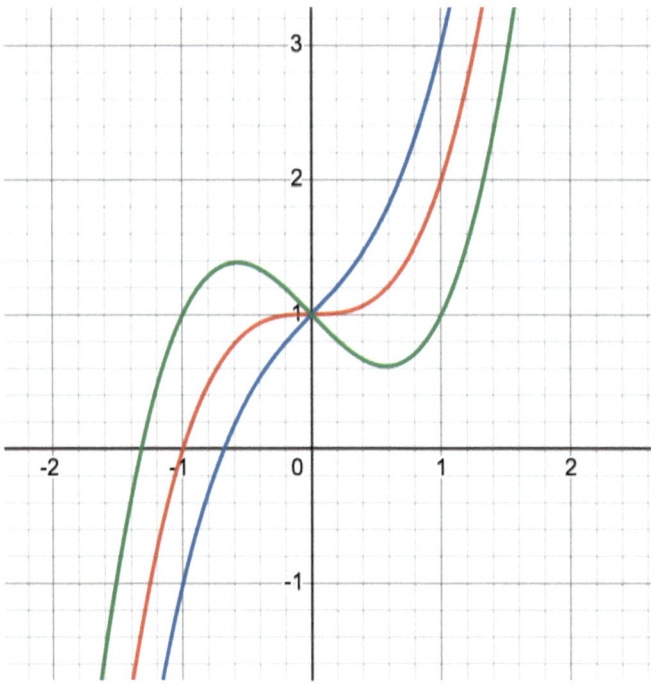

Figure 1.13: Some $f_t(x)$ for the family $F(x,t) = x^3 + tx + 1$: $t = -1$ (green), $t = 0$ (red), $t = 1$ (blue).

Example 1.2.3. Suppose $f, g : X \to \mathbb{R}^n$ for some n. The *straight-line homotopy* from f to g is the homotopy $F : X \times I \to \mathbb{R}^n$ defined by

$$F(x,t) = (1 - t)f(x) + tg(x).$$

Since f and g are continuous, so is F, and we have $F(x,0) = f(x), F(x,1) = g(x)$. Pictorially, one sees that this family of maps connects each point $f(x)$ to $g(x)$ with a straight line.

Example 1.2.4. Let X be the punctured plane $\mathbb{R}^2 \setminus \{0\}$ and consider the map $f : X \to X$ defined by $f(x) = x/\|x\|$. The straight-line homotopy from f to id_X is well-defined since the line segment joining x and $f(x)$ does not pass through 0. It follows that $f \simeq \mathrm{id}_X$.

Definition 1.2.5. Suppose $f : X \to Y$ is continuous. We say that f is *nullhomotopic* if it is homotopic to a constant map $c : X \to Y, c(x) = y_0$ for some fixed $y_0 \in Y$.

Example 1.2.6. Any map $f : X \to \mathbb{R}^n$ is nullhomotopic. We leave this as an exercise for the reader.

Example 1.2.7. The map $i : S^1 \to S^2$ defined by $i(x,y) = (x,y,0)$, which embeds the circle as the equator on the sphere, is nullhomotopic.

We leave the following result as an exercise.

Proposition 1.2.8. *Suppose (X,A) is a topological pair. Then homotopy of maps relative to A is an equivalence relation. That is, if $f, g, h : (X,A) \to (Y,B)$ are continuous, then*
(1) *$f \simeq f$ rel A;*
(2) *if $f \simeq g$ rel A, then $g \simeq f$ rel A; and*
(3) *if $f \simeq g$ rel A and $g \simeq h$ rel A, then $f \simeq h$ rel A.*

Simplicial approximation

Naturally, when working with simplicial complexes, we want to use simplicial maps. However, there may be interesting continuous maps on the underlying spaces that are not simplicial. For example, if we triangulate S^2 as the boundary of a 3-simplex with one vertex at the north pole, then a rotation by a small angle about this axis is not simplicial, and there is no way to make it so. However, all is not lost.

Theorem 1.2.9. *If K is a finite simplicial complex, then any continuous $f : K \to L$ is homotopic to a map that is simplicial with respect to an iterated barycentric subdivision of K.*

Proof. If σ is a simplex of K, denote by $\mathrm{St}(\sigma)$ (the *star of σ*) the subcomplex consisting of all simplices of K containing σ. The *open star*, $\mathrm{st}(\sigma)$, is the union of all interiors of simplices containing σ. The *link*, $\mathrm{lk}(\sigma)$ is $\mathrm{St}(\sigma) - \mathrm{st}(\sigma)$. These are simplicial analogues of a closed neighborhood, an open neighborhood, and the boundary of a closed neighborhood, respectively. See Figure 1.14 for an example.

Lemma 1.2.10. *Suppose $v_1, \ldots, v_n$ are vertices of K. Then $\mathrm{st}(v_1) \cap \cdots \cap \mathrm{st}(v_n) = \emptyset$ unless $v_1, \ldots, v_n$ are vertices of a simplex σ of K, in which case $\mathrm{st}(v_1) \cap \cdots \cap \mathrm{st}(v_n) = \mathrm{st}(\sigma)$.*

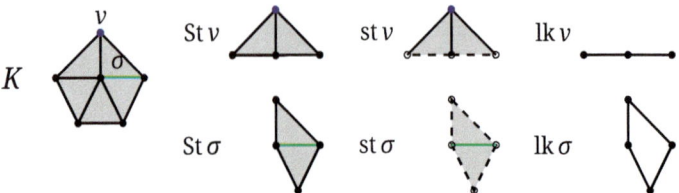

Figure 1.14: The star, open star, and link of a simplex.

Proof. The intersection consists of all interiors of simplices whose vertex set contains $\{v_1,\ldots,v_n\}$. If it is nonempty, such a τ exists and contains $\sigma = \langle v_1,\ldots,v_n \rangle \in K$. The simplices τ containing $\{v_1,\ldots,v_n\}$ are just the simplices containing σ and so $\mathrm{st}(v_1) \cap \cdots \cap \mathrm{st}(v_n) = \mathrm{st}(\sigma)$. □

Now, to prove Theorem 1.2.9, choose a metric on K restricting to the standard euclidean metric on each simplex. Let ε be the Lebesgue number of the open cover $\{f^{-1}(\mathrm{st}(w)) \mid w$ is a vertex of $L\}$. After iterated barycentric subdivision of K, we may assume each simplex has diameter less than $\varepsilon/2$. The closed star of each vertex v of K then has diameter less than ε and so this maps under f to the open star of some vertex $g(v)$ of L. This defines a map $g : K^{(0)} \to L^{(0)}$ satisfying $f(\mathrm{st}(v)) \subset \mathrm{st}(g(v))$ for all vertices v of K.

We claim that g extends to a simplicial map $g : K \to L$. Take a simplex $\langle v_1,\ldots,v_n\rangle$ of K. An interior point lies in $\mathrm{st}(v_i)$ for each i and so $f(x)$ lies in $\mathrm{st}(g(v_i))$ for each i. Thus $\mathrm{st}(g(v_1)) \cap \cdots \cap \mathrm{st}(g(v_n)) \neq \emptyset$ and so $\langle g(v_1),\ldots,g(v_n)\rangle$ is a simplex of L by Lemma 1.2.10. Thus g extends linearly over $\langle v_1,\ldots,v_n\rangle$. Both $f(x)$ and $g(x)$ lie in the same simplex of L. The straight line homotopy $(1-t)f(x) + tg(x), 0 \leq t \leq 1$ is then a well-defined homotopy from f to g. □

Exercises

Exercise 1.2.1. Prove that any map $f : X \to \mathbb{R}^n$ is nullhomotopic.

Exercise 1.2.2. Prove that the map $f(x) = x/\|x\|$ on the punctured plane is homotopic to the identity.

Exercise 1.2.3. Prove that the map $i(x,y) = (x,y,0)$ is nullhomotopic.

Exercise 1.2.4. Prove that homotopy rel A is an equivalence relation.

1.2.2 Homotopy equivalence of spaces

Homotopy equivalence is a fundamental notion in algebraic topology. It is too much to ask to be able to classify spaces up to homeomorphism; that is much too rigid. Homotopy, on the other hand, is "squishy" enough to allow us to make some progress. Moreover, the algebraic invariants we will develop in subsequent chapters do not distinguish spaces that are homotopy equivalent, so it is a good idea to get used to this notion.

Definition 1.2.11. Suppose that X and Y are topological spaces. We say that a continuous map $f : X \to Y$ is a *homotopy equivalence* if there exists a continuous map $g : Y \to X$ such that $g \circ f \simeq \mathrm{id}_X$ and $f \circ g \simeq \mathrm{id}_Y$. If such a map exists, we say that X and Y are *homotopy equivalent* and write $X \simeq Y$.

Note that if $f : X \to Y$ is a homeomorphism, then f is a homotopy equivalence. Indeed, the inverse function f^{-1} is continuous and we have $f^{-1} \circ f = \mathrm{id}_X$ and $f \circ f^{-1} = \mathrm{id}_Y$ on the nose. Homotopy equivalence requires only that each composite be homotopic to the requisite identity map; this is often easier to arrange.

Definition 1.2.12. A space X is *contractible* if there is a point $x_0 \in X$ such that $\mathrm{id} : X \to X$ is homotopic to the constant map at x_0.

Definition 1.2.13. Suppose A is a subspace of X. A map $r : X \to A$ is a *retraction* if the restriction of r to A is the identity; that is, denoting the inclusion of A into X by i, we have $r \circ i = \mathrm{id}_A$.

Definition 1.2.14. Suppose A is a subspace of X and denote by $i : A \to X$ the inclusion map. A *deformation retraction* of X onto A is a homotopy $F : X \times I \to X$ such that $F(x, 0) = x$, $F(x, 1) \in A$ for all $x \in X$ and $F(a, 1) = a$ for all $a \in A$. Note that the map $F(-, 1)$ is a retraction of X onto A.

Definition 1.2.15. A *strong deformation retraction* is a deformation retraction $F : X \times I \to X$ such that $F(a, t) = a$ for all $a \in A$ and $t \in I$.

Example 1.2.16. Let X be the punctured plane $\mathbb{R}^2 \setminus \{0\}$ and let S^1 be the unit circle embedded in X. We claim that these two spaces are homotopy equivalent. In fact, S^1 is a strong deformation retract of X. Let $j : S^1 \to X$ be the inclusion map and let $f : X \to S^1$ be the map $f(x) = x/\|x\|$. We must show that the composites $j \circ f$ and $f \circ j$ are homotopic to the identity. One of these is easy: $f(j(x)) = f(x) = x/\|x\| = x$ for all $x \in S^1$ so that $f \circ j = \mathrm{id}_{S^1}$. For the other direction, note that $j \circ f(x) = f(x)$ and this is homotopic to id_X by an exercise in the previous section.

The mapping cylinder

Suppose X and Y are spaces and that $f : X \to Y$ is continuous. Construct a new space, M_f, by

$$M_f = (X \times I) \amalg Y / \sim,$$

where $\sim$ is defined by $(x, 1) \sim f(x)$ for all $x \in X$. That is, we adjoin the cylinder $X \times I$ to Y by gluing the copy of X at $t = 1$ to its image in Y. See Figure 1.15.

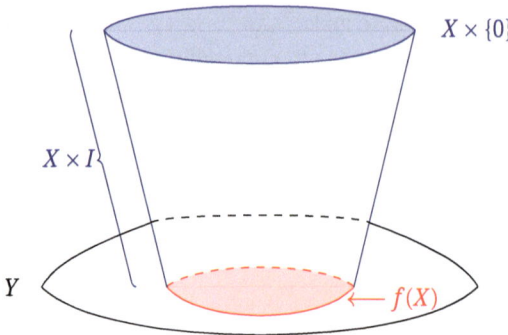

$X \times \{0\}$

$X \times I$

Y

$f(X)$

Figure 1.15: The Mapping Cylinder.

Proposition 1.2.17. *Given $f : X \to Y$, the mapping cylinder M_f is homotopy equivalent to Y. In fact, Y is a strong deformation retract of M_f.*

Proof. Denote by $j : Y \to M_f$ the inclusion of Y into M_f. Define a map $\pi : M_f \to Y$ by

$$\pi(y) = y, \text{ if } y \in Y$$
$$\pi(x, t) = f(x), \text{ if } x \in X, t \in I$$

The map π is continuous. Consider the composite $\pi \circ j : Y \to Y$. It satisfies $\pi(j(y)) = \pi(y) = y$; i. e., $\pi \circ j = \mathrm{id}_Y$. For the other composite, note that $j \circ \pi$ is defined as

$$j(\pi(y)) = y, \text{ if } y \in Y$$
$$j(\pi(x, t)) = f(x), \text{ if } x \in X, t \in I.$$

We must show that $j \circ \pi \simeq \mathrm{id}_{M_f}$. Define $F : M_f \times I \to M_f$ by

$$F(y, s) = y, \text{ if } y \in Y$$
$$F((x, t), s) = (x, (1 - s)t), \text{ if } (x, t) \in X \times I$$

Then F fixes Y for all s and $F((x, t), 0) = (x, t)$ so that $F_0 = \mathrm{id}_{M_f}$, and $F_1 = j \circ \pi$. □

Observe that there is an inclusion $i : X \to M_f$ defined by $i(x) = (x, 0)$ and since $M_f \simeq Y$, we see that we may replace any continuous $f : X \to Y$ by an inclusion up to homotopy. This is sometimes useful.

Definition 1.2.18. Suppose $f : X \to Y$ is continuous. The *mapping cone* of f is the space C_f defined by

$$M_f/(X \times \{0\});$$

that is, we collapse the copy of $X \times \{0\}$ in the mapping cylinder to a point.

We have already met an example of a mapping cone. Indeed, if $f : S^{n-1} \to X$, then since the cone on S^{n-1} is the disc D^n, the mapping cone C_f is the space $X \cup_f D^n$.

1.2.3 Homotopy extension

Definition 1.2.19. Suppose that A is a subspace of X. We say that the pair (X, A) has the *homotopy extension property* (HEP) *with respect to* Y if for any continuous $f_0 : X \to Y$ and any homotopy $F : A \times I \to Y$ with $F(a, 0) = f_0(a)$, there is a homotopy $\tilde{F} : X \times I \to Y$ extending F. If (X, A) has the HEP with respect to any space Y, we say that (X, A) has the HEP.

This may seem a bit esoteric, but this property will be very important later during our study of homology.

Proposition 1.2.20. *Suppose A is a closed subspace of X. Then the pair (X, A) has the homotopy extension property if and only if $(X \times \{0\}) \cup (A \times I)$ is a retract of $X \times I$.*

Proof. Suppose (X, A) has the homotopy extension property. Then the identity map

$$(X \times \{0\}) \cup (A \times I) \to (X \times \{0\}) \cup (A \times I)$$

extends to a map

$$X \times I \to (X \times \{0\}) \cup (A \times I).$$

Thus, this space is a retract of $X \times I$. Conversely, suppose $(X \times \{0\}) \cup (A \times I)$ is a retract of $X \times I$. Since A is closed, any two maps $X \times \{0\} \to Y$ and $A \times I \to Y$ that agree on $A \times \{0\}$ can be glued to give a continuous map $(X \times \{0\}) \cup (A \times I) \to Y$. Composing this map with a retraction $r : X \times I \to (X \times \{0\}) \cup (A \times I)$ yields the required extension. $\qquad\square$

Since we are mainly concerned with cell complexes in this text, the next result is crucial.

Proposition 1.2.21. *If (X, A) is a pair of cell complexes, then (X, A) has the homotopy extension property.*

Proof. First note that the pair (D^n, S^{n-1}) has the HEP. Indeed, there is a retraction $r : D^n \times I \to (D^n \times \{0\}) \cup (S^{n-1} \times I)$ given by radial projection from some point above the cylinder $D^n \times I$ (e. g. $(0, 2)$) (see Figure 1.16). Then the straight line homotopy from r to the identity gives a deformation retraction of $D^n \times I$ onto $(D^n \times \{0\}) \cup (S^{n-1} \times I)$. In turn, this yields a deformation retraction of $X^{(n)} \times I$ onto $(X^{(n)} \times \{0\}) \cup ((X^{(n-1)} \cup A^{(n-1)}) \times I$ since the former is obtained from the latter by attaching copies of $D^n \times I$ along $(D^n \times \{0\}) \cup (S^{n-1} \times I))$. Performing this deformation retraction along the t-interval $[1/2^{n+1}, 1/2^n]$, we obtain a deformation retraction of $X \times I$ onto $(X \times \{0\}) \cup A \times I$. Continuity at $t = 0$ is achieved since X has the weak topology. $\qquad\square$

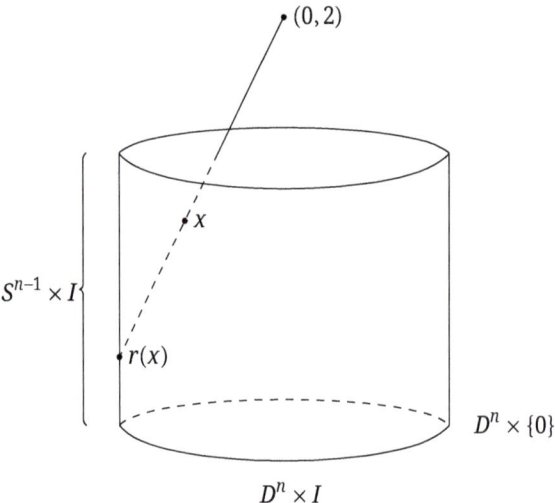

$S^{n-1} \times I$

$(0, 2)$

x

$r(x)$

$D^n \times \{0\}$

$D^n \times I$

Figure 1.16: The retraction of $D^n \times I$ onto $(D^n \times \{0\}) \cup (S^{n-1} \times I)$.

While the next result is intuitively clear (that collapsing a contractible subspace to a point should not change the homotopy type), it relies on the HEP.

Proposition 1.2.22. *Suppose* (X, A) *satisfies the homotopy extension property and that A is contractible. Then the quotient map* $q : X \to X/A$ *is a homotopy equivalence.*

Proof. We leave this for the reader. It is a good exercise in diagram chasing and the use of universal properties of quotient maps. ☐

Cofibrations
Related to the HEP is the notion of a cofibration.

Definition 1.2.23. A map $f : X' \to X$ is called a *cofibration* if, given maps $g : X \to Y$ and $G : X' \times I \to Y$ (where Y is any space) such that $g(f(x')) = G(x', 0)$ for $x' \in X'$, there is a map $F : X \times I \to Y$ such that $F(x, 0) = g(x)$ for $x \in X$, and $F(f(x'), t) = G(x', t)$ for $x' \in X'$ and $t \in I$.

Example 1.2.24. If A is a subspace of X, then the inclusion $i : A \to X$ is a cofibration if and only if (X, A) has the HEP.

Cellular approximation
It should come as no surprise, in light of Theorem 1.2.9, that if $f : X \to Y$ is a map of cell complexes, then f is homotopic to a *cellular map* $g : X \to Y$. The proof is a bit more complicated, however, as one might imagine. We first recall a result from Exercise 1.1.4 earlier in the chapter. If $q : X \to Y$ is a quotient map and Z is a compact space, then the

map $q \times \mathrm{id}_Z : X \times Z \to Y \times Z$ is a quotient map. In particular, if X is a cell complex, then the n-skeleton $X^{(n)}$ is obtained from the $(n-1)$-skeleton by attaching n-cells; that is, we have a quotient map

$$q_n : X^{(n-1)} \coprod \{e_a^n\} \to X^{(n)}.$$

Taking the product of this map with the identity map on the interval I, we have a quotient map

$$q_n \times \mathrm{id}_I : X^{(n-1)} \times I \coprod \{e_a^n \times I\} \to X^{(n)} \times I.$$

Theorem 1.2.25. *Let (X, A) be a CW pair and let Y be a CW complex. Suppose $f : X \to Y$ is a continuous map. If $f|_A : A \to Y$ is cellular, then f is homotopic to a cellular map $g : X \to Y$ relative to A. In particular, any map of cell complexes is homotopic to a cellular map.*

We first prove the following technical result.

Lemma 1.2.26. *Suppose Y is obtained from B by attaching an n-cell. Then any map $f : (D^m, \partial D^m) \to (Y, B)$ with $m < n$ is homotopic relative to ∂D^m to a map g satisfying $g(D^m) \subset B$.*

Proof. We proceed by induction on n, beginning with $n = 1$. In this case, $m = 0$ and D^m is a point with $\partial D^m = \emptyset$. A map $f : (*, \emptyset) \to (Y, B)$ is just a point $y \in Y$. There is a path $\gamma : I \to Y$ with $\gamma(0) = y$ and $\gamma(1) = b \in B$. This gives the homotopy from f to the map $g : * \mapsto b$.

The inductive step is much more complicated. Here is the idea. When we attach an n-cell, the interior of the cell is mapped homeomorphically onto its image in Y. The plan is to show that we can construct a homotopy from f to a map h relative to ∂D^m such that h omits the origin of the attached disc. Removing that origin (call it z), we see that $i : B \to Y - \{z\}$ is the inclusion of a strong deformation retraction r (collapse the punctured disc $e^n - \{z\}$ to S^{n-1}). Part of this is a homotopy $\mathrm{id}_{Y-\{z\}} \simeq i \circ r$ rel B, inducing a relative homotopy $h = \mathrm{id} \circ h \simeq i \circ r \circ h = g$ relative to ∂D^m. Putting these together gives a homotopy from f to g relative to ∂D^m.

So, assume inductively that the result holds for $n - 1$. Note that, as a consequence of this assumption, we know that any map $S^k \to S^{n-1}$ and any map $S^k \to S^{n-1} \times (a, b)$ for $k < n - 1$ is homotopic to a constant map. Moreover, any map $S^k \to S^{n-1} \times (a, b)$, $k < n - 1$, can be extended to a map on D^{k+1}. So we need only construct a homotopy $f \simeq h : D^m \to Y$ relative to ∂D^m such that h does not hit the origin of the attached e^n.

Let $\chi : e^n \to Y$ be the characteristic map for the cell e^n, with $\chi|_{\partial e^n}$ the attaching map. Consider the following two open sets in e^n:

$$A = \{x \in e^n \mid \|x\| < 2/3\} \quad \text{and} \quad B = \{x \in e^n \mid \|x\| > 1/3\}.$$

We then obtain an open cover of Y by setting $U = \chi(A)$ and $V = B \cup_{\partial e^n} \chi(B)$. We want to construct a relative homotopy $f \simeq h$ with the image of h lying entirely in V. Note that

$$U \cap V \approx S^{n-1} \times (1/3, 2/3).$$

Choose a homeomorphism $(I^m, \partial I^m) \approx (D^m, \partial D^m)$ and pull back the open cover $\{U, V\}$ to obtain an open cover of the compact metric space I^m. The Lebesgue lemma implies that there is a positive integer N such that the image of each m-cube

$$I^m_{k_1,\ldots,k_m} = [k_1/N, (k_1 + 1)/N] \times \cdots \times [k_m/N, (k_m + 1)/N], \quad 0 \le k_i < N,$$

under f lies in either U or V. Define a filtration on I^m:

$$\partial I^m \subset X_{-1} \subset X_0 \subset \cdots \subset X_m = I^m$$

as follows. Let J_{-1} be an index set for all sub ℓ-cubes, $0 \le \ell \le m$, of $I^m_{k_1,\ldots,k_m}$, $0 \le k_i < N$, which are mapped into V by f. Denote such an ℓ-cube for the index a by I^ℓ_a. Set

$$X_{-1} = \bigcup_{a \in J_{-1}} I^\ell_a.$$

By assumption, $\partial D^m \subset X_{-1}$. We now proceed by induction on the dimension of the remaining subcubes. For each $0 \le k \le m$, let J_k be an index set of all k-dimensional subcubes I^k_a satisfying $f(I^k_a) \not\subset V$. Inductively define

$$X_k = X_{k-1} \cup \bigcup_{a \in J_k} I^k_a.$$

We now want to construct maps $h_k : X_k \to Y, k \ge -1$, such that
(1) h_{-1} is obtained from f by restriction.
(2) h_k sends the cubes I^k_a to $U \cap V$ for all $a \in J_k, k \ge 0$.
(3) $h_k|_{X_{k-1}} = h_{k-1}$ for all $k \ge 0$.

For h_0, note that X_0 is obtained from X_{-1} by possibly adding some vertices that are mapped into U. For each vertex, choose a path to a point in $U \cap V$. These target points, together with h_{-1} define the map h_0. For the inductive step, assume that h_{k-1} has been constructed. For each $a \in J_k$, we have $h_{k-1}(\partial I^k_a) \subset U \cap V \approx S^{n-1} \times (1/3, 2/3)$. Our initial inductive assumption then implies that we can extend h_{k-1} to all of I^k_a, and it is easy to see that these can be assembled together to define a map $h_k : X_k \to Y$ with the desired properties. Setting $h = h_m$, we have $h(I^m) \subset V$. It remains to show that $f \simeq h$ relative to ∂I^m.

In fact, such a homotopy exists relative to X_{-1}. By construction, f and h agree on X_{-1}. Also, the restriction of both maps to $X \setminus X_{-1}$ can be considered as maps taking values in U. But U is homeomorphic to an open n-disc and via a straight-line homotopy the two restrictions are homotopic. This homotopy, with the constant homotopy on X_{-1} can be assembled to give the homotopy $f \simeq h$ relative to X_{-1}. □

Proof of Theorem 1.2.25. We have a CW-pair (X, A) and so we have a filtration

$$A = X^{(-1)} \subset X^{(0)} \subset X^{(1)} \subset \cdots \subset X$$

such that $X^{(n)}$ is obtained from $X^{(n-1)}$ by attaching n-cells, and X is the union of the $X^{(n)}$ endowed with the weak topology. We are given $f : X \to Y$ continuous with $f|_A$ cellular. We will construct maps $g_n : X \to Y$ and homotopies $H_n : g_{n-1} \simeq g_n$ such that g_n sends relative n-cells into $Y^{(n)}$, and the homotopy H_n is relative to $X^{(n-1)}$. We have the map $g_{-1} = f|_A$. Assume that we have the map g_{n-1} and let's construct g_n and H_n. Denote the set of relative n-cells by J_n and consider the diagram

$$
\begin{array}{ccc}
\coprod_{a \in J_n} \partial e_a^n & \longrightarrow & X^{(n-1)} \\
\downarrow & & \downarrow \\
\coprod_{a \in J_n} e_a^n & \xrightarrow[(\chi_a)_a]{} & X^{(n)}
\end{array}
$$

Assume there is $a \in J_n$ such that $g_{n-1}(e_a^n)$ is not contained in $Y^{(n)}$ (otherwise we could set $g_n = g_{n-1}$ and take H_n to be the constant homotopy). For each such e_a^n there is a finite relative subcomplex Y' with $Y^{(n)} \subseteq Y' \subseteq Y$ such that $g_{n-1}(e_a^n) \subseteq Y'$. Consider a cell of maximal dimension in Y' having nontrivial intersection with $g_{n-1}(e_a^n)$. Lemma 1.2.26 implies that this cell can be avoided up to relative homotopy. Repeating this finitely many times and gluing the relative homotopies, we obtain a homotopy $H_{n,a} : g_{n-1} \simeq g_{n,a} : e_a^n \to Y$ relative to ∂e_a^n, with $g_{n,a}(e_a^n) \subseteq Y^{(n)}$. As noted above, $X^{(n)} \times I$ has the quotient topology with respect to the map

$$X^{(n-1)} \times I \coprod_{a \in J_n} \{e_a^n \times I\} \to X^{(n)} \times I.$$

It follows that we can glue the homotopies $H_{n,a}$, the constant homotopies on $g_{n-1} : e_a^n \to Y$ for all n-cells with $g_{n-1}(e_a^n) \subseteq Y^{(n)}$, and the constant homotopy on $g_{n-1}|_{X^{(n-1)}}$ to obtain a homotopy

$$\tilde{H}_n : g_{n-1}|_{X^{(n)}} \simeq \tilde{g}_n : X^{(n)} \times I \to Y.$$

We therefore have an extension problem

$$
\begin{array}{ccc}
X \times \{0\} \cup X^{(n)} \times I & \xrightarrow{(g_{n-1}, \tilde{H}_n)} & Y \\
\downarrow & \nearrow & \\
X \times I & {}_{H_n} &
\end{array}
$$

which admits a solution since $X^{(n)} \to X$ is a cofibration. Set $g_n = H_n(-, 1)$. This completes the inductive step.

If X is finite-dimensional, then we are done: $H : f \simeq g_n = g$ is the required homotopy with $g : X \to Y$ a cellular map. In the infinite-dimensional case, since the homotopies H_n are relative to $X^{(n-1)}$, we see that every H_k is stationary on $X^{(k)}$ for $k \geq n$. So if we define $H : X \times I \to Y$ on $X^{(n-1)}$ by running H_k at 2^{k+1} speed, $k < n$, we get a continuous map $H : X \times I \to Y$. The map $g = H(-, 1) : X \to Y$ is the required cellular map. $\square$

Exercises

Exercise 1.2.5. Prove that a space X is contractible if and only if for any space Y any two maps of Y into X are homotopic.

Exercise 1.2.6. Let $CX = X \times I/(X \times \{0\})$ be the cone on X. Embed X in CX by $x \mapsto (x, 1)$. Prove that $f : X \to Y$ is nullhomotopic if and only if f extends to $\bar{f} : CX \to Y$.

Exercise 1.2.7. Suppose Y is contractible to $y_0 \in Y$. Define $f : X \to X \times Y$ by $f(x) = (x, y_0)$. Prove that f and the projection $p : X \times Y \to X$ are homotopy equivalences.

Exercise 1.2.8. Prove Proposition 1.2.22. (Hint: A contraction of A may be thought of as a homotopy $A \times I \to X$ extending the identity map on $X \times \{0\}$. Compose this homotopy with the quotient map and then use the universal property of quotient maps to get a map $X/A \to X$.)

Exercise 1.2.9. Prove the assertion in Example 1.2.24.

1.3 Lifting properties and fibrations

1.3.1 Liftings of continuous maps

Suppose $f : X \to Y$ is a continuous map and that $p : Z \to Y$ is another continuous map. If the map p is surjective, a natural question to ask is if one can "lift" the map f up to a map $X \to Z$ whose composition with p agrees with f. Even without the surjectivity assumption, this is still a valid question.

Definition 1.3.1. Suppose $f : X \to Y$ and $p : Z \to Y$ are continuous. A *lift* of f is a continuous map $\tilde{f} : X \to Z$ such that $p \circ \tilde{f} = f$.

Lifts need not exist, of course. Nor do they have to be unique if they do exist.

Example 1.3.2. Let $f : S^1 \to \mathbb{C} \setminus \{0\}$ be the map $f(z) = z^2$ (view S^1 as the set of unit modulus complex numbers) and let $p : \mathbb{C} \to \mathbb{C} \setminus \{0\}$ be the map $p(z) = \exp(z)$. There is no continuous lift of the map f since any such map would have to satisfy $\exp(\tilde{f}(z)) = z^2$

and therefore would be a multiple of a logarithm. Such maps cannot be defined on sets in $\mathbb{C} \setminus \{0\}$ that contain a curve around the origin.

Example 1.3.3. Let $f : S^1 \to S^1$ be the map $z \mapsto z^6$ and let $p : S^1 \to S^1$ be the map $z \mapsto z^3$. Then $\tilde{f} : S^1 \to S^1$ defined by $\tilde{f}(z) = z^2$ is a lift of f.

Example 1.3.4. Let $f : I \to S^1$ be the map $t \mapsto (\cos(4\pi t), \sin(4\pi t))$ and let $p : \mathbb{R} \to S^1$ be the map $s \mapsto (\cos(2\pi s), \sin(2\pi s))$ (as usual, we view S^1 as embedded in the plane). Define a map $\tilde{f} : I \to \mathbb{R}$ by $t \mapsto 2t$. Then $p(\tilde{f}(t)) = p(2t) = (\cos(4\pi t), \sin(4\pi t))$ and so $\tilde{f}$ is a lift of f. But if n is any integer, then the map $\tilde{f}_n$ defined by $t \mapsto 2(t + n)$ is also a lift of f.

Homotopy lifting

A natural question to ask is the following. Suppose $f : X \to Y$ and $p : Z \to Y$ are maps and that f lifts to Z. Suppose $g : X \to Y$ is a map homotopic to f. Must g also lift to Z? One would guess the answer is "yes", but this is not obvious.

Definition 1.3.5. A map $p : E \to B$ has the *homotopy lifting property* (HLP) *with respect to X* if, given maps $f' : X \to E$ and $F : X \times I \to B$ such that $F(x, 0) = pf'(x)$ for all $x \in X$, there is a map $F' : X \times I \to E$ such that $F'(x, 0) = f'(x)$ for $x \in X$ and $p \circ F' = F$.

Lemma 1.3.6. *Suppose $p : E \to B$ has the HLP with respect to X and that $f_0, f_1 : X \to B$ are homotopic. Then f_0 can be lifted to E if and only if f_1 can be lifted to E.*

Proof. Let $F : X \times I \to B$ be a homotopy with $F(x, 0) = f_0(x)$ and $F(x, 1) = f_1(x)$. If f_0 lifts to $f'_0 : X \to E$, then, since p has the HLP with respect to X, there is a lift $F' : X \times I \to E$ with $F'(x, 0) = f'_0(x)$. But then $pF'(x, 1) = F(x, 1) = f_1(x)$, so that $F'(x, 1)$ is a lift of f_1. Conversely, if f_1 lifts to E, then use the homotopy $G(x, t) = F(x, 1 - t)$ and the HLP to get a lift of f_0. $\square$

Definition 1.3.7. A map $p : E \to B$ is called a *fibration* if p has the HLP with respect to every space. In this case E is called the *total space* and B the *base space*. If $b \in B$, the space $p^{-1}(b)$ is called the *fiber* over b.

Example 1.3.8. Let F be any space and let $p : B \times F \to B$ be the projection $p(b, f) = b$. Then p is a fibration and for any $b \in B$ the fiber $p^{-1}(b)$ is homeomorphic to F.

Definition 1.3.9. Suppose $p : E \to B$ is a fibration and $f : X \to B$ is continuous. The *pullback* of p by f is the fibration $f^*E \to X$, defined by

$$f^*E = \{(x, e) \in X \times E \mid f(x) = p(e)\}.$$

One checks easily that f^*E is in fact a fibration.

1.3.2 Locally trivial fiber bundles

The product $p : B \times F \to B$ is often called a *trivial bundle* over B. There are many examples of fibrations that are not just a product space, and in this section we describe the class of *locally trivial* fibrations. As you might expect, this means that locally the fibration is just a product, but these local pieces might be glued together to yield something that is globally not just a product.

Example 1.3.10. A simple example of a nontrivial fibration is the *Möbius strip*. This is the quotient space

$$M = [0, 2\pi] \times I / \sim,$$

where we set $(0, t) \sim (2\pi, 1 - t)$. That is, if we imagine $[0, 2\pi] \times I$ as a strip of paper, we are gluing one pair of opposite sides together after giving one end a twist. There is a map $p : M \to S^1$ given by $p(x, t) = e^{ix}$ (viewing S^1 as the set of unit modulus complex numbers). Note that p is continuous since it is induced by the same map $[0, 2\pi] \times I \to S^1$ and $(0, t)$ and $(2\pi, 1 - t)$ both map to $1 \in S^1$. Given $z \in S^1$, the fiber $p^{-1}(z)$ is a copy of I, but the space M is *not* just $S^1 \times I$. See Figure 1.17.

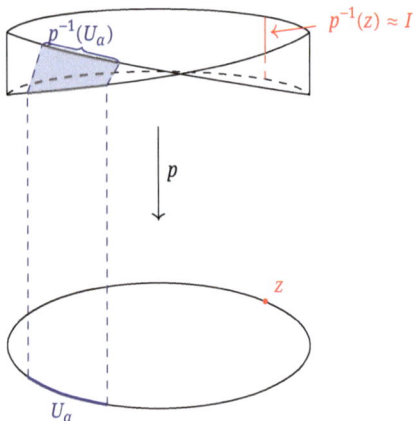

Figure 1.17: The Möbius strip as a bundle over S^1.

Note that if we choose a $z \in S^1$ and take a small neighborhood U around it, then the inverse image $p^{-1}(U)$ is homeomorphic to $U \times I$. So this fibration is locally trivial.

Definition 1.3.11. A *locally trivial fiber bundle* consists of the following.
(1) A continuous map $\pi : E \to B$ between spaces E and B. E is called the *total space* and B is called the *base space*.
(2) A topological space F called the *fiber*.

These are required to satisfy the *local triviality condition*. For each $x \in B$, there is an open neighborhood $U \subset B$ and a homeomorphism $\varphi_U : \pi^{-1}(U) \to U \times F$ such that $p_1 \circ \varphi_U = \pi|_{\pi^{-1}(U_a)}$, where $p_1 : U \times F \to U$ is the projection on the first factor. The set of all $\mathcal{U} = \{(U_i, \varphi_{U_i})\}$ is called a *local trivialization* of the bundle. Note that if $x \in B$, the preimage $F_x = \pi^{-1}(x)$ is homeomorphic to F.

If we have a local trivialization $\mathcal{U}$, consider two sets U_i and U_j with $U_i \cap U_j \neq \emptyset$. We shall use the notation φ_i for φ_{U_i}. If $x \in U_i$, define a map $\varphi_{i,x} : F \to \pi^{-1}(x)$ by $\varphi_{i,x}(y) = \varphi_i^{-1}(x, y)$. Then if $x \in U_i \cap U_j$ we have a homeomorphism

$$\varphi_{i,x}^{-1} \circ \varphi_{j,x} : F \to F.$$

Definition 1.3.12. The *group of the bundle* $\pi : E \to B$ is an effective group G of homeomorphisms of the fiber F such that for all i, j with $U_i \cap U_j \neq \emptyset$ the maps $\varphi_{i,x}^{-1} \circ \varphi_{j,x}$ are elements of G. Here, *effective* means that if $g \in G$ and $gy = y$ for some $y \in F$, then g is the identity map. We also insist that for each pair of indices i, j the map

$$g_{ij} : U_i \cap U_j \to G$$

given by $g_{ij}(x) = \varphi_{i,x}^{-1} \circ \varphi_{j,x}$ is continuous.

Note that Definition 1.3.12 uses the word *the*. By this we mean that G is the smallest subgroup of the group of homeomorphisms of the fiber that contains all the $\varphi_{i,x}^{-1} \circ \varphi_{j,x}$. For example, the group of the Möbius strip bundle is the cyclic group $\mathbb{Z}/2$, generated by the map $I \to I$ given by $t \mapsto 1 - t$. The group of any product bundle is the trivial group.

Example 1.3.13. An *n-dimensional vector bundle* is a locally trivial fiber bundle $E^n \to B$ in which the fiber is an *n*-dimensional vector space over $\mathbb{R}$ or $\mathbb{C}$. The group in this case is GL_n over the appropriate field. A primary example is obtained as follows. Let M be a (real) *n*-manifold. At each point $x \in M$, we have a tangent space T_xM, which is canonically a vector space of dimension n. Define

$$TM = \{(x, v) \mid x \in M, v \in T_xM\};$$

this is the *tangent bundle* of M. The *associated sphere bundle* consists of the subspace

$$S(TM) = \{(x, v) \mid x \in M, v \in T_xM, \|v\| = 1\}$$

obtained by replacing the fiber T_xM by the unit sphere in T_xM.

Example 1.3.14. Suppose G is a Lie group acting as a transitive group of maps on a manifold M. Fix a base point $x_0 \in M$ and define $\pi : G \to M$ by $\pi(g) = g(x_0)$. Denote by H the subgroup of G fixing x_0. The fibers of π are then the left cosets of H in G. The group of this bundle is H: if $g \in \pi^{-1}(x)$, then the map $y \mapsto gy$ is a homeomorphism of H with $\pi^{-1}(x)$ and any two such maps (for g and g') differ by left translation by $g^{-1}g'$.

Proposition 1.3.15. *Suppose $\pi : E \to B$ is a locally trivial fiber bundle. Then $\pi : E \to B$ has the HLP with respect to any CW pair (X, A).*

Proof. Because of the way CW complexes are defined, homotopy lifting for CW pairs is equivalent to homotopy lifting for discs, and therefore equivalent to lifting for cubes. Suppose we have a homotopy $F : I^n \times I \to B$, $F_t(x) = F(x, t)$ and a given lift $\tilde{F}_0$ of F_0. Choose an open cover $\{U_\alpha\}$ of B with local trivializations $\varphi_\alpha : \pi^{-1}(U_\alpha) \to U_\alpha \times F$. Since $I^n \times I$ is compact, we may subdivide I^n into cubes C and I into intervals $I_j = [t_j, t_{j+1}]$ such that each $C \times I_j$ maps into a single U_α. We may assume by induction on n that $\tilde{F}_t$ has been defined on ∂C for each of the cubes C. To extend this over C, we proceed by building $\tilde{F}_t$ in each successive I_j. But this then reduces to the case where we do not need to subdivide $I^n \times I$ and F maps all of $I^n \times I$ into a single U_α. Then we have $\tilde{F}(I^n \times \{0\} \cup \partial I^n \times I) \subset \pi^{-1}(U_\alpha)$, and composing this with φ_α reduces this to the case of the product bundle $U_\alpha \times F$. The first coordinate of a lift $\tilde{F}_t$ is the given F_t. To construct the second coordinate, use the composition $I^n \times I \to I^n \times \{0\} \cup \partial I^n \times I \to F$, where the first map is a retraction and the second is the given map. $\qquad\square$

Remark 1.3.16. Locally trivial fiber bundles over a rather general class of spaces (e. g., normal, locally compact, Lindelöf spaces) have the HLP with respect to all spaces. The proof of this fact is rather involved, however, and is not really needed here. See Section 11 of [6] for more details. Thus, for us, locally trivial fiber bundles are fibrations.

Example 1.3.17. Homotopy lifting has the following interesting and important consequence. Suppose $\pi : E \to B$ is a locally trivial fiber bundle. Choose a point $x_0 \in B$ and let $\gamma : I \to B$ be a path in B with $\gamma(0) = x_0$. Let $e_0 \in F_{x_0}$ be any point in the fiber over x_0. We may view γ as a homotopy of the form $X \times I$ where $X = \{0\}$. We can define $\tilde{\gamma}(0) = e_0$ and then by HLP we get a lift $\tilde{\gamma} : I \to E$ with $\pi \circ \tilde{\gamma} = \gamma$. The image of $\tilde{\gamma}$ is a path in E lying over the path γ. Note that $\tilde{\gamma}(1)$ lies in the fiber $F_{\gamma(1)}$. In particular, if γ is a loop in B $(\gamma(0) = x_0 = \gamma(1))$, then $\tilde{\gamma}(1)$ lies in F_{x_0}, but $\tilde{\gamma}$ is not necessarily a loop in E.

Example 1.3.18. A *covering space* is a fibration of the form $\pi : E \to B$, where the fiber $F = \pi^{-1}(x)$ is a discrete subspace of E for each $x \in B$. It is then possible to choose a sufficiently small neighborhood U of x such that $\pi^{-1}(U)$ is a disjoint union of copies of U. Since the fibers are discrete, the group of the bundle is a subgroup of the permutation group of the fiber. Path lifting plays a role here; see the project at the end of the chapter.

Example 1.3.19. Let X be a space and let x_0 be a basepoint. The *path space* $\mathcal{P}(X, x_0)$ is the set of all paths $\omega : I \to X$ with $\omega(0) = x_0$, equipped with the compact-open topology (a subbasis for this topology consists of sets $\langle K, U \rangle = \{\omega : I \to X \mid \omega(K) \subset U\}$, $K \subset I$ compact, $U \subset X$ open). Define $\pi : \mathcal{P}(X, x_0) \to X$ by $\pi(\omega) = \omega(1)$.

Proposition 1.3.20. *The map $\pi : \mathcal{P}(X, x_0) \to X$ is a fibration.*

Proof. Given a path $g : I \to X$ and $\tilde{g}_0 \in \mathcal{P}(X, x_0)$ with $\pi(\tilde{g}_0) = g(0)$ (i. e., given a path g in X and a path beginning at x_0 and ending at $g(0)$), define $\tilde{g}_t \in \mathcal{P}(X, x_0)$ by

$$\tilde{g}_t(s) = \begin{cases} \tilde{g}_0(s(1+t)) & 0 \le s \le \frac{1}{1+t} \\ g(s(1+t)-1) & \frac{1}{1+t} \le s \le 1. \end{cases}$$

Then $\pi(\tilde{g}_t) = g(t)$, and $t \mapsto \tilde{g}_t$ is a continuous map of I into $\mathcal{P}(X, x_0)$. This proves homotopy lifting for points. It is easy to see that this varies continuously in the original data and so this gives homotopy lifting for all spaces. $\qquad\qquad\square$

The fiber $\pi^{-1}(x_0) \subset \mathcal{P}(X, x_0)$ is denoted $\Omega(X, x_0)$ and is called the *loop space* of X based at x_0.

Proposition 1.3.21. *For any space X, the path space $\mathcal{P}(X, x_0)$ is contractible.*

Proof. Define $F : \mathcal{P}(X, x_0) \times I \to \mathcal{P}(X, x_0)$ by $F(\omega, t)(s) = \omega(ts)$. Then $F(\omega, 0) = \eta_{x_0}$, the constant path at x_0, and $F(\omega, 1) = \omega$. $\qquad\qquad\square$

We have seen that we can replace any map by an inclusion, up to homotopy, via the mapping cylinder construction. We can also replace any map by a fibration, up to homotopy. Indeed, if $f : X \to Y$ is a continuous map, consider the set

$$\tilde{X} = \{(x, \omega) \mid \omega(0) = f(x)\} \subset X \times \mathcal{P}(Y),$$

where $\mathcal{P}(Y)$ is the space of *all* paths in Y. Define $\pi : \tilde{X} \to Y$ by $\pi(x, \omega) = \omega(1)$ and define $i : X \hookrightarrow \tilde{X}$ by $i(x) = (x, \eta_{f(x)})$. Then

$$\pi \circ i(x) = \pi(x, \eta_{f(x)}) = \eta_{f(x)}(1) = f(x).$$

Proposition 1.3.22. *The map $i : X \to \tilde{X}$ is a homotopy equivalence.*

We leave the proof of this as an exercise.

Exercises

Exercise 1.3.1. Prove the assertions in Example 1.3.8

Exercise 1.3.2. Prove that if $\pi : E \to B$ is a locally trivial fiber bundle, then π is an open map. Conclude that π is a quotient map.

Exercise 1.3.3. Prove the assertion in Example 1.3.14 that the fibers of π are the left cosets of H in G.

Exercise 1.3.4. Show that the map $p : S^{2n+1} \to \mathbb{C}P^n$ is a fibration. What is the fiber?

Exercise 1.3.5. Prove that the pullback of a fibration is in fact a fibration.

Exercise 1.3.6. Prove Proposition 1.3.22.

1.4 Basic category theory

Category theory provides a convenient language for discussing many of the constructions that follow. To that end, we include a quick introduction to categories in this section.

Definition 1.4.1. A *category* C consists of
(1) a collection $\text{Obj}(C)$ of *objects*;
(2) for each pair $A, B \in \text{Obj}(C)$ a set $\text{Hom}(A, B)$ of *morphisms* whose elements are denoted $f : A \to B$; and
(3) for each triple A, B, C of objects a function

$$\text{Hom}(B, C) \times \text{Hom}(A, B) \to \text{Hom}(A, C)$$

called *composition* and denoted by $(g, f) \mapsto g \circ f$.

These are required to satisfy the following properties.
(1) Composition is associative: if $f : A \to B$, $g : B \to C$, and $h : C \to D$ are morphisms then $h \circ (g \circ f) = (h \circ g) \circ f$.
(2) Identity: for each object B there is a morphism $1_B : B \to B$ such that for any $f : A \to B$ and $g : B \to C$, $1_B \circ f = f$ and $g \circ 1_B = g$.

Example 1.4.2. Let <u>Set</u> be the category whose objects are the class of all sets and whose morphisms $\text{Hom}(A, B)$ consist of all set maps $f : A \to B$.

Example 1.4.3. Denote by <u>Top</u> the collection of all topological spaces where the morphisms $\text{Hom}(X, Y)$ are all continuous maps $f : X \to Y$. Then <u>Top</u> is a category. There are various *subcategories*. For example, the collection of CW complexes with cellular maps is a subcategory of <u>Top</u>, as is the collection of simplicial complexes with simplicial maps.

Example 1.4.4. Denote by <u>Grp</u> the collection of all groups with morphisms $\text{Hom}(G, H)$ the set of all group homomorphisms $f : G \to H$. The subcategory <u>Ab</u>, consisting of abelian groups, has further interesting properties that we will discuss below.

Example 1.4.5. More generally, let R be a commutative ring with identity and denote by R-mod the collection of all (left) R-modules with morphisms $\text{Hom}(M, N)$ the set of all R-module homomorphisms $f : M \to N$. In the case $R = \mathbb{Z}$ we recover <u>Ab</u>.

Example 1.4.6. Let G be a group with its operation written multiplicatively. Then G may be viewed as a category with one object, namely G itself, with morphisms $\text{Hom}(G, G) = G$. Composition of morphisms is simply given by the group multiplication.

Definition 1.4.7. Let C and $\mathcal{D}$ be categories. A *covariant functor* $F : C \to \mathcal{D}$ is an assignment

$$F : \mathrm{Obj}(\mathcal{C}) \to \mathrm{Obj}(\mathcal{D}); \quad A \mapsto F(A),$$

such that for any morphism $f : A \to B$ and $g : B \to C$ in $\mathcal{C}$ there are morphisms $F(f) :$ $F(A) \to F(B)$ and $F(g) : F(B) \to F(C)$ satisfying $F(g \circ f) = F(g) \circ F(f)$. A *contravariant* functor "reverses" arrows: given $f : A \to B$ and $g : B \to C$ in $\mathcal{C}$ there are morphisms $F(f) : F(B) \to F(A)$ and $F(g) : F(C) \to F(B)$ such that $F(g \circ f) = F(f) \circ F(g)$. In both cases, we also require that $F(1_A) = 1_{F(A)}$ for all $A \in \mathrm{Obj}(\mathcal{C})$.

Example 1.4.8. Let $\mathcal{C}$ be a category whose objects are sets. The *forgetful functor* is the map $\mathcal{C} \to \underline{\mathrm{Set}}$, taking an object of $\mathcal{C}$ to its underlying set.

One of the main uses of category theory is *diagram chasing*. That is, we often have a collection of objects in a category and we wish to build a new object related to them in some way. It may be convenient to organize this into a diagram consisting of the objects and various morphisms. The properties of categories then imply that our construction is unique, if it exists. Or, we may wish to take a diagram of objects and apply a functor; the resulting transformed objects then fit into a similar diagram. We now present a few examples of this idea.

Example 1.4.9. Let $\mathcal{C}$ be a category and let $\{A_i \mid i \in I\}$ be a family of objects in $\mathcal{C}$. A *product* for the family is an object P in $\mathcal{C}$ together with a family of morphisms $\{\pi_i : P \to A_i \mid i \in I\}$, such that for any object Q and any family of morphisms $\{\varphi_i : Q \to A_i \mid i \in I\}$ there is a unique morphism $\varphi : Q \to P$ such that $\pi_i \circ \varphi = \varphi_i$ for all $i \in I$. In terms of diagrams, we have, for each i a *commutative diagram*

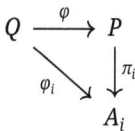

Here, *commutative* means that beginning at Q, the two paths from Q to A_i are equal.

Proposition 1.4.10. *If $(P, \{\pi_i\})$ and $(P', \{\pi_i'\})$ are products for a family $\{A_i \mid i \in I\}$, then P and P' are equivalent objects in $\mathcal{C}$.*

Proof. Since P and P' are both products there are morphisms $f : P \to P'$ and $g : P' \to P$ such that we have a pair of commutative diagrams for each $i \in I$:

Composing these gives a commutative diagram for each $i \in I$:

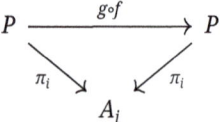

The definition of the product asserts that there is a unique morphism $P \to P$ commuting with π_i for all $i \in I$. Since the identity clearly satisfies this, we must have $g \circ f = 1_P$. Similarly, $f \circ g = 1_{P'}$. It follows that $f : P \to P'$ is an equivalence. □

Example 1.4.11. Dual to the notion of product is the *coproduct*. Given a family $\{A_i \mid I \in I\}$ of objects in $\mathcal{C}$, a coproduct for the family is an object S, together with a family of morphisms $\{\iota_i : A_i \to S \mid i \in I\}$, such that for any object T and any family of morphisms $\{\psi_i : A_i \to T \mid i \in I\}$ there is a unique morphism $\psi : S \to T$ such that $\psi \circ \iota_i = \psi_i$ for all $i \in I$. In terms of diagrams, we have, for each i a commutative diagram

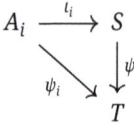

Definition 1.4.12. Let $\mathcal{C}$ be a category. An object I is an *initial object* if for any object A in $\mathcal{C}$ there is a unique morphism $I \to A$. An object T is a *terminal object* if for any object A there is a unique morphism $A \to T$.

Example 1.4.13. In Top a one-point space $\{*\}$ is terminal, but not initial, since $\{*\}$ may map to any point in a space. In Ab the trivial group $\{e\}$ is both initial and terminal.

Exercises

Exercise 1.4.1. Prove that the assignment $G \mapsto G^{ab}$ taking a group G to its abelianization is a functor Grp $\to$ Ab.

Exercise 1.4.2. Prove that the product of topological spaces, with the usual product topology, is a product in Top.

Exercise 1.4.3. Prove that in Set the coproduct is given by the disjoint union of sets.

Exercise 1.4.4. Prove that in Ab the coproduct is given by the *direct sum*,

$$\bigoplus_{i \in I} A_i = \{(a_i) : a_i = e \text{ for all but finitely many } i \in I\}.$$

Exercise 1.4.5. Prove that coproducts, if they exist, are unique up to equivalence.

Exercise 1.4.6. Prove that any two initial objects are equivalent. Similarly, any two terminal objects are equivalent.

1.5 The fundamental group

Let X be a topological space and fix a point $x_0 \in X$. Recall that a *loop* at x_0 is a path $\gamma : I \to X$ with $\gamma(0) = x_0 = \gamma(1)$. If we have two loops γ and μ, define a new loop $\gamma \cdot \mu$ by

$$\gamma \cdot \mu = \begin{cases} \gamma(2t) & 0 \le t \le 1/2 \\ \mu(2t - 1) & 1/2 \le t \le 1 \end{cases}$$

The relation of homotopy relative to the subset $\{0, 1\} \subset I$ is an equivalence relation on the set of loops at x_0; denote the homotopy class of the loop γ by $[\gamma]$. Define a set $\pi_1(X, x_0)$ by

$$\pi_1(X, x_0) = \{[\gamma] \mid \gamma \text{ is a loop at } x_0\}.$$

Proposition 1.5.1. $\pi_1(X, x_0)$ *is a group under path composition.*

Proof. Denote the homotopy class of a loop α by $[\alpha]$. Since composition respects homotopy (i. e. if $F(-, t) = \alpha_t : \alpha_0 \simeq \alpha_1$ and $G(-, t) = \beta_t : \beta_0 \simeq \beta_1$ are homotopies, then $\alpha_t \cdot \beta_t$ is a homotopy from $\alpha_0 \cdot \beta_0$ to $\alpha_1 \cdot \beta_1$), the operation $[\alpha] \cdot [\beta] = [\alpha \cdot \beta]$ is well-defined. We need only check the group axioms.

By a *reparametrization* of γ, we mean a composition $\gamma \circ \varphi$, where $\varphi : I \to I$ is a map with $\varphi(0) = 0$ and $\varphi(1) = 1$. Since any such φ is homotopic to id_I via the straight line homotopy, we see that $\gamma \circ \varphi \simeq \gamma$ for any γ.

To prove that the operation is associative, note that path composition is *not* associative in general (see Figure 1.18). However, $(\alpha \cdot \beta) \cdot \gamma$ and $\alpha \cdot (\beta \cdot \gamma)$ do differ by a reparametrization and so

$$([\alpha] \cdot [\beta]) \cdot [\gamma] = [\alpha] \cdot ([\beta] \cdot [\gamma]).$$

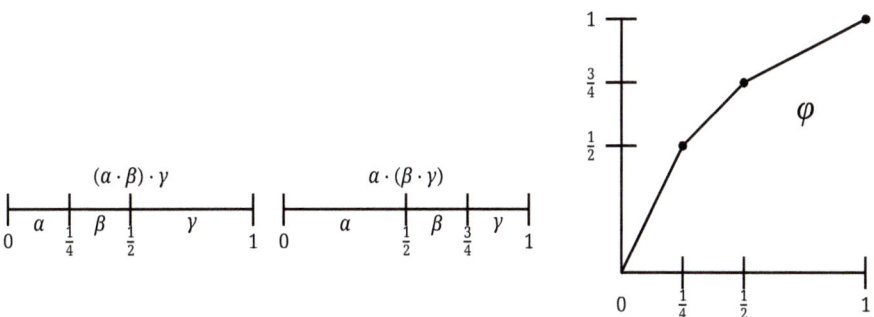

Figure 1.18: Path composition is associative up to homotopy via the reparametrization φ shown on the right.

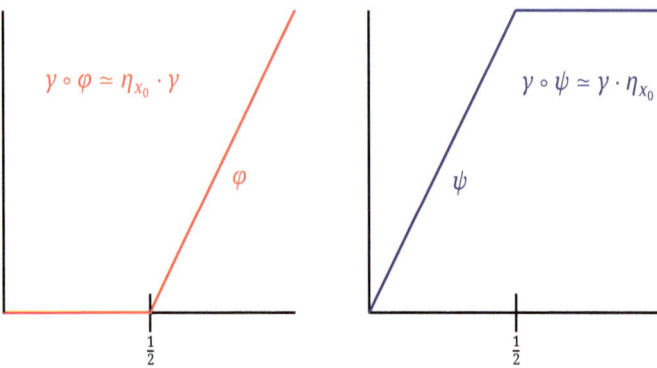

Figure 1.19: The homotopy showing that $[\eta_{x_0}]$ is the identity element.

Define $\eta_{x_0} : I \to X$ by $\eta_{x_0}(s) = x_0$ for all $s \in I$. Then $\eta_{x_0} \cdot \gamma \simeq \gamma \simeq \gamma \cdot \eta_{x_0}$ (see Figure 1.19). It follows that $[\eta_{x_0}]$ is the identity element.

Finally, if $\gamma : I \to X$ is a path from x_0 to x_1, then $\gamma^{-1} : I \to X$ defined by $\gamma^{-1}(s) = \gamma(1-s)$ is a path from x_1 to x_0. Note that $\gamma \cdot \gamma^{-1} \simeq \eta_{\gamma(0)}$ and $\gamma^{-1} \cdot \gamma \simeq \eta_{\gamma(1)}$. To see this, define a map γ_t by

$$\gamma_t = \begin{cases} \gamma(s) & 0 \le s \le 1-t \\ \gamma(1-t) & 1-t \le s \le 1 \end{cases}$$

Let $h_t = \gamma_t \cdot \gamma_t^{-1}$. Then h_t is a homotopy from $\gamma \cdot \gamma^{-1}$ to $\eta_{\gamma(0)}$. So, if we have a loop α at x_0, then $\alpha \cdot \alpha^{-1} \simeq \eta_{x_0}$ and so $[\alpha] \cdot [\alpha^{-1}] = [\eta_{x_0}]$ and $[\alpha^{-1}] \cdot [\alpha] = [\eta_{x_0}]$. ☐

The group $\pi_1(X, x_0)$ is called the *fundamental group* of X based at x_0. A connected space X with $\pi_1(X, x_0) = \{[\eta_{x_0}]\}$ is called *simply connected*.

Example 1.5.2. Suppose X is a star-shaped subset of $\mathbb{R}^n$; that is, there is an $x_0 \in X$ such that the line segment from x_0 to y lies in X for every $y \in X$. We claim that X is simply connected. To see this, if $y \in X$, let ℓ_y be the line segment from x_0 to y, parameterized as $\ell_y(t) = (1-t)x_0 + ty$. If γ is a loop at x_0, define $F : I \times I \to X$ by $F(s,t) = \ell_{\gamma(s)}(t)$. Then F is continuous and satisfies $F(s,0) = \ell_{\gamma(s)}(0) = x_0$, $F(s,1) = \ell_{\gamma(s)}(1) = \gamma(s)$, and $F(0,t) = \ell_{\gamma(0)}(t) = \ell_{x_0}(t) = x_0$ for all t. Thus, $F : \gamma \simeq \eta_{x_0}$.

Example 1.5.3. Let $X = \mathbb{C} - \{\pm 1\}$ and consider the two loops α and β in X shown in Figure 1.20. Let F_2 be the free group with generators $\{a, b\}$ and define $\theta : F_2 \to \pi_1(X, x_0)$ by $\theta(a) = [\alpha]$ and $\theta(b) = [\beta]$. We will see later that θ is an isomorphism.

The induced homomorphism

Suppose $f : (X, x_0) \to (Y, y_0)$ is continuous. Define $f_* : \pi_1(X, x_0) \to \pi_1(Y, y_0)$ by $f_*([\alpha]) = [f \circ \alpha]$. This is well-defined, since if $F : \alpha_0 \simeq \alpha_1$ rel$\{0,1\}$, then $f \circ F : f \circ \alpha_0 \simeq f \circ \alpha_1$ rel$\{0,1\}$. It

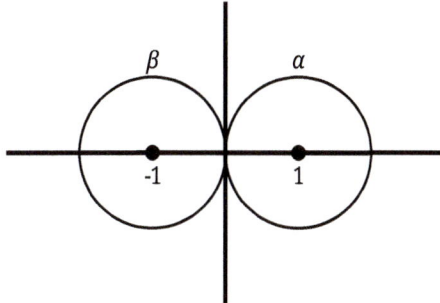

Figure 1.20: Two loops in the twice-punctured plane.

is easy to see that if α and β are composable paths in X, then $f \circ \alpha$ and $f \circ \beta$ are composable paths in Y and that $f \circ (\alpha \cdot \beta) = (f \circ \alpha) \cdot (f \circ \beta)$. It follows that f_* is a homomorphism. Moreover, $\mathrm{id}_* = \mathrm{id}_{\pi_1(X,x_0)}$ and $(g \circ f)_* = g_* \circ f_*$ for $f : X \to Y$ and $g : Y \to Z$. Thus, the fundamental group is a *functor* from $\underline{\mathrm{Top}}^*$ to $\underline{\mathrm{Grp}}$.

Theorem 1.5.4. $\pi_1(S^1, 1) \cong \mathbb{Z}$

Proof. View S^1 as sitting inside $\mathbb{C}$ as the unit circle. Given a loop $\gamma : (I, \{0, 1\}) \to (S^1, 1)$, write $\gamma(t) = \exp(2\pi i \tilde{\gamma}(t))$, where $\tilde{\gamma}(t)$ is the counterclockwise angle from 1 to $\gamma(t) \in S^1$. Thus we have the following commutative diagram

$$
\begin{array}{ccc}
 & & \mathbb{R} \\
 & \overset{\tilde{\gamma}}{\nearrow} & \downarrow {\scriptstyle \exp(2\pi i-)} \\
[0,1] & \overset{\gamma}{\longrightarrow} & S^1
\end{array}
$$

Thus, $\tilde{\gamma}$ is a lift of γ. Moreover, it is easy to see that the map $\exp : \mathbb{R} \to S^1$ is a locally trivially fiber bundle. Note that $\tilde{\gamma}(1) \in \mathbb{Z}$ since $\gamma(1) = 1$. Define the *winding number map*

$$\pi_1(S^1, 1) \longrightarrow \mathbb{Z}$$

$$[\gamma] \longmapsto \tilde{\gamma}(1)$$

We claim that this is a well-defined group isomorphism.

Lemma 1.5.5. *Suppose* $\gamma, \mu : ([0,1], \{0,1\}) \to (S^1, 1)$ *are loops. Then* $\tilde{\gamma}(1) = \tilde{\mu}(1)$ *if and only if* $\gamma \simeq \mu \, \mathrm{rel}\{0,1\}$.

Proof. Suppose $\tilde{\gamma}(1) = \tilde{\mu}(1) = n$. Define $\Phi(s,t) = s\tilde{\gamma}(t) + (1-s)\tilde{\mu}(t)$. Then $\Phi(s,0) = 0$ for all s and $\Phi(s,1) = n$ for all s so that $\Phi : \tilde{\gamma} \simeq \tilde{\mu}$. Define $F(s,t) = \exp(2\pi i \Phi(s,t))$. Then $F : \gamma \simeq \mu \, \mathrm{rel}\{0,1\}$. Conversely, if $F : \gamma \simeq \mu \, \mathrm{rel}\{0,1\}$, then we can lift the homotopy to $\tilde{F} : I \times I \to \mathbb{R}$, $\tilde{F}(s,t) = \tilde{\gamma}_t(s)$. Then $\tilde{F}(s,0) = \tilde{\gamma}_0(s)$, $\tilde{F}(s,1) = \tilde{\mu}(s)$, $\tilde{F}(0,t) = \tilde{\gamma}_t(0) = 0$, and $\tilde{F}(1,t) = \tilde{\gamma}_t(1) = \tilde{\gamma}(1)$ for all t. But this equals $\tilde{\mu}(1)$ since $\tilde{\gamma}_1 = \tilde{\mu}$. $\qquad \square$

Lemma 1.5.5 shows that the winding number map is well defined and injective. For surjectivity, note that if we set $\omega_n(t) = \exp(2\pi i n t)$, then $\tilde{\omega}_n(t) = nt$ and $\tilde{\omega}_n(1) = n$. Thus $[\omega_n] \mapsto n \in \mathbb{Z}$. Finally, this map is a homomorphism, as the diagram in Figure 1.21 shows. □

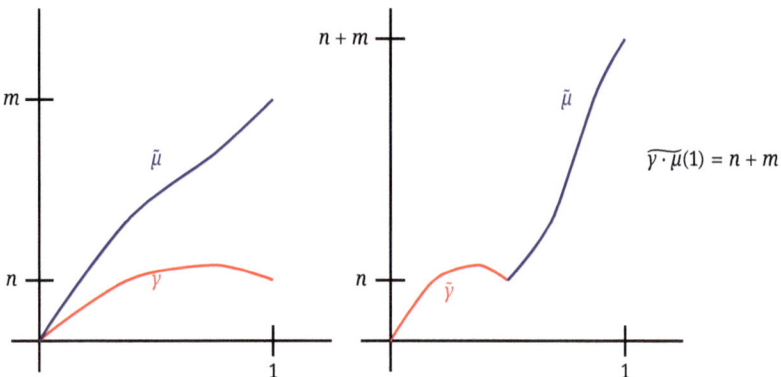

Figure 1.21: The winding number map is a homomorphism.

Suppose (X, x_0) and (Y, y_0) are pointed spaces. We have the following pair of diagrams.

$$
\begin{array}{ccc}
(X, x_0) & & \pi_1(X, x_0) \\
p \uparrow & & p_* \uparrow \quad\searrow^{i_X} \\
(X \times Y, (x_0, y_0)) & \Longrightarrow & \pi_1(X \times Y, (x_0, y_0)) \longrightarrow \pi_1(X, x_0) \times \pi_1(Y, y_0) \\
q \downarrow & & q_* \downarrow \quad\nearrow_{i_Y} \\
(Y, y_0) & & \pi_1(Y, y_0)
\end{array}
$$

We therefore obtain a homomorphism

$$
p_* \times q_* : \pi_1(X \times Y, (x_0, y_0)) \longrightarrow \pi_1(X, x_0) \times \pi_1(Y, y_0)
$$

$$
[\gamma] \longmapsto (p_*([\gamma]), q_*([\gamma]))
$$

Proposition 1.5.6. *The map $p_* \times q_*$ is an isomorphism.*

Proof. Note that products of spaces satisfy a universal mapping property: $\varphi : Z \to X \times Y$ is continuous if and only if $p \circ \varphi : Z \to X$ and $q \circ \varphi : Z \to Y$ are continuous. With this in mind, we can prove surjectivity. Suppose $([\gamma], [\mu]) \in \pi_1(X, x_0) \times \pi_1(Y, y_0)$. This is represented by a pair of loops $\gamma : ([0, 1], \{0, 1\}) \to (X, x_0)$ and $\mu : ([0, 1], \{0, 1\}) \to (Y, y_0)$.

By the universal mapping property, the map $\gamma \times \mu : ([0,1],\{0,1\}) \to (X \times Y, (x_0, y_0))$ is continuous and $(p_* \times q_*)([\gamma \times \mu]) = ([\gamma],[\mu])$. For injectivity, suppose that $(p_* \times q_*)([\alpha]) = (p_* \times q_*)([\beta])$, where α, β are loops in $X \times Y$ at (x_0, y_0). Let $\alpha_X = p \circ \alpha$, $\alpha_Y = q \circ \alpha$, $\beta_X = p \circ \beta$, and $\beta_Y = q \circ \beta$. Then

$$(p_* \times q_*)([\alpha]) = ([\alpha_X],[\alpha_Y])$$
$$(p_* \times q_*)([\beta]) = ([\beta_X],[\beta_Y])$$

By assumption, $\alpha_X \simeq \beta_X \text{ rel}\{0,1\}$ and $\alpha_Y \simeq \beta_Y \text{ rel}\{0,1\}$. Let $F_X : I \times I \to X$ and $F_Y : I \times I \to Y$ be these homotopies. Let $F : I \times I \to X \times Y$ be the corresponding continuous map $F_X \times F_Y$. Then $F : \alpha \simeq \beta \text{ rel}\{0,1\}$. □

Applications of the fundamental group

Theorem 1.5.7 (Fundamental Theorem of Algebra). *Every nonconstant polynomial with complex coefficients has a root.*

Proof. Suppose $p(z) = z^n + a_1 z^{n-1} + \cdots + a_n$ has no roots in $\mathbb{C}$. If $r \geq 0$, then the map

$$f_r(s) = \frac{p(re^{2\pi i s})/p(r)}{|p(re^{2\pi i s})/p(r)|}$$

defines a loop in S^1 based at 1. As r varies, f_r is a homotopy of loops in S^1 based at 1. Note that f_0 is the trivial loop and so $[f_r] \in \pi_1(S^1, 1) \cong \mathbb{Z}$ is the zero element for all $r \geq 0$. Now, fix $r \gg 0$, larger than $|a_1| + \cdots + |a_n| + 1$. Then for $|z| = r$ we have

$$|z^n| > (|a_1| + \cdots + |a_n|)|z^{n-1}|$$
$$> |a_1 z^{n-1}| + \cdots + |a_n|$$
$$\geq |a_1 z^{n-1} + \cdots + a_n|$$

It follows that $p_t(z) = z^n + t(a_1 z^{n-1} + \cdots + a_n)$ has no roots on the circle $|z| = r$ when $0 \leq t \leq 1$. Replacing p by p_t in the formula for f_r and letting t go from 1 to 0, we obtain a homotopy from f_r to $\omega_n(s) = e^{2\pi i n s}$. But $[\omega_n] = n \in \pi_1(S^1, 1)$, and since $[\omega_n] = [f_r] = 0$, we have $n = 0$. Thus, p is constant. □

Theorem 1.5.8 (Brouwer Fixed-Point Theorem for the disc). *Every continuous map $h : D^2 \to D^2$ has a fixed point.*

Proof. Suppose not. Define $r : D^2 \to S^1$ by setting $r(x)$ to be the point of intersection of the ray from $h(x)$ to x with S^1 (see Figure 1.22). Note that $r|_{S^1} = \text{id}_{S^1}$ and r is clearly

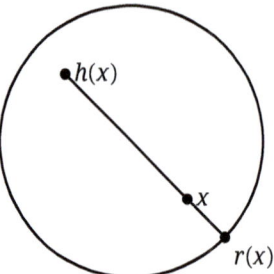

Figure 1.22: The retraction $r : D^2 \to S^1$.

continuous. But then we have a commutative diagram

$$\pi_1(S^1, 1) \xrightarrow{\ i_* \ } \pi_1(D^2, 1) = 0$$

$$\searrow \text{id} \qquad \swarrow r_*$$

$$\pi_1(S^1, 1)$$

which would imply that id $: \pi_1(S^1) \to \pi_1(S^1)$ is the zero map, a contradiction. $\qquad\square$

Theorem 1.5.9 (Borsuk–Ulam Theorem in dimension 2). *If $f : S^2 \to \mathbb{R}^2$ is continuous, then there is an $x \in S^2$ with $f(x) = f(-x)$.*

Proof. Suppose not and define $g : S^2 \to S^1$ by

$$g(x) = \frac{f(x) - f(-x)}{|f(x) - f(-x)|},$$

where we consider $\mathbb{R}^2$ as the complex plane. Define a loop η on S^2 by $\eta(s) = (\cos(2\pi s), \sin(2\pi s), 0)$ and let $h = g \circ \eta$. Since $g(-x) = -g(x)$ we have $h(s + 1/2) = -h(s)$ for all $s \in [0, 1/2]$. Lift h to a map $\tilde{h} : I \to \mathbb{R}$. Since $h(s + 1/2) = -h(s)$, we have $\tilde{h}(s + 1/2) = \tilde{h}(s) + q/2$ for some odd integer q. The integer q might depend on s, but it is easy to see that it does not, as it must depend continuously on s and so must be constant since $q \in \mathbb{Z}$. In particular, $\tilde{h}(1) = \tilde{h}(1/2) + q/2 = \tilde{h}(0) + q$. So h represents q times a generator of $\pi_1(S^1)$ and since q is odd, h is not nullhomotopic. But η is obviously nullhomotopic in S^2 and so $h = g \circ \eta$ is nullhomotopic in S^1, a contradiction. $\qquad\square$

Exercises

Exercise 1.5.1. Prove that if y is a path in X from x_0 to x_1, then the map

$$\varphi : \pi_1(X, x_0) \to \pi_1(X, x_1)$$

defined by

$$[a] \mapsto [\gamma^{-1} \cdot a\gamma]$$

is an isomorphism. Deduce that if X is path connected, then $\pi_1(X, x)$ is independent of $x \in X$.

Exercise 1.5.2. Prove that if $f : (X, x_0) \to (Y, y_0)$ is a homotopy equivalence, then $f_* : \pi_1(X, x_0) \to \pi_1(Y, y_0)$ is an isomorphism.

Exercise 1.5.3. Prove that if G is a simply connected topological group and H a discrete normal subgroup, then $\pi_1(G/H, 1) \cong H$.

Exercise 1.5.4. Prove that the fundamental group of the torus T is $\mathbb{Z} \times \mathbb{Z}$.

Exercise 1.5.5. Prove that a contractible space is simply connected.

Exercise 1.5.6. Suppose $X = U \cup V$, where U and V are simply connected open subsets such that $U \cap V$ is nonempty and path connected. Prove that X is simply connected. Deduce that the n-sphere S^n is simply connected for $n \geq 2$.

Exercise 1.5.7. Prove that every 3×3 matrix A with positive real entries has an eigenvector with positive eigenvalue. (Hint: Consider the triangle $T = \{(x, y, z) \mid x + y + z = 1\}$ and define a map f as the composition of the linear transformation associated to A with central projection onto T. Now apply Theorem 1.5.8.)

Exercise 1.5.8. Prove that if S^2 is realized as the union of three closed sets A_1, A_2, A_3, then at least one of these sets must contain a pair of antipodal points.

Exercise 1.5.9. Prove the Ham Sandwich Theorem: if A_1, A_2, and A_3 are compact sets in $\mathbb{R}^3$, then there is a plane P in $\mathbb{R}^3$ that divides each A_i into two pieces of equal measure. (Hint: the Borsuk–Ulam Theorem is useful here.)

1.6 The Seifert–Van Kampen Theorem

If a space X is written as a union of subspaces, one would like to know how the fundamental group of X is related to those of the subspaces. The Seifert–Van Kampen Theorem provides an answer, but we need to review an idea from group theory first.

Let $\{G_\alpha\}_{\alpha \in A}$ be a collection of groups. How can we build a group containing each G_α as a subgroup? One might try the direct product or some similar construction, but the "right" thing to do is to build the free product.

Definition 1.6.1. The *free product* $*_{\alpha \in A} G_\alpha$ is the set consisting of all finite words $g_1 g_2 \cdots g_m$, $m \geq 0$, where $g_\ell \in G_{\alpha_\ell}$ and $g_\ell \neq e \in G_{\alpha_\ell}$, and adjacent $g_\ell, g_{\ell+1}$ belong to different groups $G_{\alpha_\ell} \neq G_{\alpha_{\ell+1}}$. This is the set of *reduced words*. Define a product by juxtaposition: $(g_1 \cdots g_m)(h_1 \cdots h_n) = g_1 \cdots g_m h_1 \cdots h_n$ and then reduce if g_m and h_1 lie in the same group.

One can check that this operation is associative and the identity element is given by the identity from any one of the groups G_α. For inverses, note that

$$(g_1 \cdots g_m)^{-1} = g_m^{-1} \cdots g_1^{-1}.$$

Example 1.6.2. The group $\mathbb{Z} * \mathbb{Z}$ is an example of a free group (on two generators in this case). The *rank* is the number of generators.

Example 1.6.3. Consider $\mathbb{Z}_2 * \mathbb{Z}_2$. One might think this is a finite group, or at least all torsion, but it is not. Let a, b be the generators of the two factors; these satisfy $a^2 = e = b^2$. Consider the word ab. It has infinite order: $(ab)(ab) = abab \neq e$, etc. It is a countable group and it is possible to enumerate all the words:

$$e, a, b, ab, ba, aba, bab, abab, baba, \ldots$$

Note that $(ab)^{-1} = b^{-1}a^{-1} = ba$. Define $\varphi : \mathbb{Z}_2 * \mathbb{Z}_2 \to \mathbb{Z}_2$ by $\varphi(w) = \ell(w)$ mod 2, where $\ell(w)$ is the length of the reduced word w. This is clearly surjective and ker φ is the set of words of even length, which is the subgroup generated by ab. This is isomorphic to $\mathbb{Z}$, so that $\mathbb{Z}_2 * \mathbb{Z}_2$ is the semidirect product of $\mathbb{Z}$, generated by ab, and $\mathbb{Z}_2$, generated by a (say). This is the *infinite dihedral group*, which is the group of symmetries of the infinite tree with a vertex at each integer on the real line and edges joining adjacent vertices. The normal subgroup $\mathbb{Z}$ is generated by the translation $n \mapsto n + 1$ and the other generator is the reflection across 0 (this has order 2).

Example 1.6.4. The free product $\mathbb{Z}_2 * \mathbb{Z}_3$ is isomorphic to the group $\mathrm{PSL}_2(\mathbb{Z}) = \mathrm{SL}_2(\mathbb{Z})/\{\pm I\}$, where $\mathrm{SL}_2(\mathbb{Z})$ is the set of 2×2 integral matrices of determinant 1.

Proposition 1.6.5. *Given a collection $\{G_\alpha\}_{\alpha \in A}$ of groups and a homomorphism $\varphi_\alpha : G_\alpha \to H$, there is a unique $\overline{\varphi} : *_\alpha G_\alpha \to H$ such that the diagram*

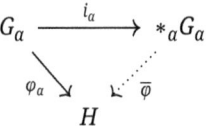

commutes, where $i_\alpha : G_\alpha \to *_\alpha G_\alpha$ is the obvious inclusion.

Proof. Define $\overline{\varphi}(g_1 \cdots g_m) = \varphi_{\alpha_1}(g_1) \cdots \varphi_{\alpha_m}(g_m)$. This must be the map and it is a homomorphism for free. □

Let us generalize this a bit. Let G_1 and G_2 be groups and suppose $i_j : A \to G_j$ is a homomorphism. Consider the following diagram (called a *pushout diagram*).

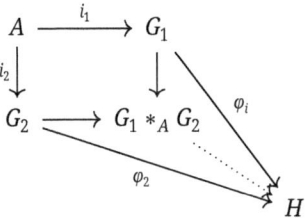

Definition 1.6.6. The *amalgamated free product* $G_1 *_A G_2$ is the group completing the diagram, satisfying the following universal mapping property. Given a group H and homomorphisms $\varphi_j : G_j \to H$ with $\varphi_1 \circ i_1 = \varphi_2 \circ i_2$, there is a unique homomorphism $\overline{\varphi} : G_1 *_A G_2 \to H$ making the diagram commute.

The amalgamated free product is unique if it exists. To construct it, consider the free product $G_1 * G_2$ and let N be the normal subgroup generated by all $i_1(a)i_2(a)^{-1}, a \in A$. We leave it as an exercise to show that the group

$$G_1 * G_2/N$$

is the amalgamated free product.

Example 1.6.7. Consider the groups $\mathbb{Z}_4$ and $\mathbb{Z}_6$ with the obvious embeddings $\mathbb{Z}_2 \to \mathbb{Z}_4$ and $\mathbb{Z}_2 \to \mathbb{Z}_6$. Then the group $\mathbb{Z}_4 *_{\mathbb{Z}_2} \mathbb{Z}_6$ is isomorphic to $SL_2(\mathbb{Z})$. This is a nontrivial fact that we leave as a project for the reader.

There is a more general amalgamated free product construction involving more groups, but this version is generally sufficient for our purposes here. We now consider the Seifert–Van Kampen Theorem.

Theorem 1.6.8 (Seifert–Van Kampen Theorem). *Suppose X is the union of path connected open sets A_α, each containing the base point x_0. If each $A_\alpha \cap A_\beta$ is path connected, then the map*

$$\Phi : *_\alpha \pi_1(A_\alpha, x_0) \to \pi_1(X, x_0)$$

is surjective. If, in addition, each $A_\alpha \cap A_\beta \cap A_\gamma$ is path connected, then $\ker \Phi$ is generated by all elements of the form $i_{\alpha\beta}(w)i_{\beta\alpha}(w)^{-1}$ for $w \in \pi_1(A_\alpha \cap A_\beta, x_0)$, where $i_{\alpha\beta}$ is the inclusion of $A_\alpha \cap A_\beta$ into A_α, and similarly for $i_{\beta\alpha}$.

Sketch. The surjectivity of Φ is fairly easy via the following fact: Under the hypotheses on the A_α, every loop in X is homotopic to a product of loops, each contained in a single A_α. To see this, suppose $\gamma : I \to X$ is a loop at x_0. Since γ is continuous, each $s \in I$ has a neighborhood V_s in I mapping into some A_α, and in fact we may assume that $\overline{V}_s$ maps into A_α. Since I is compact, a finite collection of these V_s cover I; say $0 = s_0 < s_1 < \cdots < s_m = 1$ is the corresponding collection of s values. Denote the A_α containing $\gamma([s_{i-1}, s_i])$

by A_i and let γ_i be the path $\gamma|_{[s_{i-1}, s_i]}$. Then $\gamma = \gamma_1 \cdot \gamma_2 \cdots \gamma_m$, with γ_i in A_i. Since $A_i \cap A_{i+1}$ is path connected, choose g_i in $A_i \cap A_{i+1}$ from x_0 to $\gamma(s_i) \in A_i \cap A_{i+1}$. Then the loop

$$(\gamma_1 \cdot g_1^{-1}) \cdot (g_1 \cdot \gamma_2 \cdot g_2^{-1}) \cdots (g_{m-1} \cdot \gamma_m)$$

is homotopic to γ and each piece lies in some A_i. This proves that Φ is surjective.

The computation of the kernel of Φ is a tedious combinatorial argument. It is not especially enlightening and so we omit it. A complete proof may be found in [2], for example. □

The most important case of Theorem 1.6.8 is when $X = A_1 \cup A_2$. We then have an isomorphism

$$\pi_1(X) \cong \pi_1(A_1) *_{\pi_1(A_1 \cap A_2)} \pi_1(A_2).$$

Example 1.6.9. Consider the circle S^1, decomposed as $U_1 \cup U_2$ as shown in Figure 1.23(a). Since $U_1 \cap U_2$ is not connected we cannot apply Van Kampen's Theorem to compute $\pi_1(S^1)$. This is a good thing, since each $\pi_1(U_i)$ is the trivial group.

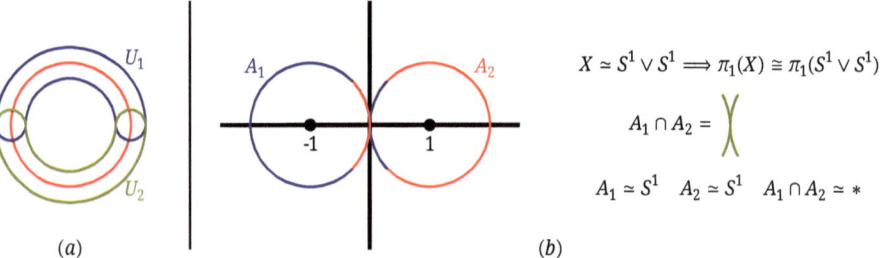

Figure 1.23: (a) The circle, and (b) $\mathbb{C} - \{\pm 1\}$.

Example 1.6.10. Let $X = \mathbb{C} - \{\pm 1\}$, as shown in Figure 1.23(b). This space is homotopy equivalent to $S^1 \vee S^1$, which we decompose as $A_1 \cup A_2$, with each $A_j \simeq S^1$ and $A_1 \cap A_2$ contractible. Then

$$\pi_1(X, 0) \cong \pi_1(A_1, 0) * \pi_1(A_2, 0) \cong \mathbb{Z} * \mathbb{Z}.$$

Example 1.6.11. Suppose A is a circle in $\mathbb{R}^3$ and let $X = \mathbb{R}^3 - A$. What is $\pi_1(X)$? Imagine enclosing A inside a large sphere S and let B be another circle inside the sphere, tangent to S at a single point and linked with A. Then X is homotopy equivalent to $S \vee B = S^2 \vee S^1$. It follows that $\pi_1(X) \cong \pi_1(S^1) * \pi_1(S^2) \cong \mathbb{Z}$. (Exercise 1.5.6 shows that $\pi_1(S^2)$ is trivial.)

Attaching cells

Suppose X is path connected with basepoint x_0 and let $\varphi_\alpha : \partial e_\alpha^2 \to X$ be attaching maps for some 2-cells e_α^2. Let $Y = X \cup_\alpha e_\alpha^2$ be the space obtained by attaching all the e_α^2. Note that φ_α determines a loop in X for each α, but the basepoints may not agree. Choose a path γ_α from x_0 to the basepoint of $\varphi_\alpha(\partial e_\alpha^2)$. Then each $\gamma_\alpha \varphi_\alpha \gamma_\alpha^{-1}$ is a loop at x_0. It might not be nullhomotopic, but it will be after e_α^2 is attached. Let N be the normal subgroup of $\pi_1(X, x_0)$ generated by the homotopy classes $[\gamma_\alpha \varphi_\alpha \gamma_\alpha^{-1}]$.

Proposition 1.6.12. (1) *The map $\pi_1(X, x_0) \to \pi_1(Y, x_0)$ is surjective with kernel N.*
(2) *If Y is obtained by attaching n-cells to X, $n > 2$, then $\pi_1(X, x_0) \to \pi(Y, x_0)$ is an isomorphism.*
(3) *For a path connected cell complex X, the inclusion of the 2-skeleton $X^{(2)} \to X$ induces an isomorphism $\pi_1(X^{(2)}, x_0) \to \pi_1(X, x_0)$.*

Proof. For the first statement, let us thicken Y a bit. Attach rectangular strips $S_\alpha = I \times I$ to Y by gluing $I \times \{0\}$ along γ_α, the edge $\{1\} \times I$ along an arc in e_α^2, and all the $\{0\} \times I$ together (see Figure 1.24). Call this space Z. Then this space clearly deformation retracts to Y. In each e_α^2, choose y_α not on the arc and let $A = Z - \bigcup_\alpha \{y_\alpha\}$. Let $B = Z - X$. Then $A \simeq X$ and B is contractible. Since $\pi_1(B) = 0$, the Seifert–Van Kampen Theorem implies that

$$\pi_1(Z) \cong \pi_1(A) / \langle \mathrm{im}(\pi_1(A \cap B) \to \pi_1(B)) \rangle.$$

What is this subgroup? Choose $z_0 \in A \cap B$ near x_0 on the segment where all the S_α meet and let δ_α be a loop in $A \cap B$ based at z_0 representing $[\gamma_\alpha \varphi_\alpha \gamma_\alpha^{-1}] \in \pi_1(A, x_0)$ (see Figure 1.24). We claim that $\pi_1(A \cap B, z_0)$ is generated by the collection of all the δ_α. To see this, cover $A \cap B$ by the collection of

$$A_\alpha = A \cap B - \bigcup_{\beta \neq \alpha} e_\beta^2.$$

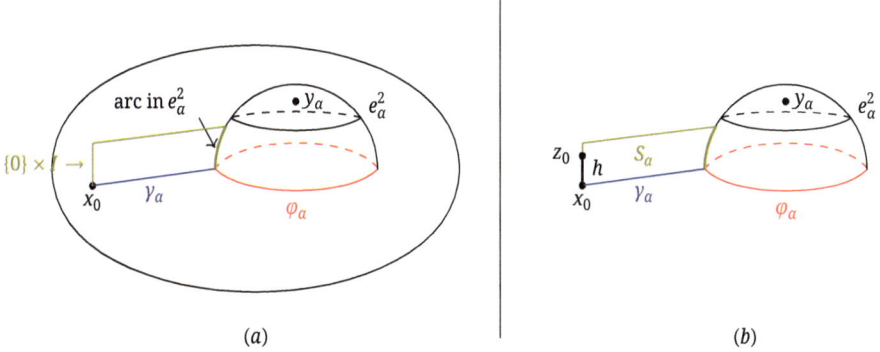

(a)   *(b)*

Figure 1.24: (a) A portion of the thickened space Z; and (b) the space A_α.

Then $A_a \simeq S^1$ in $e_a^2 - \{y_a\}$ and $\pi_1(A, z_0) \cong \mathbb{Z}$, generated by δ_a. This completes the proof of (1).

For (2), the same argument works, except we replace e_a^2 with e_a^n, $n > 2$. Then the set A_a retracts to S^{n-1} so that $\pi_1(A_a) = 0$ for all a and $\pi_1(A \cap B) = 0$.

Statement (3) follows from (2) by induction if X is finite dimensional. In general, use the fact that any loop $\gamma : I \to X$ must lie in some $X^{(n)}$ by compactness and then note that (2) implies that $\gamma \simeq \mu : I \to X^{(2)}$ so that $\pi_1(X^{(2)})$ surjects onto $\pi_1(X)$. For injectivity, if γ is nullhomotopic in X, then the image of the homotopy lies in some $X^{(n)}$ (again, by compactness of $I \times I$) with $n > 2$. Since (2) implies that $\pi_1(X^{(2)}) \to \pi_1(X)$ is injective, we are done. □

Example 1.6.13. Let M_g be the orientable surface of genus g. The 1-skeleton of M_g is homotopic to a wedge of $2g$ circles. Denote these circles by $a_1, b_1, \ldots, a_g, b_g$; they are the images of the boundary of the $4g$-gon surjecting to M_g. The 2-cell is attached by the loop

$$\gamma = [a_1, b_1][a_2, b_2] \cdots [a_g, b_g],$$

where $[a_i, b_i] = a_i b_i a_i^{-1} b_i^{-1}$. The loop γ is a map $S^1 \to \bigvee_{2g} S^1$ and the kernel of the map $\pi_1(S^1) \to \pi_1(M_g)$ is generated by $[\gamma]$. It follows that

$$\pi_1(M_g) \cong \langle a_1, b_1, \ldots, a_g, b_g \mid [a_1, b_1][a_2, b_2] \cdots [a_g, b_g] = 1 \rangle.$$

Corollary 1.6.14. *If $g \neq h$, then M_g is not homotopy equivalent to M_h.*

Proof. Recall the *abelianization* of a group G: it is the largest abelian quotient G^{ab} of G. It is constructed as the quotient $G/[G, G]$, where $[G, G]$ is generated by all $[g, h], g, h \in G$. It has the obvious universal mapping property: if $\varphi : G \to H$ is a homomorphism with H abelian, there is a unique $\overline{\varphi} : G^{ab} \to H$ making the following diagram commute:

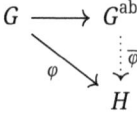

Note that

$$\pi_1(M_g)^{ab} \cong \mathbb{Z}\{a_1, b_1, \ldots, a_g, b_g\}/\langle(a_1 + b_1 - a_1 - b_1) + \cdots + (a_g + b_g - a_g + b_g)\rangle$$
$$\cong \mathbb{Z}^{2g}$$

So if $M_g \simeq M_h$, we would have $\pi_1(M_g) \cong \pi_1(M_h)$, and hence their abelianizations would be isomorphic: $\mathbb{Z}^{2g} \cong \mathbb{Z}^{2h}$. But this implies $2g = 2h$; that is, $g = h$. □

Exercises

Exercise 1.6.1. Prove that the group $G_1 * G_2/N$ is the amalgamated free product.

Exercise 1.6.2. Compute $\pi_1(\mathbb{R}P^2)$. Use this to compute $\pi_1(\mathbb{R}P^n)$ for all $n \geq 2$.

Exercise 1.6.3. Compute $\pi_1(K)$, where K is the Klein bottle.

Exercise 1.6.4. Prove that if G is a group, then there is a 2-dimensional cell complex X_G with $\pi_1(X_G) \cong G$.

Exercise 1.6.5. Prove that there are infinitely many distinct homotopy types of spaces with one 0-cell, one 1-cell, and one 2-cell.

Exercise 1.6.6. Prove that if Y is obtained from a path-connected space X by attaching a cell e^n with $n \geq 2$, then the inclusion $X \to Y$ induces a surjection on π_1. Deduce that $\pi_1(S^1 \vee S^2) = \mathbb{Z}$. Also, prove that the inclusion of the 1-skeleton $X^{(1)} \to X$ induces a surjection on π_1.

Exercise 1.6.7. Let X be the union of n lines through the origin in $\mathbb{R}^3$. Compute $\pi_1(\mathbb{R}^3 - X)$.

Exercise 1.6.8. Let X be the quotient of S^2 obtained by identifying two antipodal points. Compute $\pi_1(X)$.

Exercise 1.6.9. Let X be the union of the set of circles C_n in $\mathbb{R}^2$ of radius n and center $(n, 0)$. Show that $\pi_1(X)$ is the free group $*_n \pi_1(C_n)$.

Exercise 1.6.10. Let X be the union of the set of circles D_n in $\mathbb{R}^2$ of radius $1/n$ and center $(1/n, 0)$. Show that X is not homotopy equivalent to a countable wedge of circles by showing that the fundamental group of X surjects onto an infinite product of copies of $\mathbb{Z}$, which is an uncountable group. (Hint: Let $r_n : X \to D_n$ be the retraction collapsing all D_i except D_n to a point. Show that r_n induces a surjection on π_1. The product of the induced homomorphisms is useful here.)

Project: covering spaces and the fundamental group

Covering spaces are a special type of fibration in which the fibers are discrete. The goal of this project is to establish some of the basic properties of these spaces and to relate them the fundamental group. We first give a few examples.

Example 1.6.15. Let n be a nonzero integer and define a map $p_n : S^1 \to S^1$ by $p_n(z) = z^n$, where, as usual, we think of S^1 as the group of unit modulus complex numbers. This is a fibration, and for any fixed $w \in S^1$, the fiber $p_n^{-1}(w)$ consists of the n distinct nth roots of w. That is, if $w = \exp(2\pi i\theta)$, then

$$p_n^{-1}(w) = \{\exp(2\pi i(\theta + k)/n) \mid 0 \leq k \leq n - 1\}.$$

These fibers are discrete, and if we take a sufficiently small neighborhood U of w, the set $p_n^{-1}(U)$ consists of n disjoint sets homeomorphic to U around the n preimages of w.

Example 1.6.16. Consider the map $p : \mathbb{R} \to S^1$ given by $p(s) = \exp(2\pi i s)$. The fiber over $1 \in S^1$ is the set $\mathbb{Z}$ of integers, and for an arbitrary $w \in S^1$, $p^{-1}(w)$ is given as $s + \mathbb{Z}$, where s is any particular preimage of w under p. These fibers are discrete, and again, if we take a sufficiently small neighborhood U of w, the preimage of U consists of an infinite number of disjoint sets homeomorphic to U around the various points in $s + \mathbb{Z}$.

Example 1.6.17. Consider the infinite tree T shown in Figure 1.25 (the figure shows only a finite iteration, but imagine that it continues indefinitely). Let X be the space $S^1 \vee S^1$. Define a map $\pi : T \to X$ by identifying all the vertices in T to the wedge point in X. Then π is a fibration, and the fiber of any point on X (aside from the wedge point) is an infinite discrete subset of T with one point on each edge in T. The fiber over the wedge point is the set of vertices of T. Note that a small neighborhood of the wedge point is just a pair

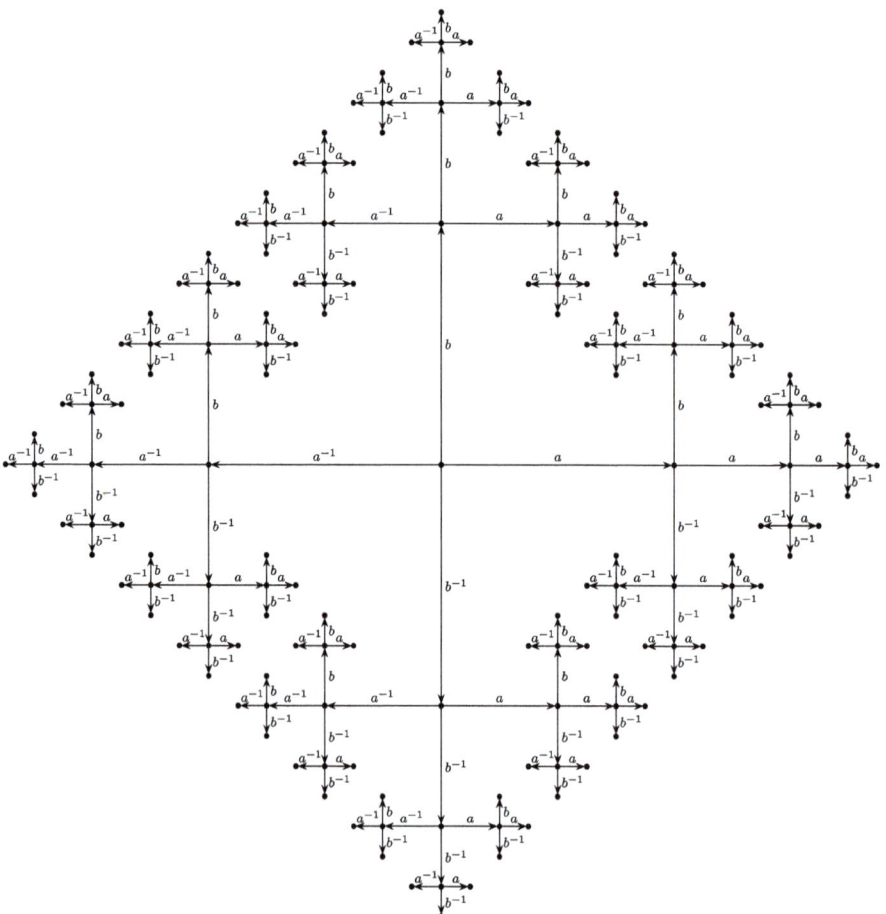

Figure 1.25: The tree T as a covering space of $S^1 \vee S^1$.

of intersecting line segments, and the preimage of such a neighborhood is the disjoint union of a collection of small intersecting line segments at each vertex of T. Away from this point, the preimage of a small neighborhood (which is just a small line segment) is a collection of line segments around the various points.

Now, since a covering space $p : E \to X$ is a fibration, we have homotopy lifting. In particular, as mentioned in Example 1.3.17, if we have a loop γ at $x_0 \in X$, and if $e_0 \in E$ lies in the fiber $p^{-1}(x_0)$, then there is a lift $\tilde{\gamma}$ with $\tilde{\gamma}(0) = e_0$. Since γ is a *loop*, we must have $\tilde{\gamma}(1) \in p^{-1}(x_0)$ as well.

(1) Prove that lifts are unique. That is, if $f : (Y, y_0) \to (X, x_0)$ is continuous, and if Y is connected, then any two lifts $f', f'' : (Y, y_0) \to (E, e_0)$ are equal. (Hint: consider the set $A = \{y \in Y \mid f'(y) = f''(y)\}$ and show that it is both open and closed.)

(2) Show that if σ and τ are paths in X with initial point x_0 with $\sigma \simeq \tau$ rel$\{0, 1\}$, then the lifts σ' and τ' are homotopic rel $\{0, 1\}$. In particular, they have the same endpoint.

The fundamental group is intimately connected with the theory of covering spaces over X. Suppose $p : (E, e_0) \to (X, x_0)$ is a covering space. If σ is a loop at x_0, then its lift σ' with $\sigma'(0) = e_0$ may not be a loop in E, but its endpoint $\sigma'(1)$ lies in $p^{-1}(x_0)$. That point depends only on the homotopy class of σ. We therefore define an action of $\pi_1(X, x_0)$ on the fiber $p^{-1}(x_0)$ by

$$ e \cdot [\sigma] = \sigma'(1); \quad \sigma'(0) = e. $$

That is, given a point $e \in p^{-1}(x_0)$, we take the lift of σ with initial point e and find its other end.

(1) Prove that $p_* : \pi_1(E, e_0) \to \pi_1(X, x_0)$ is injective.

(2) Prove that the stabilizer of the point e_0 under the action of $\pi_1(X, x_0)$ on the fiber $p^{-1}(x_0)$ is the subgroup $p_*(\pi_1(E, e_0)) \subset \pi_1(X, x_0)$. Deduce that the subgroups $p_*(\pi_1(E, e))$ as e varies in $p^{-1}(x_0)$ are conjugate in $\pi_1(X, x_0)$.

(3) Prove that if E is path connected, the map $[\sigma] \mapsto e_0 \cdot [\sigma]$ induces a bijection of the set of all cosets $p_*(\pi_1(E, e_0))[\sigma]$ onto the fiber $p^{-1}(x_0)$. In particular, if $p^{-1}(x_0)$ is finite, the number of points in the fiber is equal to the index $[\pi_1(X, x_0) : p_*(\pi_1(E, e_0))]$.

(4) Prove that if E is path connected, then all fibers have the same cardinality.

Given a covering space $p : (E, e_0) \to (X, x_0)$, the group G of *deck transformations* is the group of all fiber-preserving homeomorphisms of E:

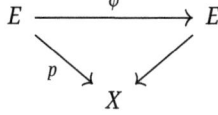

(1) Prove that if G is the group of deck transformations of $p : (E, e_0) \to (X, x_0)$, with E simply connected and locally path connected, then G is isomorphic to $\pi_1(X, x_0)$.

(Hint: Since E is simply connected, all paths from e_0 to $\varphi(e_0)$ are homotopic rel $\{0, 1\}$. So given such a path σ', the element $[p \circ \sigma'] \in \pi_1(X, x_0)$ depends only on e_0 and $\varphi(e_0)$. This defines a homomorphism $G \to \pi_1(X, x_0)$. Show that this is an isomorphism. It is injective because a deck transformation is uniquely determined by its action on one point (why?). The local path connectivity of E gives surjectivity.)

(2) More generally, if E is connected and locally path connected, show that if N is the normalizer of $p_*(\pi_1(E, e_0))$ in $\pi_1(X, x_0)$ then G is isomorphic to $N/p_*(\pi_1(E, e_0))$. If $p_*(\pi_1(E, e_0))$ is normal in $\pi_1(X, x_0)$, then the covering space is called *normal*. Prove that the covering space is normal if and only if G acts transitively on the fiber $p^{-1}(x_0)$.

(3) Suppose E is connected and locally path connected and that G is a group of homeomorphisms of E that acts properly discontinuously (that is, for any $e \in E$ there is a neighborhood V of e such that $V \cap gV = \emptyset$ for any $g \neq 1$ in G). Let $X = E/G$ be the space of orbits and let $p : E \to X$ be the quotient map $e \mapsto Ge$. Show that $p : E \to X$ is a covering space, G is the group of deck transformations, and $p_*(\pi_1(E, e_0))$ is a normal subgroup of $\pi_1(X, x_0)$ for all $e_0 \in E$. Deduce that if we know a simply connected covering space E of X and its group of deck transformations, then we know $\pi_1(X) \cong G$ and that X is homeomorphic to E/G.

(4) Prove that for $n \geq 2$, $\pi_1(\mathbb{R}P^n) \cong \mathbb{Z}/2$.

A simply connected covering space $p : E \to X$ is called (the) *universal cover* of X. It need not exist in general, but we can say a few things.

(1) Prove the following *lifting criterion*: Let $p : (E, e_0) \to (X, x_0)$ be a covering space and consider the following diagram

$$
\begin{array}{ccc}
 & & (E, e_0) \\
 & {\scriptstyle f'} \nearrow & \downarrow {\scriptstyle p} \\
(Y, y_0) & \xrightarrow{\ f\ } & (X, x_0)
\end{array}
$$

where f is any map. Then there exists a lift f' of f if and only if

$$f_*(\pi_1(Y, y_0)) \subseteq p_*(\pi_1(E, e_0)).$$

Deduce that if Y is simply connected the lift f' always exists.

(2) Prove that if, in the situation in the previous exercise, the map $f : Y \to X$ is also a covering space and if f' exists, then $f' : Y \to E$ is also a covering space.

(3) Prove that if E and E' are both simply connected covering spaces of X, then E and E' are homeomorphic. This justifies calling a simply connected covering space *the* universal cover.

(4) Fact: If X is locally simply connected (i. e., every point has a simply connected neighborhood), connected, and locally path connected, then X has a universal cover. Assuming this, prove that for every subgroup H of $\pi_1(X, x_0)$ there exists a

covering space $p : (E, e_0) \to (X, x_0)$, unique up to homeomorphism, such that $H = p_*(\pi_1(E, e_0))$.

This last result is a sort of *Galois correspondence* analogous to the theory of field extensions.

Project: SL$_2$($\mathbb{Z}$) as an amalgamated free product

We mentioned above that the group of 2×2 integer matrices of determinant 1, SL$_2$($\mathbb{Z}$), is an amalgamated free product $\mathbb{Z}_4 *_{\mathbb{Z}_2} \mathbb{Z}_6$. This project aims to prove this statement in two ways: algebraically and geometrically.

Let us start with the algebra. Consider the following two elements of SL$_2$($\mathbb{Z}$):

$$ S = \begin{pmatrix} 0 & -1 \\ 1 & 0 \end{pmatrix}, \quad T = \begin{pmatrix} 1 & 1 \\ 0 & 1 \end{pmatrix}. $$

Denote the identity matrix by I. Let G be the subgroup generated by S and T.

(1) Show that $S^4 = I$ (and $S^2 = -I$), while T has infinite order.

(2) Show that ST has order 6 (and $(ST)^3 = -I$).

(3) Show that $G = $ SL$_2$($\mathbb{Z}$) as follows. First determine the effect of left multiplication by S or T on an arbitrary element $X = \begin{pmatrix} a & b \\ c & d \end{pmatrix}$ of SL$_2$($\mathbb{Z}$). If $c = 0$, then $X = \begin{pmatrix} \pm 1 & m \\ 0 & \pm 1 \end{pmatrix}$. This matrix is either T^m or $-T^{-m}$ and since $-I \in G$, such an element lies in G as well. If $c \neq 0$, then if $|a| \geq |c|$, divide a by c: $a = cq + r$ with $0 \leq r \leq |c|$. Show that multiplying X by T^{-q} yields a matrix with a smaller lower left entry (in absolute value). Iterate this process. If $|a| < |c|$, multiply X by S to rectify this and then proceed as before.

Note that SL$_2$($\mathbb{Z}$) is also generated by S and ST, elements of orders 4 and 6 respectively, and that $S^2 = -I = (ST)^3$. Denote by PSL$_2$($\mathbb{Z}$) the quotient SL$_2$($\mathbb{Z}$)/{$\pm I$}. Consider the finite cyclic groups $\mathbb{Z}_4 = \langle a | a^4 = e \rangle$ and $\mathbb{Z}_6 = \langle b | b^6 = e \rangle$ and consider the obvious inclusions $\mathbb{Z}_2 \to \mathbb{Z}_4$ and $\mathbb{Z}_2 \to \mathbb{Z}_6$. Define $f : \mathbb{Z}_4 *_{\mathbb{Z}_2} \mathbb{Z}_6 \to $ SL$_2$($\mathbb{Z}$) by $f(a) = S$ and $f(b) = ST$. This is a well-defined homomorphism. Since $f(a^2) = f(b^3) = -I$, we obtain an induced homomorphism $\bar{f} : \mathbb{Z}_2 * \mathbb{Z}_3 \to $ PSL$_2$($\mathbb{Z}$).

(1) Prove that f is injective (resp. surjective) if and only if $\bar{f}$ is injective (resp. surjective).

(2) Show that f is surjective.

(3) Show that $\bar{f}$ is injective. (Hint: Denote the generator of $\mathbb{Z}_2$ by s and the generator of $\mathbb{Z}_3$ by t. Any word in $\mathbb{Z}_2 * \mathbb{Z}_3$ has a reduced expression of the following form: $w = s^{\pm 1} t s^{\pm 1} t \cdots t s^{\pm 1}$, wt, tw, twt, or t. Note that t is not in the kernel of $\bar{f}$. Since $twt = twt^{-1}$ is a conjugate of w and tw is a conjugate of wt, it suffices to check that $\bar{f}(w) \neq I$ and $\bar{f}(wt) \neq I$. Compute the image of st and $s^{-1}t$ under $\bar{f}$ and conclude that no product of these can give the identity in PSL$_2$($\mathbb{Z}$).)

Now for the geometry. This is more fun, and we will actually do something more general. Recall that *graph* Γ consists of a set V of vertices and E of edges, along with two maps

$$E \to V \times V, \quad e \mapsto (o(e), t(e)); \quad E \to E, \quad e \mapsto \bar{e}$$

such that for any $e \in E$, $\bar{\bar{e}} = e$, $\bar{e} \neq e$ and $o(e) = t(\bar{e})$. In other words, we think of each edge e as having an *origin* $o(e)$ and a *terminus* $t(e)$, and the map $e \mapsto \bar{e}$ reverses these. This is often referred to as a *directed graph*. There is an obvious notion of morphism for these objects. An *orientation* of the graph Γ is a subset E_+ of E such that E is the disjoint union of E_+ and $\bar{E}_+$.

Of course, we can draw graphs as one normally does, with each vertex corresponding to a node and each edge in E corresponding to an oriented edge. A *circuit* of length n in Γ is a subgraph isomorphic to the standard circuit C_n, which is the graph having n vertices, $0, 1, \ldots, n-1$ with an oriented edge $(i, i+1)$ for all $0 \le i \le n-1$ (values taken modulo n). Note that a circuit of length 1 is a loop with a single vertex. A *path* of length n in Γ is a subgraph isomorphic to the standard path P_n, which is the graph having n vertices $0, 1, \ldots, n-1$ with an oriented edge $(i, i+1)$ for all $0 \le i \le n-1$.

Now, given a group G and a subset S of G, define an oriented graph $\Gamma(G, S)$ by taking G to be the set of vertices and $E_+ = G \times S$ with $o(g, s) = g$, $t(g, s) = gs$.

(1) Show that left multiplication by elements of g defines an action of G on Γ that preserves orientation.

(2) Show that G acts *freely* on $\Gamma(G, S)$.

(3) Show that $\Gamma(G, S)$ is connected if and only if S generates G.

A *tree* is a nonempty connected graph with no circuits. A *geodesic* in a graph Γ is a path without backtracking.

(1) If p and q are vertices in a tree T, show that there is exactly one geodesic from p to q, and that this geodesic is an injective path (i. e., it does not pass through the same vertex twice).

(2) Define the *distance* from p to q, $\ell(p, q)$, to be the length of the geodesic from p to q in a tree T. Fix a vertex p and for each $n \ge 0$, let X_n be the set of vertices q of T with $\ell(p, q) = n$. Show that if $q \in X_n, n \ge 1$, there is a unique vertex q' at distance $< n$ from p to which q is adjacent. This defines a map $f_n : X_n \to X_{n-1}, q \mapsto q'$. We therefore have an inverse system

$$\cdots \to X_n \overset{f_n}{\to} X_{n-1} \to \cdots \to X_1 \to X_0 = \{p\}.$$

Show that knowledge of this system yields a reconstruction of the tree T.

(3) Consider the geometric realization of a tree T, $|T|$. It is the simplicial complex with one vertex for each vertex of T and one 1-cell for each oriented edge. Prove that $|T|$ is contractible.

Now, given a nonempty graph Γ, the set of subgraphs that are trees is a directed set, ordered by inclusion. By Zorn's Lemma it has a maximal element T, which we call the *maximal tree* of Γ.

(1) Show that the maximal tree of Γ contains all the vertices of Γ.
(2) If S is any subtree of Γ, prove that the quotient space $|\Gamma|/|S|$ is homotopy equivalent to $|\Gamma|$.
(3) Prove that $|\Gamma|$ has the homotopy type of a wedge of circles.
(4) Prove that Γ is a tree if and only if $|\Gamma|$ is contractible.

We say that a group G acts (on the left) on a graph X *without inversion* if $ge \neq \bar{e}$ for all $g \in G$ and edges e. Denote the quotient graph by $G\backslash X$. Prove that if X is connected and G acts on X without inversion, then every subtree T' of $G\backslash X$ lifts to a subtree of X. (Hint: consider the set of subtrees of X that project injectively into T'. Let T_0 be a maximal element of this set, let T_0' be its image in T' and suppose that $T_0' \neq T'$. Then there is an edge in $T' \setminus T_0'$. Lift this to X and deduce a contradiction.)

Given an action of G on X without inversion, a *tree of representatives* of X mod G is a subtree T of X that is a lift of a maximal tree in $G\backslash X$.

Proposition 1.6.18. *If $X = \Gamma(G, S)$ is the graph defined by a group G and subset S, then X is a tree if and only if G is a free group with basis S.*

Proof. Suppose G is a free group on the set S. Every element of G has the form

$$g = s_1^{\epsilon_1} s_2^{\epsilon_2} \cdots s_n^{\epsilon_n}$$

with each $\epsilon_i = \pm 1$ and $\epsilon_i = \epsilon_{i+1}$ if $s_i = s_{i+1}$. The integer n is called the *length* of the word g and is denoted by $\ell(g)$. Let G_n be the set of elements of length n. If $g \in G_n$, then g is adjacent to a unique element of G_{n-1}, namely $s_1^{\epsilon_1} \cdots s_{n-1}^{\epsilon_{n-1}}$. This gives a map $G_n \to G_{n-1}$ and we see that X is the graph defined by the inverse system

$$\cdots \to G_n \to G_{n-1} \to \cdots \to G_1 \to G_0 = \{e\}$$

so that X is a tree.

Conversely, if X is a tree, then since X is connected, S generates G. Moreover, $S \cap S^{-1} = \emptyset$ (exercise). If S is not a free family, then there is some nontrivial element g of the free group $F(S)$ with basis S whose image in G equals e. Choose such a g of minimal length n and let $g = s_1^{\epsilon_1} \cdots s_n^{\epsilon_n}$ be its reduced decomposition in $F(S)$. Since $S \cap S^{-1} = \emptyset$, $n \geq 3$. Let p_i, $0 \leq i \leq n$ be the image of $s_1^{\epsilon_1} \cdots s_i^{\epsilon_i}$ in G. The minimality of g implies that $p_0, \ldots, p_{n-1}$ are all distinct and p_i is adjacent to p_{i+1} in X. Moreover $p_n = e = p_0$. Since $n \geq 3$, the edges $\{p_i, p_{i+1}\}$ ($0 \leq i \leq n-1$) and $\{p_n, p_0\}$ are all distinct. Thus, $p_0, \ldots, p_{n-1}$ form a cycle of length n in X, contrary to the assumption that X is a tree.   □

For example, if $S = \{a, b\}$, with $G = F(S)$, then the tree X is the tree in Figure 1.25.

Recall that a group G acts freely on a graph X if it acts without inversion and no element $g \neq e$ of G leaves a vertex of X fixed. The simplest example is the group G acting on $\Gamma(G, S)$ by left multiplication. Proposition 1.6.18 shows that if G is free, then there is a tree on which G acts freely. The converse is also true.

Theorem 1.6.19. *A group which acts freely on a tree is a free group.*

Proof. Choose a tree T of representatives of X mod G and an orientation E_+ preserved by G. Let S be the set of nonidentity elements in G for which there is an edge $e \in E_+$ with origin in T and terminus in gT. We will show that S is a basis for G. Since G acts freely and since $T \to G\backslash X$ is injective, the map $g \mapsto gT$ is a bijection of G onto the set of translates of G. These translates are pairwise disjoint. Form the graph $X' = G\backslash X \cdot T$ by contracting each tree gT to a single vertex, denoted by (gT). This is a tree since each gT is contractible. The inverse of the bijection $g \mapsto (gT)$ can be viewed as a bijection $\alpha : V(X') \to V(\Gamma(G, S)) = G$. Extend α to an isomorphism $X' \to \Gamma(G, S)$ as follows. We have $E(X') = E(X) \backslash E(G \times T)$. Give X' the orientation $E'_+ = E_+ \cap E(X')$ induced from X. It suffices to define $\alpha : E'_+ \to G \times S = E(\Gamma(G, S))_+$. Given $e \in E'_+$, let $(gT) = o(e)$, $(g'T) = t(e)$. Since, in X, the edge e connects gT to $g'T$, we conclude that $s = g^{-1}g'$ belongs to S. Set $\alpha(e) = (g, s)$. The surjectivity of $\alpha : E'_+ \to G \times S$ follows from the definition of S and the injectivity follows from the fact that X' is a tree and that α is injective on the vertex set. $\square$

Corollary 1.6.20. *Every subgroup of a free group is free.*

Proof. Let G be a free group. Then G acts freely on a tree X. If H is a subgroup of G, then H also acts freely on X and so is free by Theorem 1.6.19. $\square$

The algebraic proof of this result (Schreier's Theorem) is much more complicated than this geometric one.

For actions that are not free, there is still something to say, and this will lead us to our desired goal of realizing $SL_2(\mathbb{Z})$ as an amalgamated free product. Suppose G acts on a graph X without inversion. A *fundamental domain* of X mod G is a subgraph T of X such that $T \to G\backslash X$ is an isomorphism. By one of the exercises above, if $G\backslash X$ is a tree then a fundamental domain exists. Prove the following converse.

Proposition 1.6.21. *Let G act on a tree X without inversion. A fundamental domain of X mod G exists if and only if $G\backslash X$ is a tree.*

We are now ready for the main results.

Theorem 1.6.22. *Let G be a group acting on a graph X, and let T be a segment of X (i. e., T is a subgraph isomorphic to a path of length 1, consisting of vertices P and Q connected by a single edge e). Let G_P, G_Q, and $G_e = G_{\bar{e}}$ be the stabilizers of the vertices and edges of T. Then X is a tree if and only if the homomorphism $G_P *_{G_e} G_Q \to G$ induced by the inclusions $G_P \to G$ and $G_Q \to G$ is an isomorphism.*

Proof. We first show that X is connected if and only if G is generated by $G_P \cup G_Q$. Let X' be the connected component containing T, let G' be the set of elements $g \in G$ with $gX' = X'$, and let G'' be the subgroup of G generated by $G_P \cup G_Q$. If $h \in G_P \cup G_Q$, then the segments T and hT have a common vertex and so $hT \subset X'$, which implies $hX' = X'$. Thus $h \in G'$ and so G' contains G''. Conversely, $G''T$ and $(G - G'')T$ are disjoint subgraphs of X whose union is X. This implies that $G''T$ contains X', so that $G' \subset G''$ and so we have equality: $G' = G''$. The graph X is connected if and only if $X = X'$; that is $G = G' = G''$.

We now show that X contains no circuit if and only if $G_P *_{G_e} G_Q \to G$ is injective. Indeed, X has a circuit if and only if there is an edge path $f_0, \ldots, f_n, n \geq 1$ with $o(f_0) = t(f_n)$. Write $f_i = h_i e_i$ with $h_i \in G$ and $e_i = e$ or $\bar{e}$. Project to $G \backslash X \cong T$ and note that $\bar{e}_i = e_{i-1}$. Let $P_i = o(y_i) = t(y_{i-1})$. Then we have $h_i = h_{i-1} g_i$ with $g_i \in P_i$. Since

$$h_i P_i = h_i o(y_i) = o(h_i y_i) = t(h_{i-1} y_{i-1}) = h_{i-1} t(y_{i-1}) = h_{i-1} P_i,$$

and $g_i \neq G_e$ since $\overline{h_i y_i} \neq h_{i-1} y_{i-1}$. The equality $o(f_0) = t(f_n)$ is equivalent to $t(y_n) = P_0$ and also to

$$h_0 P_0 = h_n P_0 = h_0 g_1 \cdots g_n P_0;$$

that is $g_1 \cdots g_n \in G_{P_0}$. Thus X contains a circuit if and only if we can find a sequence of vertices $P_0, \ldots, P_n$ in T with $\{P_{i-1}, P_i\} = \{P, Q\}$ for all i and a sequence of elements $g_i \in G_{P_i} - G_e$, $0 \leq i \leq n$, such that $g_0 g_1 \cdots g_n = 1$. This is equivalent to saying that $G_P *_{G_e} G_Q$ is not injective. $\square$

Conversely, amalgamated free products act on trees with a segment as fundamental domain:

Theorem 1.6.23. *Let $G = G_1 *_A G_2$ be an amalgamated free product. Then there is a tree X (unique up to isomorphism) on which G acts, with fundamental domain a segment T having vertices P and Q and edge e, with $G_P = G_1$, $G_Q = G_2$ and $G_e = A$.*

Proof. Exercise. Show that, up to isomorphism, X must be the graph with vertex set $(G/G_1) \coprod (G/G_2)$ and edge set $(G/A) \coprod (\overline{G/A})$. Show that X is a tree. $\square$

We are now ready for some examples.

$\mathbb{Z}_2 * \mathbb{Z}_2$

This is the infinite dihedral group, which acts on the infinite path shown below.

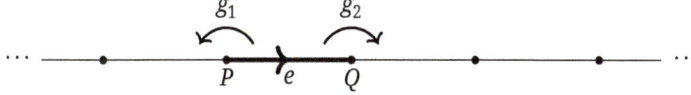

$\mathbb{Z}_n * \mathbb{Z}_m$

Generalize the previous example to construct a tree on which the free product $\mathbb{Z}_n * \mathbb{Z}_m$ acts.

$SL_2(\mathbb{Z})$

The group $SL_2(\mathbb{Z})$ acts on the upper half-plane $\mathbb{H} = \{z \mid \mathrm{Im}(z) > 0\}$ via linear fractional transformations:

$$\begin{pmatrix} a & b \\ c & d \end{pmatrix} : z \mapsto \frac{az + b}{cz + d}.$$

Let e be the circular arc consisting of the points $z = e^{i\theta}$ for $\pi/3 \le \theta \le \pi/2$. The origin of e is $P = e^{\pi i/3}$ and the terminus is $Q = i$. Let X be the union of translates of e by $SL_2(\mathbb{Z})$. This is a tree on which $SL_2(\mathbb{Z})$ acts with fundamental domain PQ; it is shown in Figure 1.26. Compute the stabilizers G_P, G_Q, and G_e and deduce the isomorphism

$$SL_2(\mathbb{Z}) \cong \mathbb{Z}_4 *_{\mathbb{Z}_2} \mathbb{Z}_6.$$

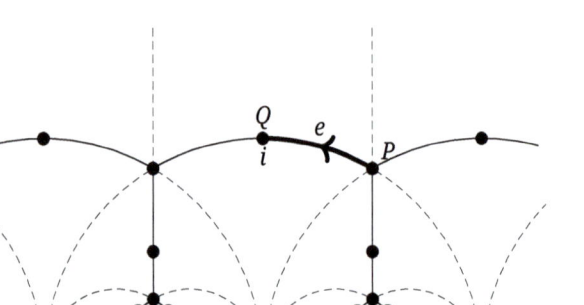

Figure 1.26: The tree X on which $SL_2(\mathbb{Z})$ acts.

Bibliographic notes

Most of the proofs in this chapter are standard and follow expositions in other texts such as Hatcher [2], Greenberg & Harper [1], and Munkres [5]. The discussion of locally trivial fiber bundles is heavily influenced by Steenrod's classic book [6]. The proof of the cellular approximation theorem follows that of [8]. The proof of the Fundamental Theorem of Algebra comes directly from Hatcher's book; it is so clean that there was little the author could do to improve it. Finally, the project on $SL_2(\mathbb{Z})$ was built out of Conrad's "blurb" [9] about this group and the book of Kassel [10] (for the algebraic portion) and Serre's book *Trees* [11] (for the geometric portion).

2 Homology

2.1 What is homology?

Our primary goal in the study of topology is to be able to distinguish among spaces. Here is a simple example: is it possible to deform the sphere S^2 into the torus T? That is, is there a homotopy equivalence $f : S^2 \to T$? Just by looking at these objects, the answer is obviously no, right? But just because we cannot see a way to do something does not mean that it is impossible.

The fundamental group provides one solution: since $\pi_1(S^2) \neq \pi_1(T)$, we cannot have such a homotopy equivalence. But fundamental groups can be difficult to compute, and what we would really like is a mechanism to make precise the following statement that is guiding our conjecture: the torus has a "hole", while the sphere does not. Moreover, if two spaces have the same fundamental group (for example, all the $\mathbb{R}P^n$ for $n \geq 2$ have the same fundamental group), we need another tool to prove that they are not all homotopy equivalent.

Homology theory makes all this hand-waving precise. Unfortunately, while this geometric intuition is fairly easy to see, defining homology actually takes quite a bit of work. We must then prove various things about it in order to conclude that the homotopy equivalences we wonder about do not exist. But this work will be worth it; the machine we construct is quite powerful.

2.1.1 Chain complexes

Throughout this book, we use the convention that a ring R is commutative with identity. Recall that an *R-module* is an abelian group M with an action of R on it:

$$R \times M \to M; \quad (r, m) \mapsto rm,$$

satisfying (1) $1.m = m$ for all $m \in M$; (2) $r(m + n) = rm + rn$; and (3) $(rs)m = r(sm)$. A group homomorphism $\varphi : M \to N$ is an *R*-module homomorphism if $\varphi(rm) = r\varphi(m)$ for all $r \in R$ and $m \in M$. Note that a $\mathbb{Z}$-module is just an abelian group. If k is a field, then a k-module is just a vector space over k.

Definition 2.1.1. A *chain complex* is a sequence of R-modules and homomorphisms

$$\cdots \xrightarrow{\partial_{i+2}} C_{i+1} \xrightarrow{\partial_{i+1}} C_i \xrightarrow{\partial_i} C_{i-1} \xrightarrow{\partial_{i-1}} \cdots$$

such that $\partial_i \circ \partial_{i+1} = 0$ for all i. We often omit the subscript on the homomorphisms and simply write $\partial^2 = 0$. A chain complex is *bounded below* if $C_i = 0$ for all $i < N$ for some $N \in \mathbb{Z}$. It is *bounded above* if $C_i = 0$ for all $i \geq N'$ for some N'. It is *nonnegative* if $C_i = 0$ for $i < 0$. We will denote a chain complex by $(C_\bullet, \partial)$.

https://doi.org/10.1515/9783111014852-002

Note that the condition $\partial^2 = 0$ implies that $\operatorname{im} \partial_{i+1} \subseteq \ker \partial_i$.

Definition 2.1.2. A sequence of R-modules and homomorphisms

$$M \xrightarrow{f} N \xrightarrow{g} P$$

is exact at N if $\ker g = \operatorname{im} f$.

Example 2.1.3. To say that the sequence

$$0 \longrightarrow N \xrightarrow{g} P$$

is exact at N is the same as saying that the map g is injective. Similarly, to say that

$$M \xrightarrow{f} N \longrightarrow 0$$

is exact at N is the same as saying that f is surjective. A *short exact sequence* is a sequence

$$0 \longrightarrow M \xrightarrow{f} N \xrightarrow{g} P \longrightarrow 0$$

that is exact at M, N, and P. That is, f is injective, g is surjective, and $\ker g = \operatorname{im} f$.

Example 2.1.4. Here are two short exact sequences:

$$0 \longrightarrow \mathbb{Z}/2 \xrightarrow{\times 2} \mathbb{Z}/4 \xrightarrow{\text{mod } 2} \mathbb{Z}/2 \longrightarrow 0$$

$$0 \longrightarrow \mathbb{Z}/2 \xrightarrow{i} \mathbb{Z}/2 \oplus \mathbb{Z}/2 \xrightarrow{\pi} \mathbb{Z}/2 \longrightarrow 0$$

where $i : \mathbb{Z}/2 \to \mathbb{Z}/2 \oplus \mathbb{Z}/2$ is the inclusion $1 \mapsto (1,0)$ and $\pi : \mathbb{Z}/2 \oplus \mathbb{Z}/2 \to \mathbb{Z}/2$ is the projection $(a,b) \mapsto b$.

2.1.2 Homology of chain complexes

Definition 2.1.5. Let $(C_\bullet, \partial)$ be a chain complex of R-modules. For each i, set $Z_i = \ker \partial_i$; this is the module of *i-cycles*. We also have $B_i = \operatorname{im} \partial_{i+1} \subseteq C_i$; this is the module of *i-boundaries*. Since $\partial^2 = 0$, we have $B_i \subseteq Z_i$. The *ith homology group* is defined to be the quotient

$$H_i(C_\bullet) = Z_i / B_i.$$

In this abstract setting, homology simply measures the failure of a chain complex to be exact. In subsequent sections we will find various geometric interpretations of this phenomenon.

Example 2.1.6. Consider the chain complex of $\mathbb{Z}$-modules

$$0 \longrightarrow \mathbb{Z} \xrightarrow{\times 6} \mathbb{Z} \xrightarrow{0} \mathbb{Z} \longrightarrow 0$$

where the final 0 is the module C_{-1}. The homology of this complex is

$$H_0 = \ker \partial_0 / \operatorname{im} \partial_1 = \mathbb{Z}$$
$$H_1 = \ker \partial_1 / \operatorname{im} \partial_2 = \mathbb{Z}/6$$
$$H_2 = \ker \partial_2 = 0.$$

Definition 2.1.7. A *chain map* $f : (C_\bullet, \partial) \to (C'_\bullet, \partial')$ is a collection of maps $f_i : C_i \to C'_i$ such that the diagram

$$\cdots \xrightarrow{\partial_{i+2}} C_{i+1} \xrightarrow{\partial_{i+1}} C_i \xrightarrow{\partial_i} C_{i-1} \xrightarrow{\partial_{i-1}} \cdots$$
$$\downarrow{f_{i+1}} \qquad \downarrow{f_i} \qquad \downarrow{f_{i-1}}$$
$$\cdots \xrightarrow{\partial'_{i+2}} C'_{i+1} \xrightarrow{\partial'_{i+1}} C'_i \xrightarrow{\partial'_i} C'_{i-1} \xrightarrow{\partial'_{i-1}} \cdots$$

commutes; that is, $f_i \circ \partial_{i+1} = \partial'_{i+1} \circ f_{i+1}$ for all i.

Note that a chain map sends cycles to cycles and boundaries to boundaries. We therefore have an induced homomorphism

$$H_i(f) : H_i(C_\bullet) \to H_i(C'_\bullet); \quad H_i(f)([z]) = [f(z)],$$

where $[z]$ denotes the homology class of the cycle z. We will often use the notation f_* for the induced map, and write $f_* : H_i(C_\bullet) \to H_i(C'_\bullet)$.

Definition 2.1.8. A *chain homotopy* between chain maps $f, g : (C_\bullet, \partial) \to (C'_\bullet, \partial')$ is a collection of maps $D_i : C_i \to C'_{i+1}$ such that

$$\partial'_{i+1} D_i + D_{i-1} \partial_i = f_i - g_i$$

for all i. That is, the following diagram commutes for all i:

$$C_{i+1} \xrightarrow{\partial_{i+1}} C_i \xrightarrow{\partial_i} C_{i-1}$$
$$\downarrow \quad {}^{D_i}\!\swarrow \quad \downarrow{f_i - g_i} \quad {}^{D_{i-1}}\!\swarrow \quad \downarrow$$
$$C'_{i+1} \xrightarrow{\partial'_{i+1}} C'_i \xrightarrow{\partial'_i} C'_{i-1}$$

Proposition 2.1.9. *Chain homotopic maps induce the same map in homology.*

Proof. Suppose D is a chain homotopy between f and g and let z be an i-cycle in $C_\bullet$. Then

$$
\begin{aligned}
(f_* - g_*)([z]) &= [(f_i - g_i)(z)] \\
&= [(\partial'_{i+1}D_i + D_{i-1}\partial_i)(z)] \\
&= [\partial'_{i+1}D_i(z)] + [D_{i-1}\partial_i(z)] \\
&= [\partial'_{i+1}D_i(z)] + [0] \text{ (since } \partial_i(z) = 0) \\
&= 0 \text{ (since the class of a boundary is 0).}
\end{aligned}
$$

Thus, $f_* - g_*$ is the zero map; that is, $f_* = g_*$. □

Our task moving forward is to put these notions into a concrete setting. Given a topological space, we will seek chain complexes that capture information about the space. Computing homology of the resulting complex will then tell us something about the space (we hope). We begin with simplicial complexes.

Exercises

Exercise 2.1.1. Suppose that $f : (C_\bullet, \partial) \to (C'_\bullet, \partial')$ and $g : (C'_\bullet, \partial') \to (C''_\bullet, \partial'')$ are chain maps. Prove that $g \circ f$ is a chain map and that $H_i(g \circ f) = H_i(g) \circ H_i(f)$ for all i.

Exercise 2.1.2. For each $n \geq 0$, let $C_n = \mathbb{Z}/8$ and let $\partial : C_n \to C_{n-1}$ be the map x mod $8 \mapsto 4x$ mod 8. Show that $C_\bullet$ is a chain complex and compute its homology.

Exercise 2.1.3. A chain complex is *acyclic* if $H_i(C_\bullet) = 0$ for all i. Assume that R is a principal ideal domain. Prove that if $C_\bullet$ is an acyclic complex of R-modules, then the identity map $1 : C_\bullet \to C_\bullet$ is chain homotopic to the zero map.

Exercise 2.1.4. Prove that a chain complex $C_\bullet$ is acyclic if and only if the map $0_\bullet \to C_\bullet$ induces an isomorphism in homology, where $0_\bullet$ is the chain complex consisting of the zero module in all degrees.

Exercise 2.1.5. Let R be a principal ideal domain and let $(C_\bullet, \partial)$ be a bounded chain complex of R-modules. Define the *Euler characteristic* of $C_\bullet$ to be

$$
\chi(C_\bullet) = \sum_i (-1)^i \operatorname{rank}(C_i).
$$

Prove that

$$
\chi(C_\bullet) = \sum_i (-1)^i \operatorname{rank}(H_i(C_\bullet)).
$$

2.2 Simplicial homology

Let us begin with a simple example. Consider the (hollow) tetrahedron shown in Figure 2.1. This is a simplicial complex with four vertices, six edges, and four 2-simplices. Topologically, this space is homeomorphic to the sphere S^2. The vertices are labeled v_0, v_1, v_2, v_3 and the order will matter for us here. Specifying an ordering of the vertices induces an *orientation* on each simplex. The edge spanned by v_0 and v_1 then has two distinct orientations: $\langle v_0, v_1 \rangle$ and $\langle v_1, v_0 \rangle$. While these are the same geometric object, they are distinct as oriented simplices.

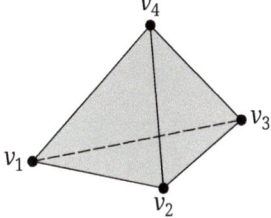

Figure 2.1: A hollow tetrahedron as a simplicial complex.

What do we mean by the boundary of a simplex? We understand what it is geometrically–it is the collection of simplices on the "border" of a simplex. So the boundary of a 1-simplex consists of its two endpoints, the boundary of a 2-simplex is its three edges, etc. But for our purposes, we need to also keep track of the orientations. Here is the convention: if $\sigma = \langle v_{i_0}, v_{i_1}, \dots, v_{i_k} \rangle$ is an oriented k-simplex with $i_0 < i_1 < \cdots < i_k$, then the jth face of σ is the $(k-1)$-simplex

$$\langle v_{i_0}, \dots, \hat{v}_{i_j}, \dots, v_{i_k} \rangle,$$

where the hat over v_{i_j} means to delete that vertex.

Definition 2.2.1. The *boundary* of the simplex $\sigma = \langle v_{i_0}, v_{i_1}, \dots, v_{i_k} \rangle$ is the formal sum

$$\partial \sigma = \sum_{j=0}^{k} (-1)^j \langle v_{i_0}, \dots, \hat{v}_{i_j}, \dots, v_{i_k} \rangle.$$

We will ignore for now where this object lives, but let us assume the linearity of this map ∂. Note that, for example

$$\partial(\partial(\langle v_0, v_1, v_2 \rangle)) = \partial(\langle v_1, v_2 \rangle - \langle v_0, v_2 \rangle + \langle v_0, v_1 \rangle)$$
$$= (v_2 - v_1) - (v_2 - v_0) + (v_1 - v_0)$$
$$= 0.$$

(Here, we drop the angle brackets around a vertex, viewed as a 0-simplex.) In fact, this is true for any simplex and we leave this as an exercise. So it would appear that we have at least some hope for building a chain complex from a simplicial complex. The reader may check that for the tetrahedron in Figure 2.1 we have the following:

$$\partial(\langle v_0, v_1, v_2 \rangle - \langle v_0, v_1, v_3 \rangle + \langle v_0, v_2, v_3 \rangle - \langle v_1, v_2, v_3 \rangle) = 0,$$

since each edge appears twice in the sum with opposite orientations. Thus the formal sum of the four triangles is in the kernel of this map ∂, and so would be a cycle in this hypothetical chain complex. Since there are no 3-simplices in sight, it would appear that this cycle would generate a homology class in degree 2. Moreover, one can check that ∂ applied to the alternating sum of three edges around a single 2-simplex is 0 and therefore these form a 1-cycle. However, in this case, ∂ of the corresponding 2-simplex equals this cycle (up to a sign) and so the corresponding homology class is 0.

As it turns out, all of this works.

2.2.1 Simplical chains

Let K be a simplicial complex and fix an ordering on the vertices of K. This then determines an orientation of each simplex in K as noted above. We will refer to this as an *ordered simplicial complex*.

Definition 2.2.2. Let K be an ordered simplicial complex. For each i, let $C_i(K)$ be the free abelian group with basis the set of oriented i-simplices of K:

$$C_i(K) = \left\{ \sum n_j \sigma_j \mid n_j \in \mathbb{Z} \right\}.$$

If $\sigma = \langle v_0, v_1, \ldots, v_k \rangle$ is an oriented k-simplex in K, define $\partial \sigma$ by

$$\partial \sigma = \sum_{j=0}^{k} (-1)^j \langle v_{i_0}, \ldots, \hat{v}_{i_j}, \ldots, v_{i_k} \rangle,$$

and extend this to a linear map $\partial_k : C_k(K) \to C_{k-1}(K)$. Since $\partial^2 \sigma = 0$ for every simplex σ, we have $\partial_k \circ \partial_{k+1} = 0$ for all k. The nonnegative chain complex $(C_.(K), \partial)$ is called the *simplicial chain complex* of K.

Definition 2.2.3. Let K be an ordered simplicial complex. The *simplicial homology* of K is the homology of the simplicial chain complex $(C_.(K), \partial)$.

Example 2.2.4. The homology of a point is easy to calculate: $H_i(\Delta^0)$ is $\mathbb{Z}$ for $i = 0$ and 0 for $i > 0$. This follows because the same is true for $(C_.(\Delta^0), \partial)$.

Example 2.2.5. Let K_n be a regular n-gon with vertices $v_1, v_2, \ldots, v_n$. There are n 1-simplices $e_i = \langle v_i, v_{i+1} \rangle$, where the indices are taken modulo n (that is, $e_n = \langle v_n, v_{n+1} \rangle = \langle v_n, v_1 \rangle$). The chain complex $C_\bullet(K_n)$ is

$$0 \longrightarrow \mathbb{Z}^n \xrightarrow{\partial_1} \mathbb{Z}^n \xrightarrow{\partial_0} 0.$$

The map ∂_0 is just the zero map, and the map ∂_1 sends the edge $\langle v_i, v_{i+1} \rangle$ to $v_{i+1} - v_i$ (indices modulo n). Note that the kernel of ∂_1 is spanned by the sum $\sum e_i$, from which it follows that $H_1(K_n) = \ker \partial_1 = \mathbb{Z}$. The image of ∂_1 consists of all 0-chains of the form $\sum n_i v_i$ with $\sum n_i = 0$. This is a submodule of rank $n - 1$, generated by the $v_{i+1} - v_i$ for $i = 1, \ldots, n$ (indices modulo n). It follows that $H_0(K_n) = C_0(K_n)/\operatorname{im} \partial_1 \cong \mathbb{Z}$.

Note that the answer in the previous example is independent of n. This is actually reassuring. As a topological space, each K_n is homeomorphic to the circle S^1 and if we want homology to be a homotopy invariant of spaces, then feeding a different triangulation of a particular simplicial complex into it should not change the answer. This is not a proof, of course, but at least it does not dash our hopes. An actual proof will be given later and will use some heavy machinery we still need to develop.

Note that if $f : K \to L$ is a simplicial map between ordered simplicial complexes, then we get an induced map

$$f_\sharp : C_k(K) \to C_k(L); \quad f_\sharp(\sigma) = f(\sigma).$$

There is one small problem: it may be that f takes a k-simplex to a simplex of lower dimension. In such a case, we set $f_\sharp(\sigma) = 0$. This actually makes good sense intuitively: the map f is defined on the set of vertices and then extended simplex-by-simplex. If f takes a k-simplex to a simplex of lower dimension, that means that f takes at least two vertices in K to the same vertex in L. We could write this "degenerate" simplex in $C_k(L)$ using these repeated vertices, but it is not an actual simplex in L. There is a more expansive point of view for dealing with this, called the theory of simplicial sets, but we will not address that here.

Observe that $f_\sharp \circ \partial = \partial \circ f_\sharp$; that is, we have a chain map. This then induces a homomorphism $f_* : H_k(K) \to H_k(L)$. If we have a second map $g : L \to M$, then by an exercise in the previous section we have $(g \circ f)_* = g_* \circ f_*$. In other words, simplicial homology is a *functor*

$$H_i : \underline{\text{Simp}} \to \underline{\text{Ab}}$$

from the category of simplicial complexes and simplicial maps to the category of abelian groups.

Exercises

Exercise 2.2.1. If σ is an oriented k-simplex, prove that $\partial^2 \sigma = 0$.

Exercise 2.2.2. Prove that if $f : K \to L$ is a simplicial map, then $f_\sharp : C_\bullet(K) \to C_\bullet(L)$ is a chain map.

Exercise 2.2.3. Let Γ be a finite graph with vertex set $V = \{v_1, \ldots, v_n\}$ and edge set $E = \{e_1, \ldots, e_m\}$. Orient the edges and let $\mathbb{I}$ be the incidence matrix of Γ: this is the $m \times n$ matrix with (i,j)-entry $+1$ if edge e_j starts at v_i, -1 if e_j ends at v_i, and 0 otherwise. Let C_0 be the free abelian group on the vertices, C_1 the free abelian group on the edges, $C_n = 0$ for all other n, and let $d : C_1 \to C_0$ be given by the matrix $\mathbb{I}$. If Γ is connected, show that $H_0(C_\bullet) \cong \mathbb{Z}$ and that $H_1(C_\bullet)$ is free abelian of rank $n - m - 1$ (this is the number of circuits in Γ).

2.2.2 Calculations

In this section, we perform a few calculations of simplicial homology. It will quickly become apparent that doing this by hand can be quite cumbersome. While simplicial complexes are often easy to describe, they generally consist of a large number of simplices. The corresponding boundary maps in the simplicial chain complex are then quite large matrices, and the linear algebra (over $\mathbb{Z}$!) is not as simple as it is over a field. Still, it is instructive to do a few examples.

Example 2.2.6. Let K be the simplicial complex shown in Figure 2.2. This space has the homotopy type of a circle and so Example 2.2.5 suggests that we should expect to see that the simplicial homology groups are $\mathbb{Z}$ in dimensions 0 and 1, and 0 otherwise. The com-

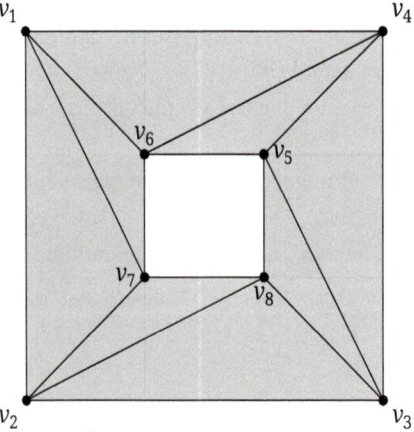

Figure 2.2: A triangulated annulus.

plex K has eight vertices, 16 edges, and eight 2-simplices. The simplicial chain complex is therefore

$$0 \longrightarrow \mathbb{Z}^8 \xrightarrow{\partial_2} \mathbb{Z}^{16} \xrightarrow{\partial_1} \mathbb{Z}^8 \longrightarrow 0$$

First note the following, which will be useful in future calculations. If $e_{ij} = \langle v_i, v_j \rangle$ is an edge in K, $\partial e_{ij} = v_j - v_i$. Since the various e_{ij} form a basis of $C_1(K)$, we see that the image of ∂_1 is equal to the submodule of $C_0(K)$ consisting of all $\sum n_i v_i$ with $\sum n_i = 0$. Define $\varepsilon : C_0(K) \rightarrow \mathbb{Z}$ by $\varepsilon(\sum n_i v_i) = \sum n_i$. The map ε is called the *augmentation map* and is clearly surjective with kernel equal to the image of ∂_1. It follows that $H_0(K) \cong \mathbb{Z}$.

Let us turn to the group $H_2(K)$. I claim that the map $\partial_2 : C_2(K) \rightarrow C_1(K)$ is injective. Indeed, if $\partial_2(\sum n_i \sigma_i) = 0$, then each edge in this boundary must occur twice with opposite signs. Note, however, that every triangle in K has exactly one edge occurring in either the outer square or the inner square, and that each of these edges is a face of exactly one triangle. It is therefore impossible for these edges to be canceled. Thus, there are no 2-cycles in K and $H_2(K) = 0$.

As for the group $H_1(K)$, we finally have to do some linear algebra over the integers. If we were working over a field, we could use the Rank–Nullity Theorem: the rank of ∂_2 is 8 and the rank of ∂_1 is 7. The nullity of ∂_1 is then 9 and so the rank of $H_1(K)$ would be $9 - 8 = 1$. This turns out to be correct over $\mathbb{Z}$ in this case, but we cannot assume that.

Order the basis elements of each $C_i(K)$ lexicographically. With respect to these, the matrices for ∂ are

$$\partial_2 = \begin{bmatrix} 1 & 0 & 0 & 0 & 0 & 0 & 0 & 0 \\ 0 & 1 & 0 & 0 & 0 & 0 & 0 & 0 \\ 0 & -1 & 1 & 0 & 0 & 0 & 0 & 0 \\ -1 & 0 & -1 & 0 & 0 & 0 & 0 & 0 \\ 0 & 0 & 0 & 1 & 0 & 0 & 0 & 0 \\ 1 & 0 & 0 & 0 & 1 & 0 & 0 & 0 \\ 0 & 0 & 0 & -1 & -1 & 0 & 0 & 0 \\ 0 & 0 & 0 & 0 & 0 & 1 & 0 & 0 \\ 0 & 0 & 0 & 0 & 0 & -1 & 1 & 0 \\ 0 & 0 & 0 & 1 & 0 & 0 & -1 & 0 \\ 0 & 0 & 0 & 0 & 0 & 1 & 0 & 1 \\ 0 & 1 & 0 & 0 & 0 & 0 & 0 & -1 \\ 0 & 0 & 0 & 0 & 0 & 0 & 0 & 1 \\ 0 & 0 & 0 & 0 & 0 & 0 & 1 & 0 \\ 0 & 0 & 1 & 0 & 0 & 0 & 0 & 0 \\ 0 & 0 & 0 & 0 & 1 & 0 & 0 & 0 \end{bmatrix}$$

$$\partial_1 = \begin{bmatrix}
-1 & -1 & -1 & -1 & 0 & 0 & 0 & 0 & 0 & 0 & 0 & 0 & 0 & 0 & 0 & 0 \\
1 & 0 & 0 & 0 & -1 & -1 & -1 & 0 & 0 & 0 & 0 & 0 & 0 & 0 & 0 & 0 \\
0 & 0 & 0 & 0 & 1 & 0 & 0 & -1 & -1 & -1 & 0 & 0 & 0 & 0 & 0 & 0 \\
0 & 1 & 0 & 0 & 0 & 0 & 0 & 1 & 0 & 0 & -1 & -1 & 0 & 0 & 0 & 0 \\
0 & 0 & 0 & 0 & 0 & 0 & 0 & 0 & 1 & 0 & 1 & 0 & -1 & -1 & 0 & 0 \\
0 & 0 & 1 & 0 & 0 & 0 & 0 & 0 & 0 & 0 & 0 & 1 & 1 & 0 & -1 & 0 \\
0 & 0 & 0 & 1 & 0 & 1 & 0 & 0 & 0 & 0 & 0 & 0 & 0 & 0 & 1 & -1 \\
0 & 0 & 0 & 0 & 0 & 0 & 1 & 0 & 0 & 1 & 0 & 0 & 0 & 1 & 0 & 1
\end{bmatrix}$$

To compute homology, we must compute the kernel of ∂_1 and the image of ∂_2. Above, we asserted that ∂_2 is injective, but to check this, we should compute the row echelon form of its matrix. Note that we must do this linear algebra over the integers. In this case, we find that the echelon form of ∂_2 is

$$\begin{bmatrix}
1 & 0 & 0 & 0 & 0 & 0 & 0 & 0 \\
0 & 1 & 0 & 0 & 0 & 0 & 0 & 0 \\
0 & 0 & 1 & 0 & 0 & 0 & 0 & 0 \\
0 & 0 & 0 & 1 & 0 & 0 & 0 & 0 \\
0 & 0 & 0 & 0 & 1 & 0 & 0 & 0 \\
0 & 0 & 0 & 0 & 0 & 1 & 0 & 0 \\
0 & 0 & 0 & 0 & 0 & 0 & 1 & 0 \\
0 & 0 & 0 & 0 & 0 & 0 & 0 & 1 \\
0 & 0 & 0 & 0 & 0 & 0 & 0 & 0 \\
0 & 0 & 0 & 0 & 0 & 0 & 0 & 0 \\
0 & 0 & 0 & 0 & 0 & 0 & 0 & 0 \\
0 & 0 & 0 & 0 & 0 & 0 & 0 & 0 \\
0 & 0 & 0 & 0 & 0 & 0 & 0 & 0 \\
0 & 0 & 0 & 0 & 0 & 0 & 0 & 0 \\
0 & 0 & 0 & 0 & 0 & 0 & 0 & 0 \\
0 & 0 & 0 & 0 & 0 & 0 & 0 & 0
\end{bmatrix}$$

confirming that ∂_2 is in fact injective. When we compute this for ∂_1, we get

$$\begin{bmatrix}
1 & 0 & 0 & 0 & 0 & -1 & 0 & -1 & 0 & 0 & 1 & 0 & -1 & 0 & 0 & 1 \\
0 & 1 & 0 & 0 & 0 & 0 & 0 & 1 & 0 & 0 & -1 & -1 & 0 & 0 & 0 & 0 \\
0 & 0 & 1 & 0 & 0 & 0 & 0 & 0 & 0 & 0 & 0 & 1 & 1 & 0 & -1 & 0 \\
0 & 0 & 0 & 1 & 0 & 1 & 0 & 0 & 0 & 0 & 0 & 0 & 0 & 0 & 1 & -1 \\
0 & 0 & 0 & 0 & 1 & 0 & 0 & -1 & 0 & -1 & 1 & 0 & -1 & -1 & 0 & 0 \\
0 & 0 & 0 & 0 & 0 & 0 & 1 & 0 & 0 & 1 & 0 & 0 & 0 & 1 & 0 & 1 \\
0 & 0 & 0 & 0 & 0 & 0 & 0 & 0 & 1 & 0 & 1 & 0 & -1 & -1 & 0 & 0 \\
0 & 0 & 0 & 0 & 0 & 0 & 0 & 0 & 0 & 0 & 0 & 0 & 0 & 0 & 0 & 0
\end{bmatrix}$$

Note that this matrix has rank 7, as we computed above. Now, over the integers, sub-modules of free modules are free. Moreover, rank is additive over short exact sequences (because we can tensor with the rational numbers $\mathbb{Q}$, which is a flat $\mathbb{Z}$-module). Thus

$$\mathrm{rank}(C_1(K)) = \mathrm{rank}(\mathrm{ker}(\partial_1)) + \mathrm{rank}(\partial_1),$$

from which we conclude that $\mathrm{ker}(\partial_1)$ has rank 9. It follows that $H_1(K)$ has rank $9 - 8 = 1$, but it could in principle have torsion as well.

We therefore need to compute a basis for $\mathrm{ker}(\partial_1)$. This is a tedious calculation, but the result is that the following cycles form a basis:

$$e_{12} - e_{17} + e_{28} - e_{78}$$
$$e_{14} - e_{17} + e_{46} + e_{67}$$
$$e_{16} - e_{17} + e_{67}$$
$$e_{23} - e_{28} + e_{38}$$
$$e_{27} - e_{28} + e_{78}$$
$$e_{34} - e_{38} + e_{46} + e_{67} + e_{78}$$
$$e_{35} - e_{38} + e_{58}$$
$$e_{45} - e_{46} + e_{58} - e_{67} - e_{78}$$
$$e_{56} - e_{58} + e_{67} + e_{78}$$

Note that many of them are obviously boundaries. Those consisting of a sum of exactly three edges are the boundary of the corresponding 2-simplex. The others among the first eight are easily seen to be boundaries as well. The last on the list, $e_{56} - e_{58} + e_{67} + e_{78}$ is *not* a boundary, however. To see this, note that a 2-boundary containing this cycle would also have to contain the 1-chain $e_{12} - e_{14} + e_{23} + e_{34}$ (up to a sign), and there is no way to cancel these edges. It follows that $H_1(K) \cong \mathbb{Z}$, generated by the cycle $e_{56} - e_{58} + e_{67} + e_{78}$. Note that we could also take $e_{12} - e_{14} + e_{23} + e_{34}$ as a generating cycle since these two differ by a boundary (take the sum of all the 2-simplices with appropriate signs, so that all the interior edges cancel).

Example 2.2.7. Suppose K is a path-connected simplicial complex. The augmentation map constructed in the previous example will allow us to prove a general fact. Note that this map exists for any simplicial complex: $\varepsilon : C_0(K) \rightarrow \mathbb{Z}$. The observation that the image of ∂_1 is equal to the kernel of ε is true in general for connected complexes. Indeed, for any simplicial complex the module $C_1(K)$ is generated by all the $\langle v_i, v_j \rangle$ for vertices v_i, v_j. If all the vertices lie in a single component of K, then there is an edge path connecting any two vertices; the boundary of this 1-cycle is then the difference of the endpoints. This shows that the image of ∂_1 spans the kernel of ε. If K were not connected, then we would not be able to conclude this. It follows that $H_0(K) \cong \mathbb{Z}$ when K is connected. Any vertex of K is a generating cycle for this group (exercise).

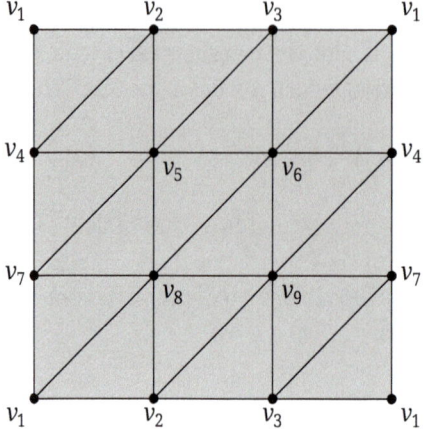

Figure 2.3: A triangulated torus.

Example 2.2.8. Consider the triangulation of the torus shown in Figure 2.3. To get the torus, one identifies the top and bottom sides of the square, and the left and right sides. The simplicial chain complex $C_*(T)$ is

$$0 \longrightarrow \mathbb{Z}^{18} \xrightarrow{\partial_2} \mathbb{Z}^{27} \xrightarrow{\partial_2} \mathbb{Z}^9 \longrightarrow 0.$$

Since T is connected, the previous example tells us that $H_0(T) \cong \mathbb{Z}$. Denote the edge $\langle v_i, v_j \rangle$ by e_{ij}, and the 2-simplex $\langle v_i, v_j, v_k \rangle$ by σ_{ijk}. There are two obvious 1-cycles:

$$a = e_{12} + e_{23} - e_{13}$$
$$b = e_{14} + e_{47} - e_{17}$$

There are other cycles, of course, such as $c = e_{18} - e_{68} - e_{16}$, but one can check that $[a + b] = [c]$. What is clear is that a and b are *not* boundaries. Consider a, for example (the argument for b is completely analogous). If $\partial_2(w) = a$ for some 2-chain w, then w must contain exactly one of σ_{124} or σ_{128}, one of σ_{235} or σ_{289}, and one of σ_{136} or σ_{137}. But then w must contain other triangles adjacent to these to cancel out the extra edges. This propagates through until we realize that no such w can exist. Moreover, there is no 2-chain that maps to $a+b$. It follows that the homology classes of a and b are independent elements in $H_1(T)$.

As for $H_2(T)$, note that the sum of all the 2-simplices, with suitable signs attached, is a 2-cycle. One can show that the rank of ∂_2 is 17, so that $H_2(T) \cong \mathbb{Z}$. Now, the Euler characteristic of $C_*(T)$ is $9 - 18 + 27 = 0$, and since this equals the alternating sum of the ranks of the homology groups, we see that $H_0(T) \cong \mathbb{Z}$, $H_1(T) \cong \mathbb{Z}^2$, and $H_2(T) \cong \mathbb{Z}$.

Exercises

Exercise 2.2.4. Let K be a connected simplicial complex. Prove that the homology class of any vertex in K generates $H_0(K)$.

Exercise 2.2.5. Let K be a finite simplicial complex with N connected components. Prove that $H_0(K) \cong \mathbb{Z}^N$.

Exercise 2.2.6. An *infinite p-chain* on K is a sum $c = \sum_{i=1}^{\infty} n_i \sigma_i$, where each σ_i is a p-simplex of K and where we do not assume all but finitely many of the n_i vanish. Note that if K is a finite complex, this condition is vacuous. Assume the following condition: for any such c and any compact subcomplex $L \subset K$, the number of n_i in c for which the corresponding σ_i intersect L is finite. We call such a chain *locally finite*. Show that if $C_p^{\mathrm{lf}}(K)$ denotes the collection of locally finite chains, then the boundary operator in K restricts to give a boundary map

$$\partial_p^{\mathrm{lf}} : C_p^{\mathrm{lf}}(K) \to C_{p-1}^{\mathrm{lf}}(K).$$

The resulting homology groups are denoted $H_p^{\mathrm{lf}}(K)$. Show that if K is finite, then $H_p^{\mathrm{lf}}(K) = H_p(K)$. If K is the complex whose underlying space is $\mathbb{R}$ with vertex set $\mathbb{Z}$, show that $H_1(K) = 0$, while $H_1^{\mathrm{lf}}(K) \cong \mathbb{Z}$.

Exercise 2.2.7. Prove that in the triangulated torus of Figure 2.3, the cycles $a + b$ and c differ by a boundary and therefore determine the same homology class.

Exercise 2.2.8. Verify the calculations of Example 2.2.8.

Exercise 2.2.9. Let K be the n-skeleton of the simplex Δ^{n+1}. This is the subcomplex consisting of the boundary of Δ^{n+1}; that is, it is the union of all the n-simplices. As a topological space, K is homeomorphic to the n-sphere S^n. Show that $H_n(K) \cong \mathbb{Z}$. What is a generating cycle?

Exercise 2.2.10. Consider the triangulation of $\mathbb{R}P^2$ shown in Figure 2.4. Use this to compute the simplicial homology of this complex.

2.3 Singular homology

As Example 2.2.6 shows, the algebra involved in computing simplicial homology becomes cumbersome very quickly. Simplicial complexes are convenient objects, and they are easily stored in a computer; it is therefore possible to compute their homology via efficient algorithms. But they are not very good for proving the general results about homology that we want, and so we must move to a more abstract realm to make progress.

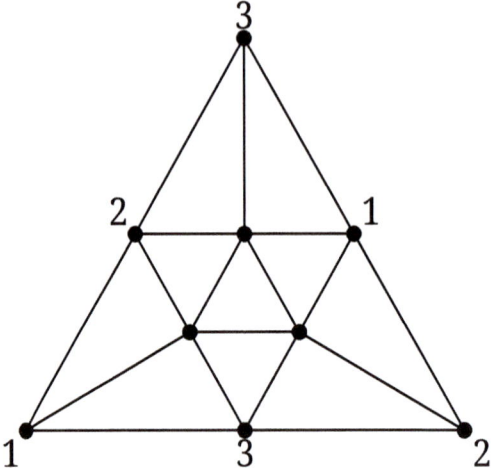

Figure 2.4: A triangulation of $\mathbb{R}P^2$.

2.3.1 Singular chains

Here is a way to think about simplicial chains that will motivate our next steps. A simplicial complex is a particular type of cell complex in which the attaching maps happen to be simplicial homeomorphisms from a k-simplex onto its image. What happens if we drop this rigid structure and just ask for continuous maps?

Definition 2.3.1. Let X be a topological space. A *singular k-simplex* is a continuous map $\sigma : \Delta^k \to X$. Let $S_k(X)$ be the free abelian group on the set of singular k-simplices and define the boundary map $\partial_k : S_k(X) \to S_{k-1}(X)$ by

$$\partial_k(\sigma) = \sum_{i=0}^{k}(-1)^i \sigma|_{\langle v_0,\dots \hat{v}_i,\dots,v_k\rangle}.$$

One checks, as in the simplicial case, that $\partial_{k-1} \circ \partial_k = 0$ so that $(S_\bullet(X), \partial)$ is a chain complex, called the *singular chain complex of X*. The nth *singular homology group of X* is the nth homology of this complex:

$$H_n(X) = \ker(\partial_n)/\operatorname{im}(\partial_{n+1}) = Z_n/B_n.$$

Note that if X is, for example, a cell complex then there are *uncountably many* $\sigma : \Delta^k \to X$. Moreover, these simplices exist for *all* $k \geq 0$. At first glance, this would make it seem that singular homology is somehow worse for computations. Even the simplest case requires some work.

Example 2.3.2. Let us compute the singular homology of a point. Let X be a one-point space. Note that for each k there is a unique map $\sigma_k : \Delta^k \to X$. The singular chain groups are therefore $C_k(X) = \mathbb{Z}$ for all $k \geq 0$. Note that the map ∂_k is then

$$\partial_k(\sigma_k) = \sum_{i=0}^{k}(-1)^i \sigma_{k-1} = \begin{cases} 0 & k \text{ odd} \\ \sigma_{k-1} & k \text{ even.} \end{cases}$$

It follows that the complex has the form

$$\cdots \xrightarrow{\text{id}} \mathbb{Z} \xrightarrow{0} \mathbb{Z} \xrightarrow{\text{id}} \mathbb{Z} \xrightarrow{0} \mathbb{Z} \longrightarrow 0.$$

We then see that $H_0(X) \cong \mathbb{Z}$ and $H_i(X) = 0$ for $i > 0$.

Proposition 2.3.3. *If $\{X_\alpha\}$ are the path components of X, then $H_n(X) \cong \bigoplus_\alpha H_n(X_\alpha)$ for all $n \geq 0$.*

Proof. If $\sigma : \Delta^n \to X$ is a singular n-simplex, then the image of σ lies in a unique X_α. It follows that

$$S_n(X) = \bigoplus_\alpha S_n(X_\alpha).$$

The maps ∂_n preserve this decomposition so that

$$\ker(\partial_n) = \bigoplus_\alpha \ker(\partial_n|_{X_\alpha}) \quad \text{and} \quad \text{im}(\partial_n) = \bigoplus_\alpha \text{im}(\partial_n|_{X_\alpha}).$$

The result follows. $\qquad\square$

Proposition 2.3.4. *If X is nonempty and path-connected, then $H_0(X) \cong \mathbb{Z}$.*

Proof. The proof is much the same as in the case of a connected simplicial complex. Define the augmentation map $\varepsilon : S_0(X) \to \mathbb{Z}$ by $\varepsilon(\sum_i n_i \sigma_i) = \sum_i n_i$. This is surjective if X is nonempty. If $\sigma : \Delta^1 \to X$ is a singular 1-simplex, then $\varepsilon\partial_1(\sigma) = \varepsilon(\sigma|_{\langle v_1 \rangle} - \sigma|_{\langle v_0 \rangle}) = 1-1 = 0$ so that $\text{im}(\partial_1) \subseteq \ker(\varepsilon)$.

Conversely, if $\varepsilon(\sum n_i \sigma_i) = 0$, then $\sum n_i = 0$. The σ_i, being 0-simplices, simply correspond to points in X. Choose a path $\tau_i : I \to X$ from a fixed $x_0 \in X$ to the point $\sigma_i(\langle v_0 \rangle) \in X$ and let σ_0 be the 0-simplex corresponding to x_0. Then $\partial_i(\sum n_i \tau_i) = \sum n_i \sigma_i$ and so $\ker(\varepsilon) \subseteq \text{im}(\partial_1)$. It follows that $H_0(X) \cong \mathbb{Z}$. $\qquad\square$

The augmentation map can be used to extend the singular chain complex

$$\cdots \longrightarrow S_n(X) \xrightarrow{\partial_n} \cdots \xrightarrow{\partial_2} S_1(X) \xrightarrow{\partial_1} S_0(X) \xrightarrow{\varepsilon} \mathbb{Z} \longrightarrow 0.$$

Definition 2.3.5. The *reduced homology groups* of X, denoted $\tilde{H}_n(X)$ are the homology groups of the augmented chain complex.

Note that $\tilde{H}_n(X) \cong H_n(X)$ for $n > 0$ and $H_0(X) \cong \tilde{H}_0(X) \oplus \mathbb{Z}$.

As in the simplicial case, singular homology is functorial. Suppose $f : X \to Y$ is a continuous map. Define $f_\sharp : S_n(X) \to S_n(Y)$ by $f_\sharp(\sigma) = f \circ \sigma$ for all $\sigma : \Delta^n \to X$ and extending linearly. Note that

$$f_\sharp(\partial\sigma) = f_\sharp\left(\sum_{i=0}^n \sigma|_{\langle v_0,\dots,\hat{v}_i,\dots,v_n \rangle} \right)$$

$$= \sum_{i=0}^n f \circ \sigma|_{\langle v_0,\dots,\hat{v}_i,\dots,v_n \rangle}$$

$$= \partial(f_\sharp(\sigma)),$$

so that $f_\sharp : S_\bullet(X) \to S_\bullet(Y)$ is a chain map. We therefore have an induced map on homology $f_* : H_n(X) \to H_n(Y)$ satisfying

(1) $\mathrm{id}_* = \mathrm{id}$, and

(2) $(g \circ f)_* = g_* \circ f_*$.

We now prove the crucial fact that singular homology is a homotopy invariant.

Theorem 2.3.6. *Suppose $f, g : X \to Y$ are homotopic maps. Then $f_\sharp, g_\sharp : S_\bullet(X) \to S_\bullet(Y)$ are chain homotopic maps.*

Corollary 2.3.7. *If $f, g : X \to Y$ are homotopic, then $f_* = g_*$.*

Corollary 2.3.8. *If $f : X \to Y$ is a homotopy equivalence, then $f_* : H_n(X) \to H_n(Y)$ is an isomorphism for all n.*

Proof. Let $g : Y \to X$ be a homotopy inverse of f. Then $f \circ g \simeq \mathrm{id}_Y$ and $g \circ f \simeq \mathrm{id}_X$. It follows that $f_* \circ g_*$ and $g_* \circ f_*$ are both the identity and are therefore inverse isomorphisms. $\square$

Corollary 2.3.9. *If X is contractible, then $H_0(X) \cong \mathbb{Z}$ and $H_n(X) = 0$ for $n > 0$.*

Proof of Theorem 2.3.6. We must construct a chain homotopy; that is, we need a map taking singular n-simplices in X to singular $(n + 1)$-simplices in Y. However, we do have a homotopy between f and g. This is a map $F : X \times I \to Y$. If σ is a singular n-simplex, then $\sigma \times \mathrm{id} : \Delta^n \times I \to X \times I$ may be composed with F to land in Y. We just need to turn $\Delta^n \times I$ into a "sum" of $(n + 1)$-simplices.

Let $\Delta^n \times \{0\} = \langle v_0, \dots, v_n \rangle$ and $\Delta^n \times \{1\} = \langle w_0, \dots, w_n \rangle$ as subsets of $\Delta^n \times I$. Iteratively divide $\Delta^n \times I$ into $(n + 1)$-simplices as follows: move $\langle v_0, \dots, v_n \rangle$ up to $\langle v_0, \dots, v_{n-1}, w_n \rangle$. Then move this up to $\langle v_0, \dots, v_{n-2}, w_{n-1}, w_n \rangle$, etc., $\langle v_0, \dots, v_i, w_{i+1}, \dots, w_n \rangle \mapsto \langle v_0, \dots, v_{i-1}, w_i, \dots, w_n \rangle$. The region between these is the $(n + 1)$-simplex $\langle v_0, \dots, v_i, w_i, \dots, w_n \rangle$ and $\Delta^n \times I$ is the union of these. See Figure 2.5 for the cases $n = 1$ and $n = 2$.

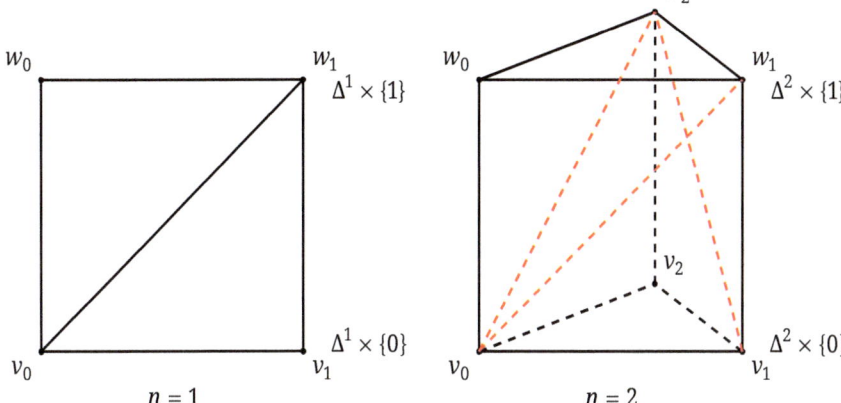

Figure 2.5: Decomposing $\Delta^n \times I$ into $(n+1)$-simplices.

Now define the *prism operator* $P : S_n(X) \to S_{n+1}(Y)$ by

$$P(\sigma) = \sum_{i=0}^{n} (-1)^i F \circ (\sigma \times \text{id})|_{\langle v_0,\dots,v_i,w_i,\dots,w_n \rangle}.$$

Note that

$$\partial P(\sigma) = \sum_{j \le i} (-1)^i (-1)^j F \circ (\sigma \times \text{id})|_{\langle v_0,\dots,\hat{v}_j,\dots,v_i,w_i,\dots w_n \rangle}$$

$$+ \sum_{j \ge i} (-1)^i (-1)^{j+1} F \circ (\sigma \times \text{id})|_{\langle v_0,\dots,v_i,w_i,\dots,\hat{w}_j,\dots,w_n \rangle}.$$

The two terms with $i = j$ cancel each other except for

(1) $F \circ (\sigma \times \text{id})|_{\langle \hat{v}_0, w_0,\dots,w_n \rangle}$, which is $g \circ \sigma = g_\#(\sigma)$; and

(2) $-F \circ (\sigma \times \text{id})|_{\langle v_0,\dots,v_n,\hat{w}_n \rangle}$, which is $-f \circ \sigma = -f_\#(\sigma)$.

The terms with $i \ne j$ are $-P\partial(\sigma)$:

$$\partial P(\sigma) = \sum_{j < i} (-1)^i (-1)^j F \circ (\sigma \times \text{id})|_{\langle v_0,\dots,v_i,w_i,\dots,\hat{w}_j,\dots w_n \rangle}$$

$$+ \sum_{j > i} (-1)^i (-1)^{j+1} F \circ (\sigma \times \text{id})|_{\langle v_0,\dots,\hat{v}_j,\dots,v_i,w_i,\dots,w_n \rangle}.$$

Thus, $\partial P + P\partial = g_\# - f_\#$. □

Note that the prism operator is very explicit if we apply it to simplicial homology. That is, the operator P takes a particular n-simplex in X to a formal sum of $(n+1)$-simplices in Y. So even though we have moved into this more general setting, the concrete nature of simplicial homology can serve as a guide of how to proceed.

Exercises

Exercise 2.3.1. Prove that $\partial^2 = 0$ in $S_\bullet(X)$.

Exercise 2.3.2. Prove that if X is homeomorphic to Y, then $H_\bullet(X) \cong H_\bullet(Y)$. (Do not simply appeal to Theorem 2.3.6.)

Exercise 2.3.3. Prove that singular homology is actually a special case of simplicial homology in the following sense. Given a space X, let $S(X)$ be the simplicial complex with one n-simplex Δ^n_σ for each singular simplex $\sigma : \Delta^n \to X$. Glue these together in the obvious way to form the complex. Prove that the simplicial homology of $S(X)$ is isomorphic to the singular homology of X.

Exercise 2.3.4. Let p, q be distinct points in $\mathbb{R}^n$. Let $\sigma : \Delta^1 \to \mathbb{R}^n$ be the line segment from p to q and let $\tau : \Delta^1 \to \mathbb{R}^n$ be the line segment from q to p. Note that $\sigma \neq -\tau$ in $S_1(\mathbb{R}^n)$. Prove that σ is homologous to $-\tau$, however.

2.3.2 Relative homology

Recall Proposition 1.2.22, which asserts that under some mild assumptions, if A is a contractible subspace of X, then the quotient map $X \to X/A$ is a homotopy equivalence. Then, by Corollary 2.3.8, this map induces an isomorphism in singular homology. Since the homology of A vanishes in positive degrees, this should perhaps not be surprising, but it does raise the following question: Given a subspace A of X, what is the relationship (if any) between $H_\bullet(A)$ and $H_\bullet(X)$? This leads to the idea of relative homology.

The chain complex $S_\bullet(A)$ is a *subcomplex* of $S_\bullet(X)$; that is, the boundary map in $S_\bullet(A)$ is simply the restriction of the boundary map of $S_\bullet(X)$. Denote the quotient $S_n(X)/S_n(A)$ by $S_n(X, A)$. The boundary map in $S_\bullet(X)$ induces a map

$$\partial_n : S_n(X, A) \to S_{n-1}(X, A)$$

and we still have $\partial^2 = 0$. Thus, $S_\bullet(X, A)$ is a chain complex; denote its homology groups by $H_n(X, A)$. These are the *homology groups of X relative to A*.

An element of $H_n(X, A)$ is represented by a chain $z \in S_n(X)$ with $\partial_n(z) \in S_{n-1}(A)$, not necessarily 0. A relative cycle z represents the trivial homology class in $H_n(X, A)$ if and only if it is a *relative boundary*: $z = \partial_{n+1}\beta + \gamma$ for some $\beta \in S_{n+1}(X)$, $\gamma \in S_n(A)$. See Figure 2.6 for an example.

Our next task is to relate all these homology groups together, and this is accomplished via the long exact sequence in homology. This is best done by generalizing the situation. To that end, consider a short exact sequence of chain complexes:

$$0 \longrightarrow A_\bullet \overset{i}{\longrightarrow} B_\bullet \overset{j}{\longrightarrow} C_\bullet \longrightarrow 0.$$

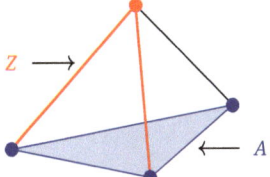

$z \longrightarrow$

$\longleftarrow A$

Figure 2.6: A relative 1-cycle.

This means that for each n we have a short exact sequence

$$0 \longrightarrow A_n \xrightarrow{\ i\ } B_n \xrightarrow{\ j\ } C_n \longrightarrow 0$$

and the maps i and j commute with the boundary maps in the respective complexes. We have induced maps in homology

$$i_* : H_n(A_\bullet) \to H_n(B_\bullet) \quad \text{and} \quad j_* : H_n(B_\bullet) \to H_n(C_\bullet).$$

Moreover, since the sequence of chain complexes is exact, we have $j_* \circ i_* = 0$. In order to push this further, we must construct the *connecting homomorphism*

$$d : H_n(C_\bullet) \to H_{n-1}(A_\bullet).$$

We will then have a sequence of homology groups:

$$\cdots \longrightarrow H_n(A_\bullet) \xrightarrow{\ i_*\ } H_n(B_\bullet) \xrightarrow{\ j_*\ } H_n(C_\bullet) \xrightarrow{\ d\ } H_{n-1}(A_\bullet) \longrightarrow \cdots$$

Lemma 2.3.10 (Snake Lemma). *Given a short exact sequence of chain complexes*

$$0 \longrightarrow A_\bullet \xrightarrow{\ i\ } B_\bullet \xrightarrow{\ j\ } C_\bullet \longrightarrow 0,$$

there is a well-defined homomorphism $d : H_n(C_\bullet) \to H_{n-1}(A_\bullet)$.

Proof. The proof is best done on a chalkboard with one's fingers, but here goes. Consider the following diagram

$$
\begin{array}{ccccc}
A_n & \xrightarrow{\ i\ } & B_n & \xrightarrow{\ j\ } & C_n \\
\downarrow{\scriptstyle\partial} & & \downarrow{\scriptstyle\partial} & & \downarrow{\scriptstyle\partial} \\
A_{n-1} & \xrightarrow{\ i\ } & B_{n-1} & \xrightarrow{\ j\ } & C_{n-1}
\end{array}
$$

Our task is to associate to a cycle in C_n a cycle in A_{n-1}, show that it is well-defined, and that the resulting map is a group homomorphism. Let $c \in Z_n(C_\bullet)$ be a cycle. Since j is surjective, there is a $b \in B_n$ with $j(b) = c$. Since

$$j(\partial b) = \partial(j(b)) = \partial(c) = 0$$

and since the sequence is exact, we have $\partial(b) = i(a)$ for some $a \in A_{n-1}$. Note that $\partial(a) = 0$:

$$i(\partial(a)) = \partial(i(a))$$
$$= \partial(\partial(b))$$
$$= 0,$$

and, since i is injective, we have $\partial(a) = 0$. Define $d : H_n(C_\bullet) \to H_{n-1}(A_\bullet)$ by $d([c]) = [a]$.

The map d is well-defined. Indeed, a is uniquely determined, since i is injective. If we choose b' instead of b as a lift of c, then $b - b' \in \ker(j) = \operatorname{im}(i)$ so that $b - b' = i(a')$ for some $a' \in A_n$; that is, $b' = b + i(a')$. But then

$$i(a + \partial(a')) = i(a) + i(\partial(a'))$$
$$= \partial(b) + \partial(i(a'))$$
$$= \partial(b + i(a'))$$
$$= \partial(b').$$

Thus, changing b to b' changes a to $a + \partial(a')$, and these are homologous. If we replace c by $c + \partial(c')$ for some $c' \in C_{n+1}$, then there is a $b' \in B_{n+1}$ with $j(b') = c'$. We then have

$$c + \partial(c') = c + \partial(j(b'))$$
$$= c + j(\partial(b'))$$
$$= j(b + \partial(b')).$$

So b is replaced by $b + \partial(b')$, leaving $\partial(b)$ and thus a unchanged.

It remains to show that d is a homomorphism. If $d([c_1]) = [a_1]$ and $d([c_2]) = [a_2]$ via b_1 and b_2, respectively, then

$$j(b_1 + b_2) = j(b_1) + j(b_2) = c_1 + c_2$$

and

$$i(a_1 + a_2) = i(a_1) + i(a_2) = \partial(b_1) + \partial(b_2) = \partial(b_1 + b_2).$$

It follows that $d([c_1] + [c_2]) = [a_1] + [a_2]$. □

Theorem 2.3.11. *The following sequence is exact.*

$$\cdots \longrightarrow H_n(A_\bullet) \xrightarrow{i_*} H_n(B_\bullet) \xrightarrow{j_*} H_n(C_\bullet) \xrightarrow{d} H_{n-1}(A_\bullet) \longrightarrow \cdots$$

Proof. We have already noted that $j_* \circ i_* = 0$ since $j \circ i = 0$. Thus, $\text{im}(i_*) \subseteq \text{ker}(j_*)$. Recalling the definition of the map d, since $\partial(j(b)) = 0$ we see that $d \circ j_* = 0$ and $\text{im}(j_*) \subseteq \text{ker}(d)$. Also, $\text{im}(d) \subseteq \text{ker}(i_*)$ since $i_*(d([c])) = [\partial(b)] = 0$. These are the easy containments.

$\text{ker}(j_*) \subseteq \text{im}(i_*)$: Suppose $j_*([b]) = 0$. Then $j(b)$ is a boundary in C_n; that is, $j(b) = \partial(c')$ for some $c' \in C_{n+1}$. Since j is surjective, $c' = j(b')$ for some $b' \in B_{n+1}$. Then $j(b - \partial(b')) = j(b) - j(\partial(b')) = j(b) - \partial(j(b')) = 0$ since $\partial(j(b')) = \partial(c') = j(b)$. Then $b - \partial(b') = i(a)$ for some $a \in A_n$. The element a is a cycle: $i(\partial(a)) = \partial(i(a)) = \partial(b - \partial(b')) = \partial(b) = 0$ and i is injective. Thus, $i_*([a]) = [b - \partial(b')] = [b]$.

$\text{ker}(d) \subseteq \text{im}(j_*)$: Suppose $[c] \in \text{ker}(d)$. Then $a = \partial(a')$ for some $a' \in A_n$. The element $b - i(a')$ is a cycle: $\partial(b - i(a')) = \partial(b) - \partial(i(a')) = \partial(b) - i(\partial(a')) = \partial(b) - i(a) = 0$. Also, $j(b - i(a')) = j(b) - j(i(a')) = j(b) = c$ so that $j_*([b - i(a')]) = [c]$.

$\text{ker}(i_*) \subseteq \text{im}(d)$: Suppose $[a] \in \text{ker}(i_*)$, $a \in A_{n-1}$. Then $i(a) = \partial(b)$ for some $b \in B_n$. Then $j(b)$ is a cycle: $\partial(j(b)) = j(\partial(b)) = j(i(a)) = 0$. Thus $d(j([b])) = [a]$. □

Now, if (X, A) is a pair of spaces, then by definition the following sequence of chain complexes is exact:

$$0 \longrightarrow S_\bullet(A) \longrightarrow S_\bullet(X) \longrightarrow S_\bullet(X, A) \longrightarrow 0.$$

We therefore have the *long exact homology sequence of the pair* (X, A):

$$\cdots \longrightarrow H_n(A) \longrightarrow H_n(X) \longrightarrow H_n(X, A) \xrightarrow{\;d\;} H_{n-1}(A) \longrightarrow \cdots$$

The map d has a more explicit description in this context. If z is a relative cycle, then $\partial(z) \in C_{n-1}(A)$. But this is a *cycle* in $C_{n-1}(A)$ since $\partial(\partial(z)) = 0$. Thus

$$d([z]) = [\partial(z)] \in H_{n-1}(A).$$

One more thing to note: we can augment the short exact sequence above with the following short exact sequence in degree -1:

$$0 \longrightarrow \mathbb{Z} \longrightarrow \mathbb{Z} \longrightarrow 0 \longrightarrow 0$$

and the long exact sequence in homology still works. This shows that $\tilde{H}_n(X, A) \cong H_n(X, A)$ for all $n \geq 0$. As a particular example, note that, taking $A = \{x_0\}$ for some point $x_0 \in X$, we see that the map $H_n(X) \to H_n(X, x_0)$ is an isomorphism for all $n > 0$ and for $n = 0$ we have $\tilde{H}_0(X) \cong H_0(X, x_0)$.

Relative homology is functorial as well. Indeed, if $f : (X, A) \to (Y, B)$ is a map of pairs, then we have an induced chain map $f_\sharp : S_\bullet(X, A) \to S_\bullet(Y, B)$ and hence an induced map in relative homology $f_* : H_n(X, A) \to H_n(Y, B)$.

Proposition 2.3.12. *If $f, g : (X, A) \to (Y, B)$ are homotopic via maps of pairs $(X, A) \to (Y, B)$, then*

$$f_* = g_* : H_n(X, A) \to H_n(Y, B).$$

Proof. The prism operator P takes $S_n(A)$ into $S_{n+1}(B)$ and so induces a relative prism operator $P : S_n(X, A) \to S_{n+1}(Y, B)$. The formula $\partial P + P\partial = g_\sharp - f_\sharp$ still holds. □

Exercises

Exercise 2.3.5. Prove that if (X, A, B) is a triple of spaces $(B \subset A \subset X)$, then there is a long exact sequence of homology groups

$$\cdots \to H_n(A, B) \to H_n(X, B) \to H_n(X, A) \to H_{n-1}(A, B) \to \cdots$$

(Hint: Consider the sequence $0 \to S_\bullet(A, B) \to S_\bullet(X, B) \to S_\bullet(X, A) \to 0$.)

Exercise 2.3.6. Relative homology makes sense in simplicial homology as well: if A is a subcomplex of K, then we have relative homology groups $H_\bullet(K, A)$, which are the homology groups of the quotient complex $C_\bullet(K)/C_\bullet(A)$. Compute the following:
(1) Let K be a complex whose underlying space is the Möbius band and let A be the boundary edge of K. Compute $H_\bullet(K)$ and $H_\bullet(K, A)$.
(2) Let K be the torus of Figure 2.3 and let A be the subcomplex consisting of the 1-chain along the top of the square (i. e., A consists of the three edges $\langle v_1, v_2 \rangle$, $\langle v_2, v_3 \rangle$, and $\langle v_3, v_1 \rangle$ and their vertices). Compute $H_\bullet(K, A)$.
(3) Let K be a 2-dimensional complex and let σ be a 2-simplex of K. Let A be the subcomplex $K \setminus \text{int}(\sigma)$. Compute $H_\bullet(K, A)$.

2.3.3 Excision

The long exact sequence is useful for computation since it allows us to build up the homology of a space X from that of a subspace A and the relative homology of (X, A). However, we need another tool to get the ball rolling, since, as it stands, the only spaces for which we know the homology are discrete (a disjoint union of points with the discrete topology). This is where *excision* comes in.

Consider the sketch in Figure 2.7. The chain α is a relative cycle in $(X - Z, A - Z)$. That is, it is a relative cycle in (X, A) that also happens to miss the subspace $Z \subset A$. Since Z does not interfere with α, we can think of it as a cycle in (X, A) and it should give us the same information. Excision makes this statement precise.

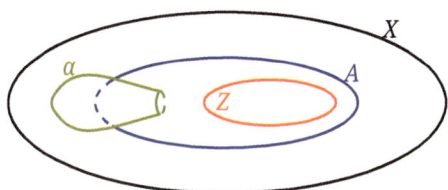

Figure 2.7: A relative cycle in $(X - Z, A - Z)$.

Theorem 2.3.13 (Excision). *Suppose $\overline{Z} \subset \text{int}(A)$. Then the inclusion $(X - Z, A - Z) \hookrightarrow (X, A)$ induces isomorphisms*

$$H_n(X - Z, A - Z) \to H_n(X, A)$$

for all $n \geq 0$.

Before proving Theorem 2.3.13, let us use it to compute the homology of the sphere S^n. This is the first nontrivial space whose singular homology we can compute. Denote the closed northern hemisphere by D^n_+ and the closed southern hemisphere by D^n_-. Note that $D^n_+ \cap D^n_-$ is the equatorial S^{n-1}. Morally, we wish to excise the open southern hemisphere

$$Z = \{x \in S^n \mid x_{n+1} < 0\},$$

but this does not satisfy the hypotheses of Theorem 2.3.13. However, if we consider the set

$$V = \{x \in S^n \mid x_{n+1} < -1/2\},$$

then via excision the inclusion $(S^n - V, D^n_- - V) \hookrightarrow (S^n, D^n_-)$ induces an isomorphism in homology. Note, however, that the pair (D^n_+, S^{n-1}) is a deformation retract of $(S^n - V, D^n_- - V)$ (move upward along the longitudes). It follows that the map $(D^n_+, S^{n-1}) \hookrightarrow (S^n, D^n_-)$ induces an isomorphism in homology.

Now, let us consider the pair (D^n, S^{n-1}). The long exact sequence then has the form

$$\cdots \to \tilde{H}_k(S^{n-1}) \to \tilde{H}_k(D^n) \to \tilde{H}_k(D^n, S^{n-1}) \xrightarrow{d} \tilde{H}_{k-1}(S^{n-1}) \to \cdots$$

Since D^n is contractible its reduced homology vanishes in all degrees. It follows that the map

$$d : \tilde{H}_k(D^n, S^{n-1}) \to \tilde{H}_{k-1}(S^{n-1})$$

is an isomorphism for all k.

On the other hand, since D^n_- is contractible, the long exact sequence of the pair (S^n, D^n_-) implies that the map $\tilde{H}_k(S^n) \to \tilde{H}_k(S^n, D^n_-)$ is an isomorphism for all k. Putting this together, we have a sequence of isomorphisms for all k and $n \geq 1$

$$\tilde{H}_k(S^n) \longrightarrow \tilde{H}_k(S^n, D^n_-) \longleftarrow \tilde{H}_k(D^n, S^{n-1}) \longrightarrow \tilde{H}_{k-1}(S^{n-1})$$

Now, since S^0 consists of two points, we have $\tilde{H}_0(S^0) \cong \mathbb{Z}$ and $\tilde{H}_k(S^0) = 0$ for $k > 0$. We therefore conclude the following.

Corollary 2.3.14. *For all $k \geq 0$ and all $n \geq 1$*

$$\tilde{H}_k(S^n) \cong \begin{cases} \mathbb{Z} & k = n \\ 0 & k \neq n. \end{cases}$$

While we are here, we may as well prove the following.

Theorem 2.3.15 (Brouwer Fixed-Point Theorem). *Any continuous map $f : D^n \to D^n$ has a fixed point.*

Proof. Suppose f does not have a fixed point. Define a map $r : D^n \to S^{n-1}$ by setting $r(x)$ to be the point at which the directed ray from $f(x)$ to x intersects S^{n-1}. Note that $r|_{S^{n-1}} = \text{id}_{S^{n-1}}$. Let $i : S^{n-1} \to D^n$ be the inclusion. Then $r \circ i$ is the identity on S^{n-1}. But then the composite

$$H_{n-1}(S^{n-1}) \xrightarrow{\ i_*\ } H_{n-1}(D^n) \xrightarrow{\ r_*\ } H_{n-1}(S^{n-1})$$

would be the identity $\mathbb{Z} \to \mathbb{Z}$, which is absurd. □

We now return to the proof of Theorem 2.3.13. This requires several new ideas. In essence, we want to "ignore" chains that live entirely inside Z. This creates two problems: (1) how do we do that? and (2) what if a chain "hits" Z but is not entirely inside of it? This suggests that we need a mechanism to "subdivide" chains in some way that allows us to kill off those that lie in Z. Luckily, our work with simplicial complexes gives us a path forward: barycentric subdivision.

Let $\mathcal{U} = \{U_i\}$ be a collection of subspaces of X such that $X = \bigcup_i \text{int}(U_i)$ and let

$$S_n^{\mathcal{U}}(X) = \left\{ \sum n_j \sigma_j \in S_n(X) \mid \text{im}(\sigma_j) \subset U_i \text{ for some } j \right\}.$$

Note that $\partial : S_n(X) \to S_{n-1}(X)$ restricts to $\partial : S_n^{\mathcal{U}}(X) \to S_{n-1}^{\mathcal{U}}(X)$. Thus, $S_\bullet^{\mathcal{U}}(X)$ is a chain complex, called the $\mathcal{U}$-*small singular chains on* X. Denote the resulting homology groups by $H_n^{\mathcal{U}}(X)$.

Proposition 2.3.16. *The inclusion $i : S_\bullet^{\mathcal{U}}(X) \to S_\bullet(X)$ is a chain homotopy equivalence. That is, there is a chain map $\rho : S_\bullet(X) \to S_\bullet^{\mathcal{U}}(X)$ with $\rho \circ i$ equal to the identity and $i \circ \rho$ chain homotopic to the identity. Hence, i induces isomorphisms $H_n^{\mathcal{U}}(X) \cong H_n(X)$ for all $n \geq 0$.*

Proof of Theorem 2.3.13. Let $\mathcal{U} = \{X - Z, A\}$. Note that $X = (X - \overline{Z}) \cup \text{int}(A)$, since $\overline{Z} \subset \text{int}(A)$. By Proposition 2.3.16 there is a chain homotopy $D : S_n(X) \to S_{n+1}(X)$ with image in $S_\bullet^{\mathcal{U}}(X)$ such that $\partial D + D\partial = \text{id} - i \circ \rho$ and $\rho \circ i = \text{id}$. All these maps take chains in A into chains in A, so they induce maps on quotients by $S_n(A)$, automatically satisfying the same formulas. It follows that the map

$$S_\bullet^{\mathcal{U}}(X)/S_\bullet(A) \longrightarrow S_\bullet(X)/S_\bullet(A)$$

induces an isomorphism on homology.

We also have a chain map

$$S_n(X - Z)/S_n(A - Z) \longrightarrow S_n^{\mathcal{U}}(X)/S_n(A).$$

This map is obviously an isomorphism since both groups are free with basis the singular simplices in $X - Z$ that do not lie in A. Thus, we have a sequence of maps

$$S_\bullet(X - Z)/S_\bullet(A - Z) \longrightarrow S_\bullet^{\mathcal{U}}(X)/S_\bullet(A) \longrightarrow S_\bullet(X)/S_\bullet(A)$$

inducing isomorphisms in homology; that is, the inclusion $(X - Z, A - Z) \to (X, A)$ induces isomorphisms

$$H_n(X - Z, A - Z) \longrightarrow H_n(X, A)$$

for all $n \geq 0$. □

As you might guess, the proof of Proposition 2.3.16 is very technical and the algebra is not especially illuminating. Here is the overall plan. Recall the barycentric subdivision of an n-simplex $\sigma = \langle v_0, \ldots, v_n \rangle$. The barycenter b is the centroid:

$$b = \sum t_i v_i \quad t_i = \frac{1}{n+1} \text{ for all } i.$$

We then use these iteratively on the faces of the n-simplex to build the barycentric subdivision. Note that the simplices of the subdivision are smaller; in fact, the diameter of any simplex in the subdivision is at most $n/(n+1)$ times the diameter of σ. It follows that if we iterate this construction r times, each simplex has diameter at most $(n/(n + 1))^r$ times the diameter of σ and hence can be made arbitrarily small.

The goal is now to define a "subdivision operator" $\text{Sd} : S_n(X) \to S_n(X)$ that is chain homotopic to the identity such that Sd^r has image lying in $S_n^{\mathcal{U}}(X)$ for some r. Suppose $\sigma : \Delta^n \to X$ is a singular n-simplex and consider the open cover $\mathcal{V} = \{\sigma^{-1}(\text{int}(U_j))\}$ of Δ^n. Since Δ^n is a compact metric space, there is an $\varepsilon > 0$ such that every set of diameter less than ε lies in some element of $\mathcal{V}$. Thus, if we iterate the barycentric subdivision enough, σ restricted to any simplex lies in some U_j and so σ is a "sum" of simplices in $S_n^{\mathcal{U}}(X)$. Of course the word "is" is doing some heavy lifting here. What we really mean is that there is a chain homotopy lying around that fixes the difference between σ and this sum of $\mathcal{U}$-small simplices.

So, this is the plan and now we have to stop beating around the bush and get on with it. Suppose $\sigma = \langle v_0, \dots, v_n \rangle$ is an n-simplex in some Euclidean space and that p is a point. We define the *join* of p and σ to be the $(n+1)$-simplex

$$p\sigma = \langle p, v_0, \dots, v_n \rangle.$$

We extend this operator linearly to linear combinations of affine simplices:

$$pc = p\left(\sum n_i \sigma_i\right) = \sum n_i p\sigma_i.$$

Note that we have the following formulas:

$$\partial pc = \begin{cases} c - p\partial c & n > 0 \\ c - (\sum n_i)p & n = 0. \end{cases}$$

We will now use functoriality to our advantage. The maps we are trying to construct are functorial with respect to maps $f : X \to Y$. It is therefore sufficient to define things for the space $X = \Delta^n$ and the singular n-simplex $\delta_n : \Delta^n \to \Delta^n$ given by the identity map. We will define Sd and the chain homotopy $D : S_n(X) \to S_{n+1}(X)$ at the same time. For $n = 0$ we set $\mathrm{Sd}(\delta_0) = \delta_0$ and $D(\delta_0) = 0$. Denote the barycenter of Δ^n by b_n. Assuming Sd and D have been defined in dimensions $< n$, we use the join operation to define them in dimension n:

$$\mathrm{Sd}(\delta_n) = b_n \mathrm{Sd}(\partial \delta_n)$$
$$D(\delta_n) = b_n(\delta_n - \mathrm{Sd}(\delta_n) - D\partial(\delta_n))$$

Note that Sd is doing what our intuition tells us to do: barycentrically subdivide the n-simplex. This is accomplished by taking the join of the barycenter with the boundary. The map D is then the correction needed to make sure the homology works out.

Lemma 2.3.17. *We have the following relations.*

$$\partial \,\mathrm{Sd} = \mathrm{Sd}\,\partial$$
$$\partial D = \mathrm{id} - \mathrm{Sd} - D\partial$$

Proof. We proceed by induction on n, with $n = 0$ being clear. For $n > 0$ it suffices by naturality to evaluate both equations on δ_n. For the first equation we have

$$\partial \,\mathrm{Sd}(\delta_n) = \partial b_n \mathrm{Sd}\partial \delta_n = \mathrm{Sd}(\partial \delta_n) - b_n \partial \,\mathrm{Sd}(\partial \delta_n),$$

and by the inductive hypothesis

$$b_n \partial \,\mathrm{Sd}(\partial \delta_n) = b_n \mathrm{Sd}(\partial^2 \delta_n) = 0.$$

For the second equation, we have

$$
\begin{aligned}
\partial D(\delta_n) &= \partial b_n(\delta_n - \mathrm{Sd}(\delta_n) - D(\partial\delta_n)) \\
&= \delta_n - \mathrm{Sd}(\delta_n) - D\partial(\delta_n) - b_n(\partial\delta_n - \partial\mathrm{Sd}(\delta_n) - \partial D(\partial\delta_n)) \\
&= \delta_n - \mathrm{Sd}(\delta_n) - D\partial(\delta_n) - b_n(\partial\delta_n - \mathrm{Sd}(\partial\delta_n) - \partial\delta_n + \mathrm{Sd}(\partial\delta_n) + D(\partial^2\delta_n)) \\
&= \delta_n - \mathrm{Sd}(\delta_n) - D\partial(\delta_n).
\end{aligned}
$$
□

Proof of Proposition 2.3.16. Let z be an n-cycle in X. Then

$$
z - \mathrm{Sd}(z) = \partial D(z) + D(\partial(z)).
$$

Since $\partial(z) = 0$, we see that z is homologous to $\mathrm{Sd}(z)$. Inductively, then, we see that z is homologous to $\mathrm{Sd}^r(z)$ for all $r \geq 0$. Since there is an r with $\mathrm{Sd}^r(z) \in S_n^{\mathcal{U}}(X)$, we see that the map

$$
H_n^{\mathcal{U}}(X) \longrightarrow H_n(X)
$$

is surjective. But it is obviously injective, since if a $\mathcal{U}$-small cycle z is a boundary in X, then we can subdivide that boundary to see that $[z] = 0$ in $H_n^{\mathcal{U}}(X)$.
□

Exercises

Exercise 2.3.7. We used functoriality to reduce the construction of Sd to consideration of the space $X = \Delta^n$ and $\sigma = \delta_n$. Make this precise as follows. First, note that a singular n-simplex $\sigma : \Delta^n \to X$ may be written $\sigma = \sigma_{\sharp}(\delta_n)$. Prove that a homomorphism $a_X : S_n(X) \to S_n(X)$ can be defined for all X using a single value $a_X(\delta_n) \in S_n(\Delta^n)$ by the equation

$$
a(\sigma) = \sigma_{\sharp}(a_X(\delta_n)).
$$

Then prove that this is *natural*: if $f : X \to Y$ is continuous, then the following diagram commutes:

$$
\begin{array}{ccc}
S_n(X) & \xrightarrow{\ a_X\ } & S_n(X) \\
\downarrow{f_{\sharp}} & & \downarrow{f_{\sharp}} \\
S_n(Y) & \xrightarrow{\ a_Y\ } & S_n(Y).
\end{array}
$$

Conversely, show that a homomorphism $a_X : S_n(X) \to S_n(X)$ defined for all X and satisfying the naturality property is determined by its value $a_{\Delta^n}(\delta_n)$.

Exercise 2.3.8. Generalize the argument we used to compute the homology of a sphere (Corollary 2.3.14) to prove the following. Let ΣX be the *suspension* of X. This is the quotient of $X \times [-1, 1]$ obtained by identifying $X \times \{-1\}$ to a point P and $X \times \{1\}$ to a point Q.

We can think of X as the subspace $X \times \{0\}$. Prove that $S^n = \Sigma S^{n-1}$ for $n \geq 1$. Prove the suspension isomorphism: $\tilde{H}_n(X) \cong \tilde{H}_{n+1}(\Sigma X)$.

Exercise 2.3.9. Let I be a homeomorphic image of $[0, 1]$ in S^2 and let $x \in I$. Show that the inclusion $(S^2 - \{x\}, I - \{x\}) \to (S^2, I)$ is not an excision.

Exercise 2.3.10 (Invariance of dimension). Prove that if $U \subseteq \mathbb{R}^m$ and $V \subseteq \mathbb{R}^n$ are homeomorphic open sets, then $m = n$.

2.3.4 Homology of spheres revisited

In the previous section, we used excision to compute the homology of a sphere. As part of that calculation, we used the long exact sequence of the pair (D^n, S^{n-1}) to show that for all k the connecting homomorphism

$$d : \tilde{H}_k(D^n, S^{n-1}) \to \tilde{H}_{k-1}(S^{n-1})$$

is an isomorphism. In this section, we show how to compute the former group more explicitly in terms of quotient spaces.

Definition 2.3.18. A *good pair* is a pair (X, A) where $A \subset X$ is a closed subspace that is a deformation retract of some neighborhood in X.

Example 2.3.19. The primary example of a good pair is (X, A), where X is a cell complex and A is a subcomplex. The space X is obtained from A by attaching cells of various dimensions. Consider the case of $X = A \cup_f e^n$, where $f : \partial e^n \to A$ is the attaching map. Let U be a neighborhood of ∂e^n in e^n and let $Y = A \cup_f U$. Then Y deformation retracts to A and $X = Y \cup (e^n)'$, where the cell $(e^n)'$ is the complement of U in e^n (homeomorphic to e^n). See Figure 2.8. Choosing such neighborhoods for all the cells in X yields the desired deformation retraction to A.

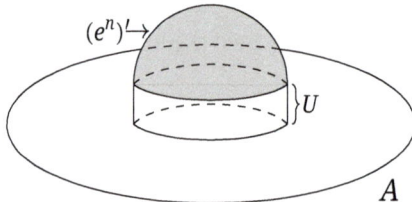

Figure 2.8: A neighborhood of A in the cell complex X.

Theorem 2.3.20. *Suppose that (X, A) is a good pair. Then there is a long exact sequence*

$$\cdots \longrightarrow \tilde{H}_n(A) \xrightarrow{\ i\ } \tilde{H}_n(X) \xrightarrow{\ j\ } \tilde{H}_n(X/A) \xrightarrow{\ d\ } \tilde{H}_{n-1}(A) \longrightarrow \cdots$$

where $i : A \hookrightarrow X$ is the inclusion and $j : X \to X/A$ is the quotient map.

Example 2.3.21. Taking $(X, A) = (D^n, S^{n-1})$, since D^n is contractible, we have that the map

$$d : \tilde{H}_k(D^n/S^{n-1}) \to \tilde{H}_{k-1}(S^{n-1})$$

is an isomorphism for all k. But D^n/S^{n-1} is homeomorphic to the sphere S^n, so proceeding by induction from the base case of $\tilde{H}_0(S^0) \cong \mathbb{Z}$, we recover the calculation of the previous section.

Proof of Theorem 2.3.20. The sequence in question is really just the long exact sequence of the pair (X, A). We simply need to show that the quotient map $q : (X, A) \to (X/A, A/A)$ induces isomorphisms on singular homology and then note that

$$H_n(X/A, A/A) \cong \tilde{H}_n(X/A)$$

for all $n \geq 0$. Suppose A is a deformation retract of the neighborhood V. Consider the commutative diagram

$$
\begin{array}{ccccc}
H_n(X, A) & \longrightarrow & H_n(X, V) & \longleftarrow & H_n(X - A, V - A) \\
\downarrow{q_*} & & \downarrow{q_*} & & \downarrow{q_*} \\
H_n(X/A, A/A) & \longrightarrow & H_n(X/A, V/A) & \longleftarrow & H_n(X/A - A/A, V/A - A/A)
\end{array}
$$

Since $(X, A) \hookrightarrow (X, V)$ is a homotopy equivalence, the two left horizontal maps are isomorphisms. The two right horizontal maps are isomorphisms by excision. The right vertical q_* is an isomorphism, since q restricts to a homeomorphism on $X - A$. Commutativity then gives that the other two q_* are isomorphisms as well. □

Corollary 2.3.22. *Suppose (X_α, x_α) is a good pair for each $\alpha \in \Lambda$ and let $X = \bigvee_{\alpha \in \Lambda} X_\alpha$. Then the inclusions $i_\alpha : X_\alpha \hookrightarrow X$ induce an isomorphism*

$$\oplus_\alpha i_\alpha : \bigoplus_{\alpha \in \Lambda} \tilde{H}_n(X_\alpha) \longrightarrow \tilde{H}_n(X)$$

for all n.

Proof. We have the following chain of isomorphisms

$$\tilde{H}_n(X) \cong H_n\left(\coprod X_\alpha, \coprod \{x_\alpha\}\right) \cong \bigoplus_\alpha H_n(X_\alpha, x_\alpha) \cong \bigoplus_\alpha \tilde{H}_n(X_\alpha). \qquad \square$$

Example 2.3.23. Let us compute the homology of the torus. Recall that T is the quotient of the square obtained by identifying opposite edges. Let Z be a small closed disc of radius ε in the center of the square and let A be a slightly larger disc of radius 2ε. Then $\bar{Z} \subset \text{int}(A)$ so that we may excise Z:

$$H_n(T - Z, A - Z) \cong H_n(T, A)$$

for all n. Since A is contractible, the relative homology $H_n(T,A)$ is isomorphic to the reduced homology $\tilde{H}_n(T)$ for all n. Now, the space $A - Z$ is an annulus and deformation retracts to a circle of radius 1.5ε. The space $T - Z$ is the quotient of the square minus Z, and this deformation retracts to its boundary square. Thus $T - Z$ has the homotopy type of $S^1 \vee S^1$. The long exact sequence of the pair $(T - Z, A - Z)$ then has the form

$$0 \longrightarrow H_2(T - Z, A - Z) \longrightarrow H_1(S^1) \longrightarrow H_1(S^1 \vee S^1) \longrightarrow H_1(T - Z, A - Z) \longrightarrow 0$$

The initial 0 is the group $H_2(S^1 \vee S^1)$, which we know vanishes thanks to Corollary 2.3.22. The group $H_1(S^1 \vee S^1)$ is $\mathbb{Z} \oplus \mathbb{Z}$ (Corollary 2.3.22 again). Thus we need to determine the map

$$H_1(S^1) \longrightarrow H_1(S^1 \vee S^1),$$

which is just a map $\mathbb{Z} \to \mathbb{Z} \oplus \mathbb{Z}$. Note that the generator of $H_1(S^1)$ corresponds to a small circle in the interior of $T - Z$ and under the deformation retraction to the boundary square it maps to the whole square. Under the quotient map, this then traverses each S^1 in the wedge twice, in opposite directions. Algebraically, then, this is the map

$$1 \mapsto (1,0) + (0,1) - (1,0) - (0,1) = (0,0);$$

that is, it is the zero map. We therefore conclude that

$$H_0(T) \cong \mathbb{Z}$$
$$H_1(T) \cong H_1(T - Z, A - Z) \cong \mathbb{Z} \oplus \mathbb{Z}$$
$$H_2(T) \cong H_2(T - Z, A - Z) \cong \mathbb{Z}$$

The copy of $S^1 \vee S^1$ embedded in T gives us a pair of generating cycles for $H_1(T)$. Indeed, the two singular simplices $\alpha, \beta : \Delta^1 \to T$, illustrated in Figure 2.9 with red and blue circles respectively, are cycles that span $H_1(T)$.

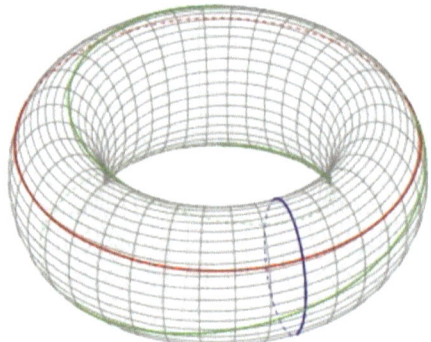

Figure 2.9: Generating cycles on the torus.

Exercises

Exercise 2.3.11. Using the fact that $(D^n, S^{n-1}) \approx (\Delta^n, \partial\Delta^n)$, show that the identity map $\delta_n : \Delta^n \to \Delta^n$ is a cycle generating $H_n(\Delta^n, \partial\Delta^n)$. (Hint: This is obviously true for $n = 0$. Proceed by induction. Let $\Lambda \subset \Delta^n$ be the union of all but one of the $(n-1)$-faces in Δ. Use the long exact sequence of the triple $(\Delta^n, \partial\Delta^n, \Lambda)$ to finish the argument.)

Exercise 2.3.12. Use the previous exercise and the fact that $S^n \approx \Delta_1^n \cup \Delta_2^n$ identified along their boundaries, preserving the order of the vertices, to show that $\Delta_1^n - \Delta_2^n$ is a cycle generating $\tilde{H}_n(S^n)$.

Exercise 2.3.13. Let X be a 2-sphere with a disc D^2 attached to the equator along its boundary. Compute the homology of X.

Exercise 2.3.14. Let X_2 be the torus T with two disjoint discs attached along curves homologous to the blue curve in Figure 2.9; that is, one disc could be attached along this curve, and another attached along a parallel translate of this curve. Compute the homology of X_2. If X_n is the analogous space with n disjoint discs attached to T, compute the homology of X_n.

Exercise 2.3.15. Consider the green curve y in Figure 2.9. Express $\bar{y}$ in terms of the homology classes of the generating cycles α and β.

2.3.5 The Mayer–Vietoris sequence

The Mayer–Vietoris sequence is an extremely useful tool for calculation. Its existence is equivalent to the excision property, as we shall see.

Theorem 2.3.24. *Suppose that $X = \mathrm{int}(A) \cup \mathrm{int}(B)$. Then there is a long exact sequence*

$$\cdots \to H_n(A \cap B) \xrightarrow{\Phi} H_n(A) \oplus H_n(B) \xrightarrow{\Psi} H_n(X) \xrightarrow{d} H_{n-1}(A \cap B) \to \cdots$$

Proof. Consider the $\mathcal{U}$-small chains $S_n^{\mathcal{U}}(X)$ for $\mathcal{U} = \{A, B\}$. Then we have a sequence of chain complexes

$$0 \longrightarrow S_n(A \cap B) \xrightarrow{\varphi} S_n(A) \oplus S_n(B) \xrightarrow{\psi} S_n^{\mathcal{U}}(X) \longrightarrow 0$$

$$x \xmapsto{\varphi} (x, -x)$$

$$(x, y) \xmapsto{\psi} x + y$$

Note that $\psi \circ \varphi = 0$. If $x \in \ker(\varepsilon)$, then $x = 0$ so that φ is injective. Suppose $\psi(x,y) = 0$. Then $x = -y$ and so x is a chain in both A and B; that is, $(x,y) = \varphi(x)$. The map ψ is clearly surjective and so the sequence is exact. We therefore have a long exact sequence in homology and since we know $H_n^{\mathcal{U}}(X) \cong H_n(X)$, this is the sequence we seek. □

An explicit description of the map d is possible. If z is an n-cycle in X, then we can write $z = x + y$ with $x \in S_n(A), y \in S_n(B)$. Since $\partial(x+y) = \partial(z) = 0$, we see that $\partial(x) = -\partial(y)$. Then $d([z]) = [\partial(x)] = -[\partial(y)]$.

Example 2.3.25. In Example 2.3.23 we computed the homology of the torus by excising a small disc. Let us compute this homology again using the Mayer–Vietoris sequence. There are any number of ways we might do this; here are two solutions.

First, as in Example 2.3.23, let A be a small disc of radius 2ε in T and let B be the complement of the disc of radius ε inside of A. Then $T = \mathrm{int}(A) \cup \mathrm{int}(B)$, $A \cap B \approx S^1$, A is contractible, and $B \simeq S^1 \vee S^1$. The Mayer–Vietoris sequence then has the form

$$0 \longrightarrow H_2(T) \xrightarrow{d} H_1(S^1) \longrightarrow H_1(S^1) \oplus H_1(S^1) \longrightarrow H_1(T) \longrightarrow 0$$

As in Example 2.3.23, the map $H_1(S^1) \longrightarrow H_1(S^1) \oplus H_1(S^1)$ is the zero map, since the S^1 in $A \cap B$ retracts to the boundary square and under the projection maps to each circle twice with opposite orientations. We therefore recover the homology of T. Note that this computation is essentially the same as Example 2.3.23, illustrating the equivalence of excision and the Mayer–Vietoris sequence.

Here is another approach. Cut the torus T along two longitudinal circles (i. e., perpendicular to the hole). Let A and B be the resulting two halves. Note that T is not quite the union of the interiors of these, so we need to thicken them a bit so that they overlap, and then $A \cap B$ is the disjoint union of two small cylinders, each homotopic to S^1. Moreover, both A and B are cylinders and are therefore homotopic to S^1. The associated Mayer–Vietoris sequence is then

$$0 \longrightarrow H_2(T) \xrightarrow{d} H_1(S^1) \oplus H_1(S^1) \longrightarrow H_1(S^1) \oplus H_1(S^1) \longrightarrow H_1(T) \;\;\rceil$$

$$\lfloor \xrightarrow{d} H_0(S^1) \oplus H_0(S^1) \longrightarrow H_0(S^1) \oplus H_0(S^1) \longrightarrow H_0(T) \longrightarrow 0$$

The map $H_1(S^1) \oplus H_1(S^1) \longrightarrow H_1(S^1) \oplus H_1(S^1)$ is the algebraic version of the "gluing" map that joins the two cylinders together. We must orient the boundary circles on the cylinders A and B compatibly and then the map is

$$(1,0) \mapsto (1,1)$$
$$(0,1) \mapsto (1,1)$$

This homomorphism has a kernel isomorphic to $\mathbb{Z}$, spanned by $(1,-1)$; since d maps $H_2(T)$ isomorphically to this kernel, we see that $H_2(T) \cong \mathbb{Z}$. The map

$$H_0(S^1) \oplus H_0(S^1) \longrightarrow H_0(S^1) \oplus H_0(S^1)$$

does something similar. If we take the generators of $H_0(A)$ and $H_0(B)$ to be points on the same circle in $A \cap B$, say corresponding to the first factor of $H_0(A \cap B)$, then the map is

$$(1,0) \mapsto (1,1)$$
$$(0,1) \mapsto (1,1)$$

so that $H_0(T)$ is isomorphic to the cokernel of this map; i. e. $H_0(T) \cong \mathbb{Z}$. Since the kernel of this map is isomorphic to $\mathbb{Z}$ and equals the image of d, we have a short exact sequence

$$0 \longrightarrow \mathbb{Z} \longrightarrow H_1(T) \overset{d}{\longrightarrow} \mathbb{Z} \longrightarrow 0$$

so that $H_1(T) \cong \mathbb{Z} \oplus \mathbb{Z}$.

Example 2.3.26. Consider the sphere S^n. Let A be the complement of the south pole and let B be the complement of the north pole. Both A and B are homeomorphic to $\mathbb{R}^n$ and are therefore contractible. The intersection $A \cap B$ deformation retracts to the equatorial S^{n-1}. The Mayer–Vietoris sequence then yields an isomorphism

$$\tilde{H}_k(S^n) \overset{d}{\longrightarrow} \tilde{H}_{k-1}(S^{n-1})$$

for all k. Proceeding by induction from $\tilde{H}_0(S^0) \cong \mathbb{Z}$, we recover our previous calculation of the homology of S^n.

Example 2.3.27. The real projective plane $\mathbb{RP}^2$ is obtained as the quotient of a square by identifying each pair of sides with opposite orientation. Let A be a small disc in the center of the square and let B be the complement of A, slightly thickened as usual. Then A is contractible, B has the homotopy type of $\mathbb{RP}^2 - A$, which is homotopic to S^1, and $A \cap B \simeq S^1$. The Mayer–Vietoris sequence then has the form

$$0 \longrightarrow H_2(\mathbb{RP}^2) \overset{d}{\longrightarrow} H_1(S^1) \longrightarrow H_1(S^1) \longrightarrow H_1(\mathbb{RP}^2) \longrightarrow 0.$$

So we need only compute the map $H_1(A \cap B) \to H_1(B)$. The circle in $A \cap B$ deforms to the boundary of the square and under the quotient maps to a loop traversing the circle twice. Thus this map is the map $\mathbb{Z} \to \mathbb{Z}$, given by multiplication by 2. This is injective and so we have

$$H_2(\mathbb{RP}^2) = 0$$
$$H_1(\mathbb{RP}^2) \cong \mathbb{Z}_2.$$

Thus, we have finally computed an example with torsion in the homology.

Example 2.3.28. Recall that the Klein bottle is the space obtained from the square by identifying one pair of edges with the same orientation and the other pair with a twist. If we again let A be a small disc in the interior of the square and let B be the thickened complement, then K is the union of the interiors of A and B. We see that A is contractible, B has the homotopy type of $S^1 \vee S^1$, and $A \cap B \simeq S^1$. The Mayer–Vietoris sequence then looks like

$$0 \longrightarrow H_2(K) \xrightarrow{\ d\ } H_1(S^1) \longrightarrow H_1(S^1 \vee S^1) \longrightarrow H_1(K) \longrightarrow 0.$$

The relevant map is the map $\mathbb{Z} \to \mathbb{Z} \oplus \mathbb{Z}$ realizing the retraction of the circle in $A \cap B$ to the boundary. In the quotient, this wraps around one circle twice with opposite orientations and around the other twice with the same orientation; that is, it is the map $1 \mapsto (0, 2)$. It follows that $H_2(K) = \ker(d) = 0$ and

$$H_1(K) \cong (\mathbb{Z} \oplus \mathbb{Z})/\langle (0, 2) \rangle \cong \mathbb{Z} \oplus \mathbb{Z}_2.$$

Exercises

Exercise 2.3.16. The *connected sum* $S_1 \# S_2$ of two surfaces S_1 and S_2 is defined as follows. Choose closed embedded discs D_j in $S_j, j = 1, 2$. Let $Y_j = S_j - \mathrm{int} D_j$. Define

$$S_1 \# S_2 = Y_1 \coprod Y_2 / \sim,$$

where the equivalence relation identifies the point θ of $\partial D_1 \cong \mathbb{R}/2\pi\mathbb{Z}$ with $-\theta$ of $\partial D_2 \cong \mathbb{R}/2\pi\mathbb{Z}$. Denote the connected sum of S^2 minus r discs with g copies of the torus T by $M_{g,r}$. Use the Mayer–Vietoris sequence to compute $H_*(M_{g,r}; \mathbb{Z})$ for all $g, r \geq 0$ and give explicit generators of the homology groups. (Note that $r = 0$ is the closed surface of genus g so that $M_{g,r}$ is M_g with r discs removed.)

Exercise 2.3.17. Compute the homology of $M_{g,r} \# \mathbb{R}P^2$. (Remark: every non-orientable compact surface with r boundary components is diffeomorphic to one of these.)

Exercise 2.3.18. Let X be the *topologist's sine curve*, which is the union of the points $\{(x, \sin(1/x)) : 0 < x \leq 1\}$ and the interval $[-1, 1]$, and an arc intersecting this union only at $(0, -1)$ and $(1, \sin(1))$. Compute the homology of X.

Exercise 2.3.19. Let X be the space obtained from the torus $S^1 \times S^1$ by attaching a Möbius band via a homeomorphism of the boundary circle of the band to the circle $S^1 \times \{x_0\}$. Compute the homology of X. What if we do the same procedure to $\mathbb{R}P^2$, attaching along the standard $\mathbb{R}P^1 \subset \mathbb{R}P^2$?

Exercise 2.3.20. The surface M_g, embedded in $\mathbb{R}^3$, bounds a compact region R. Glue two copies of R together along their boundary surfaces M_g via the identity map to form a closed 3-manifold X. Compute the homology of X.

Exercise 2.3.21. Let D_k be the space obtained by removing k disjoint small discs from the unit disc D^2 and let M_k be the space obtained by gluing two copies of D_k together along their boundaries. Compute $H_*(M_k;\mathbb{Z})$. What space is this?

Exercise 2.3.22. Repeat the previous exercise with the n-disc D^n replacing D^2, removing k disjoint small n-discs.

Exercise 2.3.23. Let $X \subset \mathbb{R}^n$ be a space homeomorphic to S^{n-1} which has a neighborhood U homeomorphic to $X \times (-1,1)$ via homeomorphism of X with $X \times \{0\}$. Show that $\mathbb{R}^n \setminus X$ has two components.

2.3.6 Homology with coefficients

All of the preceding theory works with an arbitrary ring of coefficients. That is, suppose R is a commutative ring with identity. If X is a topological space, then we have the singular chains on X with coefficients in R:

$$S_n(X;R) = \text{free } R\text{-module on the singular } n\text{-simplices in } X.$$

The boundary map still satisfies $\partial^2 = 0$ and we therefore have the resulting homology groups, which we denote by $H_n(X;R)$. Excision and the Mayer–Vietoris sequence work in this context since the proofs of these facts had nothing to do with the coefficients. The coefficient ring can influence the homology groups, however.

Example 2.3.29. Consider $\mathbb{R}P^2$ and $R = \mathbb{Z}_2$. Using the Mayer–Vietoris sequence, as in Example 2.3.27, we have

$$0 \longrightarrow H_2(\mathbb{R}P^2;\mathbb{Z}_2) \overset{d}{\longrightarrow} H_1(S^1;\mathbb{Z}_2) \longrightarrow H_1(S^1;\mathbb{Z}_2) \longrightarrow H_1(\mathbb{R}P^2;\mathbb{Z}_2) \longrightarrow 0$$

Note that the homology of any sphere S^n is the same for any coefficient ring R: $\tilde{H}_n(S^n;R) \cong R$ and $\tilde{H}_k(S^n;R) = 0$ for $k \neq n$. The map $H_1(S^1;\mathbb{Z}_2) \to H_1(S^1;\mathbb{Z}_2)$ is the map $\mathbb{Z}_2 \to \mathbb{Z}_2$ given by multiplication by 2, just as in the integral case. Over $\mathbb{Z}_2$, however, this is the zero map. It follows that

$$H_2(\mathbb{R}P^2;\mathbb{Z}_2) \cong \mathbb{Z}_2$$
$$H_1(\mathbb{R}P^2;\mathbb{Z}_2) \cong \mathbb{Z}_2.$$

Example 2.3.30. Consider the Klein bottle K and $R = \mathbb{Z}_2$. Using the Mayer–Vietoris sequence, we have

$$0 \rightarrow H_2(K;\mathbb{Z}_2) \xrightarrow{d} H_1(S^1;\mathbb{Z}_2) \rightarrow H_1(S^1 \vee S^1;\mathbb{Z}_2) \rightarrow H_1(K;\mathbb{Z}_2) \rightarrow 0$$

The relevant map is still $1 \mapsto (0,2)$, but with $\mathbb{Z}_2$ coefficients this is the zero map. It follows that

$$H_2(K;\mathbb{Z}_2) \cong \mathbb{Z}_2$$
$$H_1(K;\mathbb{Z}_2) \cong \mathbb{Z}_2 \oplus \mathbb{Z}_2.$$

Observe that the $\mathbb{Z}_2$-homology of the torus T is the same as that of K.

Exercises

Exercise 2.3.24. Prove that the homology of S^n with any coefficient ring R is R in dimensions 0 and n and 0 otherwise.

Exercise 2.3.25. Compute $H_*(T;\mathbb{Z}_2)$.

Exercise 2.3.26. Compute the homology of $M_{g,r}\#\mathbb{R}P^2$ with $\mathbb{Z}_2$ coefficients.

2.4 Cellular homology

Simplicial complexes admit study via simplicial homology, and arbitrary spaces have singular homology. It is natural to ask if cell complexes admit a homology theory based on their basic building blocks. The answer is yes, and we study this in this section. We begin with a short detour, however.

Suppose that $f : S^n \rightarrow S^n$ is a continuous map. This induces a homomorphism

$$f_* : H_n(S^n) \longrightarrow H_n(S^n).$$

Since this is just a map $\mathbb{Z} \rightarrow \mathbb{Z}$, it is given by multiplication by some integer d. We call this integer the degree of f. For example, if we consider the map $f : S^1 \rightarrow S^1$ given by $z \mapsto z^d$, then the degree of f is d. This is most easily proved using simplicial homology (which will be shown to be equivalent to singular homology later). Take the triangulation of S^1 with d vertices corresponding to the dth roots of unity. Then the map f takes each edge and wraps it around the entire circle. Since the chain $z = \sum e_i$ is a generating cycle for $H_1(S^1)$, we see that $f_*([z]) = d[z]$. This works for $d \geq 3$. For $d = 2$, this is not a simplicial complex, but the principle is the same and the calculation goes through (the cellular homology developed below works here). For $d < 0$, the map f reverses the orientation of the edges and so we still see that the degree of f is d.

The degree enjoys various important properties.

(1) The degree of the identity map on S^n is 1, clearly.

(2) Suppose f is not surjective. Then $\deg f = 0$. This follows because if the image of f omits $x_0 \in S^n$, then it factors as $S^n \to S^n - \{x_0\} \to S^n$ and $S^n - \{x_0\}$ is contractible.

(3) If f is homotopic to g, then $\deg f = \deg g$, since $f_* = g_*$.

(4) $\deg(f \circ g) = (\deg f)(\deg g)$ since $(f \circ g)_* = f_* \circ g_*$.

(5) If f is a reflection of S^n, then $\deg f = -1$. To see this, note that f fixes an equator of S^n. Then $S^n = \Delta_1^n \cup \Delta_2^n$, glued along this fixed equator. Then $f_*(\Delta_1^n) = -\Delta_2^n$ and so $f_*(\Delta_1^n - \Delta_2^n) = \Delta_2^n - \Delta_1^n$.

(6) If $a : S^n \to S^n$ is the antipodal map $a(x) = -x$, then $\deg a = (-1)^{n+1}$. This follows since a is the composition of $(n+1)$ reflections.

(7) If $f : S^n \to S^n$ has no fixed points, then $\deg f = (-1)^{n+1}$. To see this, we will prove that f is homotopic to the map a. Define $F : S^n \times I \to S^n$ by

$$F(x, t) = \frac{(1-t)f(x) - tx}{\|(1-t)f(x) - tx\|}.$$

This is simply the straight-line homotopy between f and a and it is well-defined since $f(x) \neq x$ for all $x \in S^n$ so that $(1-t)f(x) - tx \neq 0$ for all x, t. It follows that $\deg f = \deg a = (-1)^{n+1}$.

(8) Let R be a commutative ring with identity. If $f : S^k \to S^k$ has degree m, then $f_* : H_k(S^k; R) \to H_k(S^k; R)$ is given by multiplication by m. To see this, note that if $\varphi : R_1 \to R_2$ is a ring homomorphism, then we get an induced map $f_\# : S_n(X, A; R_1) \to S_n(X, A; R_2)$, which commutes with the boundary map and so we get an induced map on homology. Given $f : S^k \to S^k$ of degree m, choose a ring homomorphism $\varphi : \mathbb{Z} \to R$ sending 1 to $r \in R$. Then we have a commutative diagram

$$
\begin{array}{ccc}
\tilde{H}_k(S^k; \mathbb{Z}) & \xrightarrow{f_*} & \tilde{H}_k(S^k; \mathbb{Z}) \\
\downarrow{\varphi_*} & & \downarrow{\varphi_*} \\
\tilde{H}_k(S^k; R) & \xrightarrow{f_*} & \tilde{H}_k(S^k; R)
\end{array}
\qquad\qquad
\begin{array}{ccc}
1 & \xrightarrow{f_*} & m \\
\downarrow{\varphi_*} & & \downarrow{\varphi_*} \\
r & \xrightarrow{f_*} & rm
\end{array}
$$

and so f_* is multiplication by m.

While we are here, we may as well prove the following interesting fact. By a *tangent vector field* on S^n, we mean a continuous map $v : S^n \to \mathbb{R}^{n+1}$ such that $x \cdot v(x) = 0$ for all x, where $\cdot$ is the usual dot product in $\mathbb{R}^{n+1}$.

Theorem 2.4.1. *S^n has a continuous nonvanishing tangent vector field if and only if n is odd.*

Proof. If $n = 2k - 1$ is odd, the map

$$v(x_1, \ldots, x_{2k}) = (-x_2, x_1, -x_4, x_3, \ldots, -x_{2k}, x_{2k-1})$$

is nonvanishing. Conversely, suppose $v(x) \neq 0$ for all $x \in S^n$. Define a map $\bar{v} : S^n \to S^n$ by $\bar{v}(x) = v(x)/\|v(x)\|$. Define $F : S^n \times [0, \pi] \to S^n$ by

$$F(x, t) = (\cos t)x + (\sin t)\bar{v}(x).$$

Then F is well-defined since

$$\left\|F(x, t)\right\|^2 = (\cos^2 t)\|x\|^2 + (\cos t \sin t)x \cdot \bar{v}(x) + (\sin^2 t)\|\bar{v}(x)\|^2 = \cos^2 t + \sin^2 t = 1.$$

Note that $F(x, 0) = x$ and $F(x, \pi) = -x$ and so F is a homotopy from the identity map to a. It follows that $1 = \deg \mathrm{id} = \deg a = (-1)^{n+1}$, so that n must be odd. $\qquad\square$

Exercises

Exercise 2.4.1. Let $f, g : S^n \to S^n$ be maps such that $f(x) \neq g(x)$ for all $x \in S^n$. Prove that $f \simeq a \circ g$.

Exercise 2.4.2. Suppose $f : S^n \to S^n$ is homotopic to a constant map. Then f has a fixed point and also a point x with $f(x) = -x$.

Exercise 2.4.3. Prove that any map $f : S^{2n} \to S^{2n}$ either has a fixed point or a point x with $f(x) = -x$.

Exercise 2.4.4. Let G be a group of homeomorphisms acting freely on S^{2n}. Prove that $|G| \leq 2$.

Exercise 2.4.5. Prove that every map $f : \mathbb{RP}^{2n} \to \mathbb{RP}^{2n}$ has a fixed point.

Exercise 2.4.6. Call two maps $f, g : S^n \to S^n$ *orthogonal* at $x \in S^n$ if $f(x) \cdot g(x) = 0$ ($\cdot$ is the usual dot product in $\mathbb{R}^{n+1}$). Prove that if $|\deg f| \neq |\deg g|$, then f and g are orthogonal at some point $x \in S^n$.

Exercise 2.4.7. Let $T : \mathbb{R}^n \to \mathbb{R}^n$ be an invertible linear transformation. Show that the induced map on $H_n(\mathbb{R}^n, \mathbb{R}^n - \{0\}) \cong \tilde{H}_{n-1}(\mathbb{R}^n - \{0\}) \cong \mathbb{Z}$ is $\pm\mathrm{id}$ according to whether $\det T$ is positive or negative. (Hint: Gauss elimination is useful here.)

We need the degree of a map $S^n \to S^n$ in the definition of cellular homology. Let us proceed.

2.4.1 Cellular chains

Let X be a cell complex and recall that $X^{(n)}$ denotes the n-skeleton of X. The $(n + 1)$-skeleton is obtained by attaching $(n + 1)$-cells to $X^{(n)}$ via attaching maps $\varphi : \partial e^{n+1} \to X^{(n)}$. Note that a cell e^n has two orientations (given by choosing a normal vector to the boundary sphere); denote by $C_n^{CW}(X)$ the free abelian group on the oriented n-cells in X with the relation that the two generators corresponding to different orientations of an n-cell are negatives of each other. This is a free abelian group with rank equal to the number of n-cells in X. Given an orientation of an n-cell e^n, then choosing an orientation-preserving homeomorphism $\Delta^n \to e^n$, we obtain an element in the relative singular homology group $H_n(X^{(n)}, X^{(n-1)})$. The resulting element depends only on the orientation and changes sign if we reverse the orientation. This determines a homomorphism

$$\varphi_n : C_n^{CW}(X) \to H_n(X^{(n)}, X^{(n-1)}).$$

Lemma 2.4.2. *The map φ_n is an isomorphism.*

Proof. Note that $(X^{(n)}, X^{(n-1)})$ is a good pair. The quotient $X^{(n)}/X^{(n-1)}$ is a wedge of n-spheres, one for each n-cell in X. It follows that

$$H_n(X^{(n)}, X^{(n-1)}) \cong H_n(X^{(n)}/X^{(n-1)}) \cong \bigoplus_a \mathbb{Z}[e_a^n],$$

where e_a^n are the n-cells in X. The map φ_n maps the generator e_a^n of $C_n^{CW}(X)$ to a generator of the factor of $\mathbb{Z}$ corresponding to e_a^n, and is therefore clearly a bijection. $\square$

Note that we also have $H_i(X^{(n)}, X^{(n-1)}) = 0$ for $i \neq n$.

Lemma 2.4.3. *Let X be a cell complex. Then $H_k(X^{(n)}) = 0$ for $k > n$ and the inclusion $i : X^{(n)} \hookrightarrow X$ induces an isomorphism $i_* : H_k(X^{(n)}) \to H_k(X)$ for $k < n$.*

Proof. The remark above implies, via the long exact sequence of the pair $(X^{(n)}, X^{(n-1)})$ that $H_k(X^{(n-1)}) \cong H_k(X^{(n)})$ for $k \neq n, n-1$. So if $k > n$, we have

$$H_k(X^{(n)}) \cong H_k(X^{(n-1)}) \cong \cdots \cong H_k(X^{(0)}) = 0.$$

Moreover, if X is finite-dimensional, then for $k < n$, we have

$$H_k(X^{(n)}) \cong H_k(X^{(n+1)}) \cong \cdots \cong H_k(X^{(n+m)}) \cong H_k(X)$$

for m sufficiently large. If X is infinite dimensional, then we use compactness of the image of any singular k-simplex to deduce that the image lies in some finite-dimensional skeleton of X. $\square$

Now, define a map $d_n : C_n^{CW}(X) \to C_{n-1}^{CW}(X)$ as follows. Denote by $\partial_n : H_n(X^{(n)}, X^{(n-1)}) \to H_{n-1}(X^{(n-1)}, X^{(n-2)})$ the map in the long exact sequence of the triple $(X^{(n)}, X^{(n-1)}, X^{(n-2)})$ (see Exercise 2.3.5). We then set

$$d_n = \varphi_{n-1}^{-1} \circ \partial_n \circ \varphi_n.$$

Proposition 2.4.4. *Let X be a cell complex. Then $(C_\bullet^{CW}(X), d)$ is a chain complex whose homology groups are denoted $H_\bullet^{CW}(X)$ and are called the cellular homology groups of X.*

Proof. Since $\partial_{n-1} \circ \partial_n = 0$ (exercise), we have $d_{n-1} \circ d_n = 0$ as well. □

Let us develop a practical formula for d_n. First note that for $n = 1$, the map $d_1 : H_1(X^{(1)}, X^{(0)}) \to H_0(X^{(0)})$ is just the boundary map $S_1(X^{(1)}, X^{(0)}) \to S_0(X^{(0)})$. In general, we have the following result.

Proposition 2.4.5. *If e_α^n is an oriented n-cell in X, then*

$$d_n(e_\alpha^n) = \sum_\beta d_{\alpha\beta} e_\beta^{n-1},$$

where $d_{\alpha\beta}$ is the degree of the map

$$S_\alpha^{n-1} \longrightarrow X^{(n-1)} \longrightarrow S_\beta^{(n-1)},$$

obtained as the composition of the attaching map $\partial e_\alpha^n = S_\alpha^{n-1} \to X$ with the quotient map collapsing $X^{(n-1)} - e_\beta^{n-1}$ to a point.

Proof. Let Φ_α be the characteristic map of e_α^n and φ_α the attaching map. Consider the commutative diagram

$$
\begin{array}{ccccc}
H_n(D_\alpha^n, \partial D_\alpha^n) & \xrightarrow{\partial}_{\cong} & \tilde{H}_{n-1}(\partial D_\alpha^n) & \xrightarrow{\Delta_{\alpha\beta*}} & \tilde{H}_{n-1}(S_\beta^{n-1}) \\
\downarrow{\Phi_{\alpha*}} & & \downarrow{\varphi_{\alpha*}} & & \uparrow{q_{\beta*}} \\
H_n(X^{(n)}, X^{(n-1)}) & \xrightarrow{\partial_n} & \tilde{H}_{n-1}(X^{(n-1)}) & \xrightarrow{q_*} & \tilde{H}_{n-1}(X^{(n-1)}/X^{(n-2)}) \\
& \searrow{d_n} & \downarrow{j_{n-1}} & & \cong\downarrow \\
& & H_{n-1}(X^{(n-1)}, X^{(n-2)}) & \xrightarrow{\cong} & H_{n-1}(X^{(n-1)}/X^{(n-2)}, *)
\end{array}
$$

Here $*$ in the lower right group is the quotient $X^{(n-2)}/X^{(n-2)}$, and $\Delta_{\alpha\beta} = q_\beta \circ q \circ \varphi_\alpha$, where $q : X^{(n-1)} \to X^{(n-1)}/X^{(n-2)}$ is the quotient map and q_β collapses all but e_β^{n-1} to a point. Note that $\Phi_{\alpha*}$ takes the generator $[D_\alpha^n]$ to the generator of $H_n(X^{(n)}, X^{(n-1)})$ corresponding to e_α^n. Then

$$d_n(e_\alpha^n) = j_{n-1}\varphi_{\alpha*}\partial([D_\alpha^n]).$$

In terms of the basis $\{e_\beta^{n-1}\}$ of $H_{n-1}(X^{(n-1)}, X^{(n-2)})$, $q_{\beta*}$ is the projection of $\tilde{H}_{n-1}(X^{(n-1)}, X^{(n-2)})$ onto the summand corresponding to e_β^{n-1}. The result follows. □

Exercises

Exercise 2.4.8. In this exercise, we will develop a formula for computing $\deg f$ for a map $f : S^n \to S^n$. Suppose that there is some $y \in S^n$ such that $f^{-1}(y) = \{x_1, \ldots, x_m\}$. Show that $\deg f = m$ as follows. Let $U_1, \ldots, U_m$ be disjoint open neighborhoods of $x_1, \ldots, x_m$, respectively, which map into some neighborhood V of y. Then $f(U_i - \{x_i\}) \subset V - \{y\}$ for all i. Consider the commutative diagram

$$
\begin{array}{ccc}
H_n(U_i, U_i - \{x_i\}) & \xrightarrow{\;f_*\;} & H_n(V, V - \{y\}) \\
\end{array}
$$

Denote by $\deg f|_{x_i}$ the degree of $f_* : H_n(U_i, U_i - \{x_i\}) \to H_n(V, V - \{y\})$. Prove that

$$\deg f = \sum_i \deg f|_{x_i}.$$

Since the local degree is usually ± 1, this is a practical formula.

Exercise 2.4.9. A polynomial $f : \mathbb{C} \to \mathbb{C}$ of degree n can be extended to a continuous map of one-point compactifications $\hat{f} : S^2 \to S^2$. Prove that $\deg \hat{f} = n$. Show that the local degree of $\hat{f}$ at a root of f is the multiplicity of the root.

Exercise 2.4.10. Prove that $\partial_{n-1} \circ \partial_n = 0$ in the proof of Proposition 2.4.4.

2.4.2 Calculations

If X is a finite chain complex, then its cellular chain groups are finitely generated and the boundary map is described in terms of degrees of maps of spheres. In principle, this allows us to compute the cellular homology efficiently. Let us do a few examples.

Example 2.4.6. The sphere S^n admits a particularly nice cell decomposition: $S^n = e^0 \cup e^n$. The cellular chain complex therefore has a copy of $\mathbb{Z}$ in degrees 0 and n and is 0 elsewhere. If $n \geq 2$, the boundary map is therefore the zero map in all degrees and we obtain

$$
H_k^{CW}(S^n) = \begin{cases} \mathbb{Z} & k = 0, n \\ 0 & \text{otherwise.} \end{cases}
$$

The same is true for $n = 1$, since the boundary map $d_1 : C_1^{CW}(S^1) \to C_0^{CW}(S^1)$ maps the 1-cell generating the domain to the generating 0-cell twice with opposite orientation.

Example 2.4.7. Along the same lines as the previous example, recall that the complex projective space $\mathbb{C}P^n$ admits a cell decomposition

$$\mathbb{C}P^n = e^0 \cup e^2 \cup \cdots \cup e^{2n}.$$

The cellular chain complex then has a copy of $\mathbb{Z}$ in each even degree and is 0 otherwise. The boundary map is then zero in all degrees and we see that

$$H_k^{CW}(\mathbb{C}P^n) = \begin{cases} \mathbb{Z} & 0 \le k \le 2n \ k \text{ even} \\ 0 & k \text{ odd or } k > 2n. \end{cases}$$

Example 2.4.8. Let M_g be the closed surface of genus g. Recall that this space has a cell decomposition

$$M_g = e^0 \cup \underbrace{e^1 \cup \cdots \cup e^1}_{2g} \cup e^2.$$

The cellular chain complex then has the form

$$0 \longrightarrow \mathbb{Z} \xrightarrow{d_2} \mathbb{Z}^{2g} \xrightarrow{d_1} \mathbb{Z} \longrightarrow 0$$

Note that d_1 is the zero map, since it maps each edge to the single vertex twice with opposite orientation. We claim that d_2 is also the zero map. To see this, denote the edges by $a_1, b_1, a_2, b_2, \ldots, a_g, b_g$ and note that the attaching map $\varphi : \partial e^2 \to M_g^{(1)}$ takes the boundary S^1 to the path $a_1 b_1 a_1^{-1} b_1^{-1} \cdots a_g b_g a_g^{-1} b_g^{-1}$. On the chain level, we see that the corresponding $\Delta_{\alpha\beta}$ are all homotopic to the identity and since each edge appears once with its inverse, $d_2 \equiv 0$. It follows that

$$H_k^{CW}(M_g) = \begin{cases} \mathbb{Z} & k = 0, 2 \\ \mathbb{Z}^{2g} & k = 1 \\ 0 & \text{otherwise.} \end{cases}$$

Example 2.4.9. The real projective space has a cell decomposition of the form

$$\mathbb{R}P^n = e^0 \cup e^1 \cup \cdots \cup e^n.$$

Recall that the attaching map for e^k is the $2 : 1$ quotient map $\varphi : S^{k-1} \to \mathbb{R}P^{k-1} = (\mathbb{R}P^n)^{(k-1)}$. We therefore need to compute the degree of the composite

$$S^{k-1} \xrightarrow{\varphi} \mathbb{R}P^{k-1} \xrightarrow{q} \mathbb{R}P^{k-1}/\mathbb{R}P^{k-2} = S^{k-1}$$

Note that $q \circ \varphi$ is a homeomorphism when restricted to each component of $S^{k-1} - S^{k-2}$ and these two homeomorphisms are obtained from each other by precomposing with the antipodal map a of S^{k-1}, which has degree $(-1)^k$. Thus,

$$\deg(q \circ \varphi) = \deg(\text{id}) + \deg(a) = 1 + (-1)^k.$$

So the cellular chain complex has the form

$$0 \to \mathbb{Z} \xrightarrow{1+(-1)^n} \mathbb{Z} \xrightarrow{1+(-1)^{n-1}} \mathbb{Z} \to \cdots \xrightarrow{0} \mathbb{Z} \xrightarrow{\times 2} \mathbb{Z} \xrightarrow{0} \mathbb{Z} \to 0$$

We therefore have the following homology groups:

$$H_k^{CW}(\mathbb{R}P^n) = \begin{cases} \mathbb{Z} & k = 0, k = n, n \text{ odd} \\ \mathbb{Z}_2 & k \text{ odd}, 0 < k < n \\ 0 & \text{otherwise.} \end{cases}$$

Exercises

Exercise 2.4.11. Compute the cellular homology of the Klein bottle.

Exercise 2.4.12. Let X be the space obtained from S^1 by attaching a 2-cell via the map $\partial e^2 = S^1 \to S^1$ given by $z \mapsto z^d$ for some $d > 0$. Compute the cellular homology of X.

Exercise 2.4.13. In the previous exercise, attach two 2-cells by maps of degrees 2 and 3, respectively. Show that the resulting space X is homotopy equivalent to S^2.

Exercise 2.4.14. Expanding a previous exercise, suppose we attach k 2-cells to S^1 via maps $\partial e_i^2 = S^1 \to S^1$ given by $z \mapsto z^{d_i}$ for some $d_i > 0$, $i = 1, \ldots, k$, to obtain a space X. Compute the cellular homology of X.

Exercise 2.4.15. Compute the cellular homology of the space $S^1 \times (S^1 \vee S^1)$.

Exercise 2.4.16. In this exercise, we explore *lens spaces*. Fix an integer $m > 1$ and integers $\ell_1, \ldots, \ell_n$ relatively prime to m. The lens space $L = L_m(\ell_1, \ldots, \ell_n)$ is the quotient $S^{2n-1}/\mathbb{Z}_m$ of the unit sphere $S^{2n-1} \subset \mathbb{C}^n$ by the action of $\mathbb{Z}_m$ generated by the rotation $\rho(z_1, \ldots, z_n) = (e^{2\pi i \ell_1/m} z_1, \ldots, e^{2\pi i \ell_n/m} z_n)$. This rotates the jth factor by $2\pi\ell_j/m$. Note that when $m = 2$ ρ is the antipodal map and so $L = \mathbb{R}P^{2n-1}$. Thus the lens spaces are generalizations of the familiar projective spaces. Show that this is a free action of $\mathbb{Z}_m$ on S^{2n-1} so that $S^{2n-1} \to L$ is an m-fold cover.

We now describe a cell structure on L. Subdivide the unit circle C in the nth factor of $\mathbb{C}^n$ by taking the points $e^{2\pi i j/m} \in C$, $j = 1, \ldots, m$ as vertices. Join the jth vertex to the unit sphere $S^{2n-3} \subset \mathbb{C}^{n-1}$ by arcs of great circles in S^{2n-1} to create a $(2n-2)$ ball D_j^{2n-2} bounded by S^{2n-3}; this consists of the points $\cos\theta(0, \ldots, 0, e^{2\pi i j/m}) + \sin\theta(z_1, \ldots, z_{n-1}, 0)$ for $0 \le \theta\pi/2$. Join the jth edge of C to S^{2n-3} to yield a ball D_j^{2n-1} founded by D_j^{2n-2} and D_{j+1}^{2n-2}, subscripts mod m. The map ρ takes S^{2n-3} to itself and rotates C by the angle $2\pi\ell_n/m$, and therefore permutes the D_j^{2n-2} and D_j^{2n-1}. A suitable power r of ρ takes each D_j^{2n-2} and D_j^{2n-1} to the next one (prove that $r\ell_n \equiv 1 \bmod m$ works). Since ρ^r has order m it also generates $\mathbb{Z}_m$ and so we obtain L as the quotient of one D_j^{2n-1} by identifying its two faces

D_j^{2n-2} and D_{j+1}^{2n-2} together by ρ^r. Now observe that the $(2n-3)$-dimensional lens space $L_m(\ell_1, \ldots, \ell_{n-1})$ sits in $L_m(\ell_1, \ldots, \ell_n)$ as the quotient of S^{2n-3}, and the latter is obtained from the former by attaching two cells, of dimensions $2n-2$ and $2n-1$. Inductively, this gives a cell structure on L with one cell e^k in each dimension $0 \le k \le 2n-1$. Prove that the boundary maps in the cellular chain complex are alternately 0 and multiplication by m, where d_{2n-1} is 0, and d_{2n-2} is multiplication by m. Compute the cellular homology of L.

Exercise 2.4.17. A map $f : S^n \to S^n$ is *even* if $f(x) = f(-x)$ for all $x \in S^n$. Prove that an even map has even degree, and if n is even, this degree must be 0. When n is odd, show that there exist even maps of any given degree. (Hint: an even map factors as a composition $S^n \to \mathbb{R}P^n \to S^n$. Use the calculation of cellular homology of $\mathbb{R}P^n$ and the corresponding induced homomorphisms.)

2.5 Comparison of homology theories

We have seen three homology theories: simplicial, singular, and cellular. It would be really unfortunate if they were giving us different answers. That is, if we have a simplicial complex, for example, then we could compute all three homologies and, a priori, we could get three different answers. In a different direction, if we had two different simplicial realizations of a space X or two different cell decompositions, we could, in principle, obtain different answers when we compute the simplicial or cellular homology using these decompositions. Luckily, that does not happen, as we will now show.

Before proceeding, we state the following result, whose proof is left as an exercise in diagram chasing.

Lemma 2.5.1 (Five Lemma). *Consider the commutative diagram of abelian groups with exact rows*

$$
\begin{array}{ccccccccc}
A & \xrightarrow{i} & B & \xrightarrow{j} & C & \xrightarrow{k} & D & \xrightarrow{\ell} & E \\
\downarrow{\alpha} & & \downarrow{\beta} & & \downarrow{\gamma} & & \downarrow{\delta} & & \downarrow{\varepsilon} \\
A' & \xrightarrow{i'} & B' & \xrightarrow{j'} & C' & \xrightarrow{k'} & D' & \xrightarrow{\ell'} & E'
\end{array}
$$

If $\alpha, \beta, \delta, \varepsilon$ are isomorphisms, then γ is an isomorphism.

2.5.1 Simplicial vs. singular

Suppose (X, A) is a simplicial pair. Each n-simplex σ in X has a characteristic map $\hat{\sigma} : \Delta^n \to X$, mapping isomorphically onto the simplex. Define a chain map $C_n(X, A) \to S_n(X, A)$ by $\sigma \mapsto \hat{\sigma}$.

Theorem 2.5.2. *This map induces isomorphisms in homology.*

Proof. First assume that X is finite-dimensional and that $A = \emptyset$. Let $X^{(k)}$ be the k-skeleton of X. We have a commutative diagram

$$H_{n+1}^{\Delta}(X^{(k)}, X^{(k-1)}) \to H_n^{\Delta}(X^{(k-1)}) \to H_n^{\Delta}(X^{(k)}) \to H_n^{\Delta}(X^{(k)}, X^{(k-1)}) \to H_{n-1}^{\Delta}(X^{(k-1)})$$
$$\downarrow \qquad\qquad \downarrow \qquad\qquad \downarrow \qquad\qquad \downarrow \qquad\qquad \downarrow$$
$$H_{n+1}(X^{(k)}, X^{(k-1)}) \to H_n(X^{(k-1)}) \to H_n(X^{(k)}) \to H_n(X^{(k)}, X^{(k-1)}) \to H_{n-1}(X^{(k-1)})$$

We claim that the first and fourth vertical maps are isomorphisms. To see this, observe that $C_n(X^{(k)}, X^{(k-1)}) = 0$ for $n \neq k$ and is free abelian with basis the k-simplices of X for $n = k$. The same is therefore true of $H_n^{\Delta}(X^{(k)}, X^{(k-1)})$. The singular homology groups were computed in the proof of Lemma 2.4.3; the answer is the same as for the simplicial homology. By induction, we may assume the second and fifth vertical maps are isomorphisms (the base case of $X^{(0)}$ follows because the simplicial and singular homology of a discrete set are clearly isomorphic). Then by Lemma 2.5.1 we conclude that the map $H_n^{\Delta}(X^{(k)}) \to H_n(X^{(k)})$ is an isomorphism for all k and therefore that $H_n^{\Delta}(X) \cong H_n(X)$.

If X is infinite dimensional, then, since the image of any singular simplex lies in a finite skeleton (by compactness), we can reduce to the finite-dimensional case to conclude that $H_n^{\Delta}(X) \cong H_n(X)$. Finally, if $A \neq \emptyset$, we may apply Lemma 2.5.1 to the long exact sequences of the pair (X, A) to conclude that $H_n^{\Delta}(X, A) \cong H_n(X, A)$. □

Since a continuous map between simplicial complexes is homotopic to a simplicial map, this isomorphism is functorial.

Exercises

Exercise 2.5.1. Prove Lemma 2.5.1. This result actually holds with weaker conditions on $\alpha, \beta, \delta, \varepsilon$. What are they?

2.5.2 Cellular vs. singular

Theorem 2.5.3. *Let X be a cell complex. Then the cellular homology $H_{\bullet}^{CW}(X)$ is isomorphic to the singular homology $H_{\bullet}(X)$.*

Proof. Recall that the boundary map in $C_{\bullet}^{CW}(X)$ was built from the long exact sequence of the triple $(X^{(n)}, X^{(n-1)}, X^{(n-2)})$. We have a commutative diagram with exact rows

$$0 \to H_n(X^{(n)}, X^{(n-2)}) \to H_n(X^{(n)}, X^{(n-1)}) \xrightarrow{\partial_n} H_{n-1}(X^{(n-1)}, X^{(n-2)})$$

$$0 \longrightarrow H_n(X^{(n)}) \longrightarrow C_n^{CW}(X) \xrightarrow{d_n} C_{n-1}^{CW}(X)$$

This shows that the group $H_n(X^{(n)})$ is precisely the subgroup of cycles in $C_n^{CW}(X)$. We also have a commutative diagram

$$C_{n+1}^{CW}(X) = H_{n+1}(X^{(n+1)}, X^{(n)}) \xrightarrow{\partial} H_n(X^{(n)})$$

$$H_{n+1}(X, X^{(n)})$$

$$0$$

This allows us to identify $d_{n+1}C_{n+1}^{CW}(X) \subseteq Z_n(C_\bullet^{CW}(X))$ with $\partial H_{n+1}(X, X^{(n)}) \subseteq H_n(X^{(n)})$. By Lemma 2.4.3 we see that the following sequence is exact

$$H_{n+1}(X, X^{(n)}) \xrightarrow{\partial} H_n(X^{(n)}) \longrightarrow H_n(X) \longrightarrow 0$$

We therefore conclude that

$$H_n^{CW}(X) = H_n(X^{(n)})/\partial H_{n+1}(X, X^{(n)}) = H_n(X). \qquad \square$$

Note further that this isomorphism is functorial. If $f : X \to Y$ is a continuous map between cell complexes, then f is homotopic to a cellular map g. The map g induces maps on all relative groups in the proof above and so we see that the isomorphism commutes with homomorphisms induced by continuous maps.

Thus, if a space X admits two different cell decompositions, then the cellular homology groups are the same, no matter which decomposition we use to compute them.

There is one more important result that we can now prove.

Theorem 2.5.4 (Borsuk–Ulam). *If $g : S^n \to \mathbb{R}^n$ is continuous, then there is a point $x \in S^n$ with $g(x) = g(-x)$.*

Proof. The bulk of the proof relies on a technical lemma.

Lemma 2.5.5. *Suppose $f : S^n \to S^n$ is an odd function ($f(-x) = -f(x)$ for all x). Then $\deg f$ is odd.*

Proof. To prove this we need a fact from the project about covering spaces at the end of Chapter 1. Suppose $p : \tilde{X} \to X$ is a 2 : 1 covering map. There is a sequence of chain modules

$$0 \longrightarrow S_n(X;\mathbb{Z}_2) \xrightarrow{\tau} S_n(\tilde{X};\mathbb{Z}_2) \xrightarrow{p_\sharp} S_n(X;\mathbb{Z}_2) \longrightarrow 0$$

The map $p_\sharp$ is surjective, since singular simplices $\sigma : \Delta^n \to X$ always lift to $\tilde{X}$ (see the project in Chapter 1). Each such σ has precisely two lifts $\tilde{\sigma}_1$ and $\tilde{\sigma}_2$. Since we are using $\mathbb{Z}_2$ coefficients, the kernel of $p_\sharp$ is generated by sums $\tilde{\sigma}_1 + \tilde{\sigma}_2$. Define $\tau(\sigma) = \tilde{\sigma}_1 + \tilde{\sigma}_2$. The map τ is obviously injective. Moreover, these maps commute with the boundary maps and so we obtain a long exact sequence

$$\cdots \to H_n(X;\mathbb{Z}_2) \xrightarrow{\tau_*} H_n(\tilde{X};\mathbb{Z}_2) \xrightarrow{p_*} H_n(X;\mathbb{Z}_2) \xrightarrow{\partial} H_{n-1}(X;\mathbb{Z}_2) \to \cdots$$

The map τ_* is called the *transfer homomorphism*.

Now, consider the quotient $p : S^n \to \mathbb{R}P^n$. An odd map $f : S^n \to S^n$ induces a map $\bar{f} : \mathbb{R}P^n \to \mathbb{R}P^n$. If we can verify that the following diagram commutes, then we get an induced map on transfer sequences.

$$
\begin{array}{ccccccccc}
0 & \longrightarrow & S_i(\mathbb{R}P^n;\mathbb{Z}_2) & \xrightarrow{\tau} & S_i(S^n;\mathbb{Z}_2) & \xrightarrow{p_\sharp} & S_i(\mathbb{R}P^n;\mathbb{Z}_2) & \longrightarrow & 0 \\
& & \downarrow{\bar{f}_\sharp} & & \downarrow{f_\sharp} & & \downarrow{\bar{f}_\sharp} & & \\
0 & \longrightarrow & S_i(\mathbb{R}P^n;\mathbb{Z}_2) & \xrightarrow{\tau} & S_i(S^n;\mathbb{Z}_2) & \xrightarrow{p_\sharp} & S_i(\mathbb{R}P^n;\mathbb{Z}_2) & \longrightarrow & 0
\end{array}
$$

The right square commutes, since $p \circ f = \bar{f} \circ p$. For the left square, note that $\sigma : \Delta^i \to \mathbb{R}P^n$ lifts to $\tilde{\sigma}_1$ and $\tilde{\sigma}_2$. The two lifts of $\bar{f} \circ \sigma$ are $f \circ \tilde{\sigma}_1$ and $f \circ \tilde{\sigma}_2$, since f takes antipodal points to antipodal points and so the left square commutes as well.

Since $H_i(S^n;\mathbb{Z}_2) = 0$ for $i \neq 0, n$, the map $\partial : H_i(\mathbb{R}P^n;\mathbb{Z}_2) \to H_{i-1}(\mathbb{R}P^n;\mathbb{Z}_2)$ is an isomorphism for all $i \leq n$ and the map $\tau_* : H_n(\mathbb{R}P^n;\mathbb{Z}_2) \to H_n(S^n;\mathbb{Z}_2)$ is also an isomorphism (it is injective by the exactness of the transfer sequence and since both groups are $\mathbb{Z}_2$, it must be an isomorphism). Since $\bar{f}_* : H_0(\mathbb{R}P^n;\mathbb{Z}_2) \to H_0(\mathbb{R}P^n;\mathbb{Z}_2)$ is an isomorphism, we deduce inductively that $\bar{f}_* : H_i(\mathbb{R}P^n;\mathbb{Z}_2) \to H_i(\mathbb{R}P^n;\mathbb{Z}_2)$ is an isomorphism for all i. In particular, we have a commutative diagram

$$
\begin{array}{ccc}
H_n(\mathbb{R}P^n;\mathbb{Z}_2) & \xrightarrow[\cong]{\tau_*} & H_n(S^n;\mathbb{Z}_2) \\
\cong \downarrow{\bar{f}_*} & & \downarrow{f_*} \\
H_n(\mathbb{R}P^n;\mathbb{Z}_2) & \xrightarrow[\cong]{\tau_*} & H_n(S^n;\mathbb{Z}_2)
\end{array}
$$

from which we deduce that $f_* : H_n(S^n;\mathbb{Z}_2) \to H_n(S^n;\mathbb{Z}_2)$ is an isomorphism. Since this map is multiplication by $\deg f$, we deduce that $\deg f$ must be odd. $\quad\square$

Now to prove Theorem 2.5.4, let $f(x) = g(x) - g(-x)$. Then f is odd. If $f(x) \neq 0$ for all x, replace $f(x)$ by $f(x)/\|f(x)\|$ to get a new $f : S^n \to S^{n-1}$, which is still odd. Restricting this to the equator $S^{n-1} \subset S^n$ gives an odd map $S^{n-1} \to S^{n-1}$ and by Lemma 2.5.5 this has odd degree. But this map is nullhomotopic (restrict f to a hemisphere) and so must have degree 0. Thus there is an $x \in S^n$ with $f(x) = 0$; that is $g(x) = g(-x)$. $\quad\square$

Exercises

Exercise 2.5.2. In the proof of Lemma 2.5.5 we used singular chains to construct a trans-fer map on the chain level. Here is an alternate approach, using cellular chains. The standard cell decomposition of $\mathbb{R}P^n$ has one cell in each dimension. There is a corre-sponding decomposition of S^n with two cells in each dimension. Use this to explicitly define $\tau : C_k^{CW}(\mathbb{R}P^n; \mathbb{Z}_2) \to C_k^{CW}(S^n; \mathbb{Z}_2)$ and show that the proof goes through with this definition. You may assume that any map $f : S^n \to S^n$ is homotopic to a cellular map.

Exercise 2.5.3. Give an alternate proof of Theorem 2.5.4 using Tucker's Lemma: Let T be a triangulation of the unit ball D^n, and assume it is antipodally symmetric on the boundary sphere S^{n-1} (i. e., S^{n-1} is a subcomplex in this triangulation, and if σ is a sim-plex in S^{n-1}, then so is $-\sigma$). Let $L : V(T) \to \{\pm1, \pm2, \ldots, \pm n\}$ be a labeling of the vertices in T which is an odd function on S^{n-1}. Then there is an edge in the triangulation whose vertices are labeled $\pm k$ for some $1 \le k \le n$. Proceed as follows: Suppose $g : S^n \to \mathbb{R}^n$ is a continuous odd function. Since S^n is compact, g is uniformly continuous and so for ev-ery $\varepsilon > 0$ there is a $\delta > 0$ such that if $\|x - y\| < \delta$ then $\|g(x) - g(y)\| < \varepsilon$ (choose whatever norm you like). Take an antipodally symmetric triangulation of S^n with all edge lengths less than δ. Define $L : V(T) \to \{\pm1, \ldots, \pm n\}$ by requiring (1) the absolute value of $L(v)$ is $\mathrm{argmax}_k(|g(v)_k|)$ and (2) the sign of $L(v)$ is the sign of g: $L(v) = \mathrm{sgn}(g(v))|L(v)|$. Note that g is odd and apply Tucker's Lemma to get an edge whose ends are labeled with opposite labels. Deduce the theorem.

2.5.3 Eilenberg–Steenrod axioms

Suppose we have a sequence $\{H_n\}_{n \in \mathbb{Z}}$ of functors from the category of pairs (X, A) to Ab together with natural transformations $\partial_n : H_n(X, A) \to H_{n-1}(A)$ satisfying the following.
(1) Homotopy. If $f \simeq g : (X, A) \to (Y, B)$, then $H_n(f) = H_n(g)$ for all n.
(2) Excision. The map $i : (X - U, A - U) \hookrightarrow (X, A)$ induces isomorphisms $H_n(i) : H_n(X - U, A - U) \to H_n(X, A)$ for all n.
(3) Dimension. $H_n(\{*\}) = 0$ for $n \ne 0$.
(4) Additivity. If $X = \coprod_\alpha X_\alpha$, then $H_n(X) = \bigoplus_\alpha H_n(X_\alpha)$ for all n.
(5) Exactness. For each pair (X, A) we have a long exact sequence

$$\cdots \longrightarrow H_n(A) \longrightarrow H_n(X) \longrightarrow H_n(X, A) \xrightarrow{\partial_n} H_{n-1}(A) \longrightarrow \cdots$$

Such a collection is called a *homology theory*.

Theorem 2.5.6 (Eilenberg–Steenrod). *Any such homology theory is isomorphic to singu-lar homology.*

In the preceding two sections we saw two specific instances of this result, along with explicit isomorphisms. The proof of Theorem 2.5.6 is too complicated to include here, so we omit it.

The upshot here is that we are free to use whichever homology theory seems most computable in any particular case. Moreover, all of the theories we have discussed satisfy excision and admit Mayer–Vietoris sequences, even though that was not immediately apparent from their definitions.

2.6 Homology of manifolds

Throughout this section, M denotes an n-dimensional manifold. We note the following fact.

Proposition 2.6.1. *Suppose M is an n-dimensional manifold and that x is a point of M. Then for all i, $H_i(M, M - \{x\}; \mathbb{Z}) \cong \tilde{H}_{i-1}(S^{n-1}; \mathbb{Z})$.*

Proof. Let U be an open neighborhood of x homeomorphic to $\mathbb{R}^n$. Then excising the closed subset $M - U$ of the open set $M - \{x\}$, we get the following chain of isomorphisms:

$$
\begin{aligned}
H_i(M, M - \{x\}; \mathbb{Z}) &\cong H_i(U, U - \{x\}; \mathbb{Z}) \text{ (excision)} \\
&\cong H_i(\mathbb{R}^n, \mathbb{R}^n - \{0\}; \mathbb{Z}) \\
&\cong \tilde{H}_{i-1}(\mathbb{R}^n - \{0\}; \mathbb{Z}) \text{ (LES of the pair)} \\
&\cong \tilde{H}_{i-1}(S^{n-1}; \mathbb{Z}) \text{ (homotopy equivalence)} \qquad \square
\end{aligned}
$$

The groups $H_i(M, M - \{x\}; \mathbb{Z})$ are called the *local homology groups of M*. Note that these vanish for $i \neq n$.

2.6.1 Orientability

You probably have some experience with the notion of orientation. Loosely speaking, an orientable manifold, such as the sphere S^n, has an "inside" and an "outside." The ur-example of a nonorientable manifold is the Möbius strip; it has only one "side." Or, even more concretely, we understand "right-handed" and "left-handed" orientations of $\mathbb{R}^n$. The standard basis $e_1, \ldots, e_n$ of $\mathbb{R}^n$ gives us the standard orientation. Rotating this basis preserves the orientation, while reflecting in some hyperplane reverses it.

This suggests a plausible orientation of $\mathbb{R}^n$ at a point x: a choice of generator for the local homology group $H_n(\mathbb{R}^n, \mathbb{R}^n - \{x\}; \mathbb{Z}) \cong \mathbb{Z}$. Since this homology group is the homology of the sphere S_x^{n-1} centered at x, and since rotations of $\mathbb{R}^n$ have degree 1 and reflections have degree −1, this object does what we want. So this is what we will do in general. Recall that R is a commutative ring with identity.

Definition 2.6.2. A *local orientation* at a point x in M is a choice of generator $\mu_x \in H_n(M, M - \{x\}; R) \cong R$.

Local orientations exist at each point. What we want is for them to "match up" coherently. That is, as we move around the manifold, we should be able to keep track of the local orientations in a consistent way. Here are two good examples to keep in mind. On the sphere S^2, we have a good sense of the inside and outside, and a convenient way of thinking about that is to choose an outward pointing normal vector at each point. As one walks around the surface of the sphere, these vectors always point out. By contrast, if one imagines the Möbius band, it is possible to choose vectors pointing up in a neighborhood of any point, and if one walks along a curve, the vectors point up for a while. However, upon returning to the original point, you would be on the other "side" of the surface and so there is no way to choose these vectors globally.

Lemma 2.6.3. *Given an element $\mu_x \in H_n(M, M - \{x\}; R)$, there is an open neighborhood U of x and $\mu \in H_n(M, M - U; R)$ such that $\mu_x = j_x^U(\mu)$, where*

$$j_x^U : H_n(M, M - U; R) \longrightarrow H_n(M, M - \{x\}; R)$$

is the homomorphism induced by inclusion.

Proof. Let z be a relative cycle representing μ_x. The support of ∂z is a compact subset of M contained in $M - \{x\}$ and so $U = M - \mathrm{supp}(\partial z)$ is an open neighborhood of x. Taking μ to be the homology class of z relative to $M - U$ provides the required element. □

Thus, we can get local orientations μ_y for y near x by taking $\mu_y = j_y^U(\mu)$. This is what we mean by "matching" local orientations. The class μ is called a *continuation* of a_x in U. This element is actually unique.

Lemma 2.6.4. *Every neighborhood W of x contains a neighborhood U of x such that for every $y \in U, j_y^U$ is an isomorphism (and hence μ_x has a unique continuation in U).*

Proof. Choose a coordinate neighborhood V of x, homeomorphic to $\mathbb{R}^n$, contained in W, and let U be a smaller open set corresponding to an open ball of radius 1. There is a commutative diagram

$$
\begin{array}{ccccc}
H_n(M, M - U) & \xleftarrow{\cong} & H_n(V, V - U) & \xrightarrow{\cong} & \tilde{H}_{n-1}(V - U) \\
{\scriptstyle j_y^U}\downarrow & & \downarrow & & \downarrow \\
H_n(M, M - \{y\}) & \xleftarrow{\cong} & H_n(V, V - \{y\}) & \xrightarrow{\cong} & \tilde{H}_{n-1}(V - \{y\})
\end{array}
$$

The left horizontal isomorphisms are excisions and the right horizontal isomorphisms are connecting homomorphisms (V is contractible). The right vertical map is an isomorphism because the inclusion $V - U \hookrightarrow V - \{y\}$ is a homotopy equivalence and so the map j_y^U is an isomorphism. □

We call an element $\mu \in H_n(M, M - U; R)$ such that $j_y^U(\mu)$ generates the local homology group at y a *local R-orientation along U*. Note that if $V \subset U$ and j_V^U is the induced homomorphism in homology, then if μ is a local R-orientation along U, then $j_V^U(\mu)$ is one along V.

Definition 2.6.5. Let $\{U_i\}$ be an open cover of M and for each i let μ_i be a local R-orientation along U_i. We call this an *R-orientation system* if the following compatibility condition holds: if $x \in U_i \cap U_j$ then $j_x^{U_i}(\mu_i) = j_x^{U_j}(\mu_j)$. In this case a local R-orientation is defined unambiguously at each point x by $\mu_x = j_x^{U_i}(\mu_i)$ for $x \in U_i$. Given another such system $\{V_k, \nu_k\}$ we say it defines the same R-orientation if $\mu_x = \nu_x$ for all $x \in M$. A *global R-orientation* is an equivalence class of R-orientation systems, the equivalence relation being this final condition. If such a system exists, we say that M is *R-orientable*. If $R = \mathbb{Z}$, we simply say that M is *orientable*.

Example 2.6.6. For any $x \in S^n$, the map $H_n(S^n) \to H_n(S^n, S^n - \{x\})$ is an isomorphism, since $S^n - \{x\}$ is contractible. Taking the open covering $\{S^n\}$ and a_x a generator of $H_n(S^n)$, we see that S^n is orientable.

Proposition 2.6.7. *An open submanifold of an R-orientable manifold is orientable. A manifold is R-orientable if and only if all its connected components are.*

Proof. Let $\{U_i, \mu_i\}$ be an R-orientation system for M and let V be an open submanifold. If $x \in V$, let $\nu_x \in H_n(V, V - \{x\}; R)$ correspond to μ_x under the excision isomorphism $H_n(V, V - \{x\}; R) \longrightarrow H_n(M, M - \{x\}; R)$. By Lemma 2.6.4, there is a neighborhood V_x such that $V_x \subset V \cap U_i$ for some i such that ν_x has a unique continuation to a local R-orientation ρ_x along V_x. We may assume that $M - V$ is contained in the interior of $M - V_x$. Then for any $y \in V_x$ the diagram

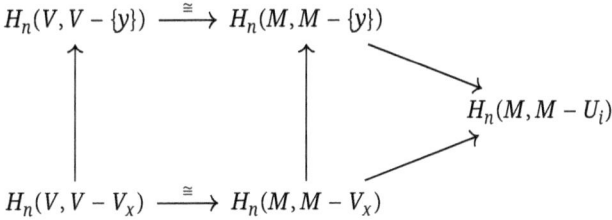

shows that the local R-orientation of V at y induced by ρ_x equals ν_y. It follows that $\{V_x, \rho_x\}$ is an orientation system for V. The second statement follows because the connected components of a manifold are open. $\square$

Example 2.6.8. $\mathbb{R}^n$ is orientable since it is homeomorphic to S^n minus a point.

Proposition 2.6.9. *If M is connected, then two R-orientations of M that agree at one point are equal.*

Proof. The set U of points in M at which the two orientations agree is open by Lemma 2.6.4. So is the set $M - U$, and since M is connected, we see that $U = M$. □

Corollary 2.6.10. *A connected orientable manifold has exactly two distinct orientations.*

Proposition 2.6.11. *Every manifold has a unique $\mathbb{Z}_2$-orientation.*

Proof. For each $x \in M$, μ_x must be the unique nonzero element of $H_n(M, M - \{x\}; \mathbb{Z}_2)$. We can choose an open neighborhood U_x in which μ_x has a unique continuation and these are clearly compatible. □

2.6.2 The top homology group and the fundamental class

Orientations are built from the homology modules $H_n(M, M - \{x\}; R)$. The long exact sequence of the pair then begins

$$0 \longrightarrow H_n(M - \{x\}) \longrightarrow H_n(M) \longrightarrow H_n(M, M - \{x\}) \longrightarrow H_{n-1}(M - \{x\}) \longrightarrow \cdots$$

Of course, there are some unproved assertions in this. For example, it is not clear that the homology of an n-dimensional manifold vanishes above degree n. This is certainly a plausible assertion, especially if one believes that every manifold can be given the structure of a simplicial complex, but it does require proof. There is also the question of whether we can compute $H_n(M - \{x\})$. This space is not compact (unless M is finite, a trivial case), so we will need to deal with this somehow.

The approach we will take is rather abstract, but it tells the whole story. It will involve the theory of covering spaces, but only a bit. The first result is that every manifold has an orientable double cover.

Theorem 2.6.12. *Suppose M is a connected nonorientable manifold. Then there is a connected double cover $p : E \to M$ such that E is orientable.*

Proof. As a set we take

$$E = \{(x, \alpha_x) : x \in M, \alpha_x \text{ a generator of } H_n(M, M - \{x\}; \mathbb{Z})\}.$$

The map p is the projection $p(x, \alpha_x) = x$. We must define a topology on E that makes p a covering map. Note that p is $2 : 1$ as a set map since the group $H_n(M, M - \{x\}; \mathbb{Z}) \cong \mathbb{Z}$ has precisely two generators.

If U is an open set in M with local orientation α_U along U, define

$$\langle U, \alpha_U \rangle = \{(x, \alpha_x) : x \in U, \alpha_x = j_x^U(\alpha_U)\}.$$

By Lemma 2.6.4 these sets form a base for a topology on E. Note that p maps $\langle U, \alpha_U \rangle$ homeomorphically to U. Moreover,

$$p^{-1}(U) = \langle U, a_U \rangle \amalg \langle U, -a_U \rangle$$

so that E is a covering space of M. To see that E is orientable, let $x \in M$ and choose a coordinate neighborhood V, which contains a smaller open set U (corresponding to the unit ball in $\mathbb{R}^n$) for which a_V exists. Set $a_U = j_U^V(a_V)$. Then we have an isomorphism

$$\tilde{H}_{n-1}(\langle V, a_V \rangle - \langle U, a_U \rangle) \xrightarrow{\cong} \tilde{H}_{n-1}(V - U).$$

This provides a local orientation of E along $\langle U, a_U \rangle$.

Finally, if E were not connected, then for each component C, $p|_C : C \to M$ is a covering space which must then be a homeomorphism since the fibers of p consist of only two points. This would imply that C is nonorientable, contrary to Proposition 2.6.7. □

Corollary 2.6.13. *A simply connected manifold is orientable.*

Proof. More is true: if $\pi_1(M)$ has no subgroup of index 2, then M is orientable. This follows because $p_*(\pi_1(E))$ would have index 2 in $\pi_1(M)$. □

While Theorem 2.6.12 is very much dependent on the fact that $R = \mathbb{Z}$, we are often interested in other coefficient rings for homology and with orientability with respect to those rings. The proof of Theorem 2.6.12 serves as a guide for the general case. Define a set

$$X = \{(x, a_x) : x \in M, a_x \in H_n(M, M - \{x\}; R)\}.$$

Note that we have dropped the requirement that a_x be a generator of R. Define $p : X \to M$ by $p(x, a_x) = x$ as before. The fiber $p^{-1}(x)$ is in one-to-one correspondence with the elements of R. If U is open in M, we may define the set $\langle U, a_U \rangle$ just as we did above, and these sets form the basis of a topology on X, where p is a covering map. The space X is called the *R-orientation sheaf of M*.

Let us specialize to the case $R = \mathbb{Z}$. Define a function $\rho : X \to \mathbb{Z}_{\geq 0}$ as follows. Given (x, a_x), the element a_x is an integer multiple of a generator of $H_n(M, M - \{x\}; \mathbb{Z})$ whose absolute value is independent of the chosen generator. Set $\rho(x, a_x)$ to be this absolute value. Note that $\rho^{-1}(1)$ is precisely the space E defined in Theorem 2.6.12.

Lemma 2.6.14. *If $m > 0$, then $\rho^{-1}(m)$ is open in X and $p : \rho^{-1}(m) \to M$ is a double cover.*

Proof. This follows from Lemma 2.6.4. □

The set $\rho^{-1}(0)$ is also open, but it is homeomorphic to M. Note that since X is the disjoint union of the various $\rho^{-1}(m)$, it is not connected.

Definition 2.6.15. Let A be a subspace of M. A continuous map $s : A \to X$ such that $p \circ s = i$, where $i : A \to M$ is the inclusion, is called a *section over A*.

If $x \in A$, let $s'(x) \in H_n(M, M - \{x\}; R)$ be the second coordinate of $s(x)$; that is, $s(x) = (x, s'(x))$. Denote the set of all sections over A by $\Gamma(A)$. If $s_1, s_2 \in \Gamma(A)$, then the map

$$x \mapsto (x, s_1'(x) + s_2'(x))$$

defines another section over A, denoted $s_1 + s_2$. If $\lambda \in R$ then the map

$$x \mapsto (x, \lambda s'(x))$$

is a section over A, denoted λs. This makes $\Gamma(A)$ an R-module. The zero element of this is the section $x \mapsto (x, 0)$. Sections over all of M are called *global sections*.

Proposition 2.6.16. *There exists a global section s mapping M into $p^{-1}(1)$ if and only if M is R-orientable. Moreover, there is a one-to-one correspondence between the different R-orientations of M and the set $\Gamma(M)$.*

Proof. This is left as an exercise. □

For any subspace A, we say that M is *R-orientable along A* if there is a section over A mapping into $p^{-1}(1)$.

Proposition 2.6.17. *M is R-orientable along A if and only if there is a homeomorphism $\varphi : p^{-1}(A) \to A \times R$ (where R has the discrete topology) such that the diagram*

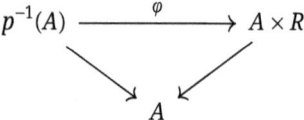

commutes. In this case, $\Gamma(A)$ is isomorphic to the module of all continuous maps $A \to R$ and if A has k connected components (k finite), then $\Gamma(A) \cong R^k$.

Proof. Suppose $s : A \to p^{-1}(1)$ is a section. Then for each $x \in A$, the element $s'(x) \in H_n(M, M - \{x\})$ is a generator. If $(x, a_x) \in p^{-1}(A)$, then there is a unique $\lambda_x \in R$ such that $a_x = \lambda_x s'(x)$. Define φ by

$$\varphi(x, a_x) = (x, \lambda_x).$$

If U is an open neighborhood of x on which a_x has a unique continuation a_U, then φ maps $\langle U, a_U \rangle$ one-to-one onto $U \times \{\lambda_x\}$ and is hence a homeomorphism. Conversely, given φ define s by $s(x) = \varphi^{-1}(x, 1)$ for all $x \in A$.

Since R has the discrete topology, the only continuous maps $A \to R$ are the locally constant maps. It follows that $\Gamma(A) \cong R^k$. □

We are now ready to compute the degree n homology. There is a canonical homomorphism $j_A : H_n(M, M - A; R) \to \Gamma(A)$ defined for $\alpha \in H_n(M, M - A; R)$ and $x \in A$ by

$$j_A(\alpha) : x \mapsto (x, j_x^A(\alpha)).$$

Lemma 2.6.18. *The map $j_A(\alpha) : A \to X$ is continuous, and so j_A is well-defined.*

Proof. Let z be a relative cycle representing α. If $U = M - \text{supp}(\partial z)$, then U is open and contains A. Let $\alpha_U \in H_n(M, M - U; R)$ be the homology class of z relative to $M - U$. Then α_U induces α under the inclusion of $M - U$ into $M - A$. If $x \in A$, let $V \subset U$ be a neighborhood of x on which $j_x^A(\alpha)$ has a unique continuation α_V over V (these exist via Lemma 2.6.4). Then $j_A(\alpha)$ maps $V \cap A$ into $\langle V, \alpha_V \rangle$, and these latter sets from a basis of neighborhoods of $(x, j_x^A(\alpha))$ in X. $\qquad\square$

Note that if $B \subset A$, we have a commutative diagram

$$
\begin{array}{ccc}
H_n(M, M - A; R) & \xrightarrow{\ j_A\ } & \Gamma(A) \\
{\scriptstyle j_B^A}\big\downarrow & & \big\downarrow{\scriptstyle r} \\
H_n(M, M - B; R) & \xrightarrow{\ j_B\ } & \Gamma(B)
\end{array}
$$

where the map r is the map restricting sections over A to sections over B.

Definition 2.6.19. A section $s \in \Gamma(A)$ has *compact support* if it agrees with the zero section outside some compact subset of A. The submodule of such sections is denoted $\Gamma_c(A)$. Note that if A is compact, then $\Gamma_c(A) = \Gamma(A)$.

Theorem 2.6.20. *Suppose A is a closed subset of M. Then*
(1) $H_k(M, M - A; R) = 0$ *for $k > n$; and*
(2) j_A *is injective with image equal to the submodule $\Gamma_c(A)$; that is, we have an isomorphism*

$$j_A : H_n(M, M - A; R) \xrightarrow{\cong} \Gamma_c(A).$$

In particular, $j_M : H_n(M; R) \to \Gamma_c(M)$ is an isomorphism, and $H_k(M; R) = 0$ for $k > n$.

Proof. The argument proceeds in various steps. First, note that if A is empty, then the result holds trivially. Now suppose the theorem holds for closed subsets A_1, A_2, and $A_1 \cap A_2$. We claim that it then holds for $A = A_1 \cup A_2$. To see this, use the Mayer–Vietoris sequence for the triple $(M, M - A_1, M - A_2)$ to see that $H_k(M, M - A; R) = 0$ for $k > n$ (assuming this is true for each A_i) and we have a commutative diagram

$$0 \to H_n(M, M - A) \to H_n(M, M - A_1) \oplus H_n(M, M - A_2) \to H_n(M, M - (A_1 \cap A_2))$$

$$\downarrow j_A \qquad\qquad \cong \downarrow j_{A_1} \oplus j_{A_2} \qquad\qquad \cong \downarrow j_{A_1 \cap A_2}$$

$$0 \longrightarrow \Gamma_c(A) \xrightarrow{\ (r_1, -r_2)\ } \Gamma_c(A_1) \oplus \Gamma_c(A_2) \xrightarrow{\ r_1 + r_2\ } \Gamma_c(A_1 \cap A_2)$$

A diagram chase shows that j_A is an isomorphism.

Now suppose A is compact, connected, and contained in a coordinate neighborhood evenly covered by p. By excision, we may replace M with the interior of D^n; we will abuse notation and still call this D^n. If A is a rectangular parallelepiped of dimension $\le n$, then the first assertion follows from the chain of isomorphisms

$$H_k(D^n, D^n - A) \xrightarrow{\cong} \tilde{H}_{k-1}(D^n - A) \xrightarrow{\cong} \tilde{H}_{k-1}(S^{n-1}).$$

Since A is connected and D^n is R-orientable the second assertion follows from Proposition 2.6.17. If A is a union of rectangular parallelepipeds $A_1, \ldots, A_m$ such that each face of each A_i is parallel to a coordinate hyperplane in $\mathbb{R}^n$, then we proceed by induction on m, the case $m = 1$ being what we just proved. Let $A' = A_1 \cup \cdots \cup A_{m-1}$. Then $A' \cap A_m$ is a set of the same type–a union of at most $m - 1$ such parallelepipeds. The inductive hypothesis then holds for both A' and $A' \cap A_m$, and therefore for $A = A' \cup (A' \cap A_m)$. Finally, if A is compact and contained in an evenly covered coordinate neighborhood U, we proceed as follows. If $s \in \Gamma(A)$, we may assume that s maps A into one sheet over U (if not, the compactness of A and the normality of U would give m disjoint open sets covering A such that it is true for each $U_i \cap A$). Then s extends to $\bar{s} \in \Gamma(U)$. For each $x \in A$, choose a rectangular parallelepiped of dimension n containing x in its interior, having its faces parallel to the coordinate hyperplanes, and contained in U. Let A' be the union of all these parallelepipeds; this is a finite union since A is compact. By the previous case and by considering the commutative diagram

$$\begin{array}{ccc} H_n(M, M - A') & \xrightarrow{\ j_{A'}\ }_{\cong} & \Gamma(A') \\ \downarrow & & \downarrow r \\ H_n(M, M - A) & \xrightarrow{\ j_A\ } & \Gamma(A) \end{array}$$

we see that s is in the image of j_A (on the right, $\bar{s}$ maps to s). Thus, j_A is surjective. Now let $\alpha \in H_k(M, M - A)$, $k \ge n$. If $k = n$, assume $j_A(\alpha) = 0$. We wish to show that $\alpha = 0$. Let z be a relative cycle representing α. Then $V = M - \mathrm{supp}(\partial z)$ is an open set containing A. Let α' be the homology class of z in $H_k(M, M - V)$. If $k = n$, since $j_x^V(\alpha') = j_x^V(\alpha) = 0$ for all $x \in A$, Lemma 2.6.4 produces an open V' with $A \subset V' \subset V$ such that $j_x^V(\alpha') = 0$ for all $x \in V'$. Construct an A' as above with $A \subset A' \subset V' \cap U$. Then $j_{A'}^V(\alpha') = 0$ by the previous case and so $\alpha = j_A^{A'}(j_{A'}^V(\alpha)) = 0$.

Now if A is compact, then A is a finite union of compact sets $A_1, \ldots, A_m$, each of which is contained in a coordinate neighborhood evenly covered by p. By induction, using the previous steps, the result is true for A.

If $A \subset U$, where U is open with compact closure $\overline{U}$, then we claim the theorem is true for A and U. To see this, use the exact homology sequence for the triple $(M, U \cup (M - \overline{U}))$, $(U - A) \cup (M - \overline{U})$. By excision we have

$$H_k(U, U - A) \cong H_k(U \cup (M - \overline{U}), (U - A) \cup (M - \overline{U}))$$

and so for $k > n$ we obtain

$$H_{k+1}(M, U \cup (M - \overline{U})) \longrightarrow H_k(U, U - A) \longrightarrow H_k(M, (U - A) \cup (M - \overline{U})).$$

The first and third modules are 0 by the previous steps applied to the manifold M and the compact subsets $\overline{U} - U$ and $\overline{A} \cup (\overline{U} - U)$. Thus, the middle term is 0 as well. For $k = n$ we have the commutative diagram

$$0 \to H_n(U, U - A) \to H_n(M, (U - A) \cup (M - \overline{U})) \to H_n(M, U \cup (M - \overline{U}))$$
$$\downarrow j_A \qquad\qquad \downarrow \qquad\qquad \downarrow$$
$$0 \longrightarrow \Gamma_c(A) \xrightarrow{\ i\ } \Gamma(\overline{A} \cup (\overline{U} - u)) \xrightarrow{\ r\ } \Gamma(\overline{U} - U)$$

The map i is defined as follows. If $s \in \Gamma_c(A)$ is zero outside a compact $K \subset A$, then $i(s)|_A = s$ and $i(s) = 0$ outside K. This is clearly injective and via diagram chase we see that j_A is an isomorphism.

Finally, in the case of a general A, let $s \in \Gamma_c(A)$ be zero outside a compact $K \subset A$. There is an open set U containing K with compact closure (since we can cover K with finitely many coordinate neighborhoods). Let $A' = A \cap U$ and let $s' = s|_{A'}$. By the previous case, applied to $j_{A'}$, the commutative diagram

$$H_n(U, U - A') \longrightarrow H_n(M, M - A)$$
$$\cong \downarrow j_{A'} \qquad\qquad \downarrow j_A$$
$$\Gamma_c(A') \xrightarrow{\ i\ } \Gamma_c(A)$$

shows that s lies in the image of j_A (since s' maps to s under i). Now suppose that $a \in H_k(M, M - A)$ and if $k = n$ assume $j_A(a) = 0$. We must show that $a = 0$. Let z be a relative cycle representing a. Applying the above to the compact set $\mathrm{supp}(z)$, we find an open set $U \supset \mathrm{supp}(z)$ such that $\overline{U}$ is compact. Let $A' = A \cap U$. The same commutative diagram finishes the argument for $k = n$ and for $k > n$ we know the class of z in $H_k(U, U - A')$ is zero by the previous step. So $a = 0$. □

This was a long and complicated argument, but it has a number of important practical corollaries.

Corollary 2.6.21. *If A is connected and non-compact, then $H_n(M, M - A; R) = 0$. In particular, if M is connected and non-compact, then $H_n(M; R) = 0$.*

Proof. If $\alpha \in H_n(M, M - A; R)$, then by connectedness, $\rho j_A(\alpha)$ is constant, and since $j_A(\alpha)$ is zero outside a compact set, $\alpha = 0$. □

Corollary 2.6.22. *Let M be a compact connected manifold. Assume that for any nonzero $a \in R$ and any unit $u \in R$, $ua = a$ implies $u = 1$ (e. g., R an integral domain). Then*

$$H_n(M; R) \cong \begin{cases} R & \text{if } M \text{ is } R\text{-orientable} \\ 0 & \text{otherwise.} \end{cases}$$

Proof. If M is R-orientable, then $\Gamma(M) \cong R$ by Proposition 2.6.17. By Theorem 2.6.20, $H_n(M; R) \cong \Gamma_c(M)$. But since M is compact, $\Gamma_c(M) = \Gamma(M)$.

Conversely, suppose there is a global section $s \in \Gamma(M)$, $s \neq 0$. Then $\rho(s(M))$ is a constant in R modulo units; that is, there is a nonzero $a \in R$ such that $s'(x)$ is a times a generator of $H_n(M, M - \{x\})$ for all $x \in M$ ($s(x) = (x, s'(x))$). The hypothesis on R implies that $s'(x)/a$ is a well-defined generator. The map $x \mapsto (x, s'(x)/a)$ is a section $M \to \rho^{-1}(1)$, so that M is R-orientable. □

Corollary 2.6.23. *If M is a compact connected manifold, then $H_n(M; \mathbb{Z}_2) \cong \mathbb{Z}_2$.*

Proof. Every manifold has a unique $\mathbb{Z}_2$ orientation. □

You may have noticed that so far the only manifolds we have shown to be orientable are the sphere S^n and euclidean space $\mathbb{R}^n$. That is because the definition of orientability involves making many local choices and then proving they are compatible; for a general manifold this can be quite daunting. However, Corollary 2.6.22 allows us to deduce that the following manifolds are orientable:

$$S^n \ (n \geq 1); \quad M_g \ (g \geq 0); \quad \mathbb{R}P^n \ (n \text{ odd}); \quad \mathbb{C}P^n \ (\text{all } n),$$

and that the following manifolds are nonorientable:

$$K \text{ (the Klein bottle)}; \quad \mathbb{R}P^n \ (n \text{ even}).$$

The fundamental class
If the compact, connected manifold M is R-orientable, then $H_n(M; R) \cong \Gamma(M) \cong R$. This leads to the following definition.

Definition 2.6.24. A generator α of $H_n(M; R)$ is called the *fundamental class* of M. We typically denote this class by $[M]$.

Note that the fundamental class corresponds to a global section of the covering space $\rho^{-1}(1) \to M$. The local R-orientation at each $x \in M$ is given by $j_x^M([M])$. Since there are two choices of generator for $H_n(M)$, we see that there are two distinct orientations of M, differing by a sign.

If we assume that M has been triangulated, then we can be more explicit about the fundamental class. Let $R = \mathbb{Z}$. We know that $[M]$ is represented by some cycle $z = \sum k_i \sigma_i$, where σ_i are the n-simplices in M, and $k_i \in \mathbb{Z}$. Since $[M]$ maps to a generator of $H_n(M, M - \{x\}; \mathbb{Z})$ for all x in the interior of σ_i, we must have $k_i = \pm 1$ for all i. Also, since z is a cycle, if σ_i and σ_j share a face, then k_i determines k_j and vice versa (the faces must cancel out via the boundary map). Since M is orientable, a choice of the k_i is possible. In the case $R = \mathbb{Z}_2$, we can always take $z = \sum \sigma_i$ as a generating cycle for the fundamental class.

Exercises

Exercise 2.6.1. Prove the assertion made in the proof of Theorem 2.6.12 that the sets $\langle U, a_U \rangle$ for a base for a topology on E.

Exercise 2.6.2. Consider the Klein bottle K. This is a nonorientable manifold. What is the orientable double cover $E \to K$?

Exercise 2.6.3. Prove Proposition 2.6.16.

Exercise 2.6.4. Let M be a compact, connected n-manifold. Prove that if M is R-orientable, then the map $H_n(M; R) \to H_n(M, M - \{x\}; R)$ is an isomorphism for all $x \in M$.

Exercise 2.6.5. Let M be a compact, connected n-manifold. Prove that if M is not R-orientable, then the map $H_n(M : R) \to H_n(M, M - \{x\}; R)$ is injective with image $\{r \in R : 2r = 0\}$.

Exercise 2.6.6. Here is an alternate definition of orientability. Suppose that $\{(U_\alpha, \varphi_\alpha)\}$ is an atlas for the compact, connected manifold M. Denote the *transition functions* $\varphi_\beta^{-1} \circ \varphi_\alpha$ by $\varphi_{\alpha\beta}$. We say that two charts U_α and U_β have compatible orientations if $\det D\varphi_{\alpha\beta} > 0$. Prove that M is orientable if and only if it admits an atlas all of whose charts have compatible orientations. Show that there are charts in the standard atlas for $\mathbb{R}P^3$ that do not have compatible orientations, yet this manifold is orientable. Alter the standard atlas to correct this issue.

2.6.3 Euler characteristic

Recall that the *Euler characteristic* of a chain complex $(C_\bullet, d)$ is defined to be the alternating sum

$$\chi(C_\bullet) = \sum_i (-1)^i \operatorname{rank}(C_i).$$

An exercise in Chapter 1 shows that

$$\chi(C_\bullet) = \sum_i (-1)^i \operatorname{rank}(H_i(C_\bullet)).$$

Definition 2.6.25. If X is a topological space and R is a commutative ring, the *ith R-Betti number of X* is

$$\beta_i(X;R) = \text{rank}(H_i(X;R)).$$

If $R = \mathbb{Z}$ we simply refer to these as the Betti numbers of X.

The Betti numbers are topological invariants, but they may depend on the ring R. For example, for $X = \mathbb{R}P^2$, the Betti number β_2 is 0 for $R = \mathbb{Z}$, but it is 1 for $R = \mathbb{Z}_2$.

Definition 2.6.26. Let X be a topological space and R a commutative ring. The *Euler characteristic of X* is

$$\chi(X;R) = \sum_i (-1)^i \beta_i(X;R).$$

When $R = \mathbb{Z}$, we omit it from the notation.

A priori, this integer depends on the ring R. In Section 3.4, however, we will see that it does not. Let us compute some examples.

Example 2.6.27. For the sphere S^n, we have $H_k(S^n;R) \cong R$ for $k = 0, n$ and 0 otherwise. Thus

$$\chi(S^n;R) = 1 + (-1)^n$$

for any ring R.

Example 2.6.28. For complex projective space $\mathbb{C}P^n$, we have that $H_i(\mathbb{C}P^n;R) \cong R$ for $0 \leq i \leq 2n$, i even. Thus

$$\chi(\mathbb{C}P^n;R) = n + 1$$

for any R.

Example 2.6.29. For the real projective space $\mathbb{R}P^n$, let's first consider $R = \mathbb{Z}$. The homology groups are all torsion except for H_0 and H_n for n odd. It follows that

$$\chi(\mathbb{R}P^n;\mathbb{Z}) = \begin{cases} 1 & n \text{ even} \\ 0 & n \text{ odd.} \end{cases}$$

For $R = \mathbb{Z}_2$, we have that $H_i(\mathbb{R}P^n;\mathbb{Z}_2) \cong \mathbb{Z}_2$ for all $0 \leq i \leq n$. It follows that

$$\chi(\mathbb{R}P^n;\mathbb{Z}_2) = \sum_{i=0}^n (-1)^i = \begin{cases} 1 & n \text{ even} \\ 0 & n \text{ odd.} \end{cases}$$

The Leftschetz Fixed Point Theorem

A map $\varphi : \mathbb{Z}^n \to \mathbb{Z}^n$ is represented by a matrix of integers $[a_{ij}]$. The *trace* of φ is

$$\mathrm{tr}(\varphi) = \sum_{i=1}^{n} a_{ii}.$$

This is independent of the choice of basis of $\mathbb{Z}^n$, since similar matrices have the same trace. If A is a finitely generated abelian group and $\varphi : A \to A$ a group homomorphism, define $\mathrm{tr}(\varphi)$ to be the trace of the induced map $\overline{\varphi} : A/\mathrm{torsion} \to A/\mathrm{torsion}$.

Definition 2.6.30. If X is a finite cell complex and $f : X \to X$ is a continuous map, the *Lefschetz number* of f is

$$\tau(f) = \sum_i (-1)^i \, \mathrm{tr}(f_* : H_i(X; \mathbb{Z}) \to H_i(X; \mathbb{Z})).$$

Example 2.6.31. If $f \simeq \mathrm{id}_X$, then

$$\tau(f) = \sum_i (-1)^i \, \mathrm{tr}(\mathrm{id} : H_i(X) \to H_i(X)) = \sum_i (-1)^i \beta_i(X) = \chi(X).$$

Theorem 2.6.32 (Lefschetz Fixed-Point Theorem). *If X is a finite simplicial complex and $f : X \to X$ satisfies $\tau(f) \neq 0$, then f has a fixed point.*

Proof. Suppose f does not have a fixed point. Then after sufficiently many barycentric subdivisions of X, we have that $f \simeq g$, with g simplicial and $g(\sigma) \cap \sigma = \emptyset$ for all simplices σ. Then $\mathrm{tr}(g_\# : C_n(X) \to C_n(X))$ is 0 for all n. But if z is an n-cycle, $z = \sum m_i \sigma_i$, then $g_\#(z) = \sum m_i g_\#(\sigma_i)$ and so $\mathrm{tr}(g_* : H_n(X) \to H_n(X))$ is 0 as well. Thus,

$$\tau(f) = \sum_i (-1)^i \, \mathrm{tr}(f_* : H_i(X) \to H_i(X)) = \sum_i (-1)^i \, \mathrm{tr}(g_* : H_i(X) \to H_i(X)) = 0. \qquad \square$$

Corollary 2.6.33. *Suppose f is a polynomial with real coefficients of odd degree $\deg(f) = 2k + 1$. Then f has a real root.*

Proof. We may assume that $f(0) \neq 0$ (otherwise 0 is a root). Let $T : \mathbb{R}^{2k+1} \to \mathbb{R}^{2k+1}$ be a linear transformation with characteristic polynomial f (e. g., take left multiplication by the companion matrix of f). Then T is invertible since $\det T = f(0) \neq 0$. The map T induces a map $t : \mathbb{RP}^{2k} \to \mathbb{RP}^{2k}$, since T takes lines in $\mathbb{R}^{2k+1}$ to lines in $\mathbb{R}^{2k+1}$. Since the integral homology of $\mathbb{RP}^{2k}$ is all torsion except in degree 0, we must have $\tau(t) = 1$ (the map t_* is the identity in degree 0). By Theorem 2.6.32, t has a fixed point. This fixed point is a line ℓ in $\mathbb{R}^{2k+1}$ with $T(\ell) = \ell$; that is, ℓ is a 1-dimensional eigenspace of T. The corresponding eigenvalue λ is a root of f. $\qquad \square$

Continuing to use our sledgehammer, we also have the following.

Corollary 2.6.34. $\mathbb{C}$ *is algebraically closed.*

Proof. Let f be a polynomial with complex coefficients of degree $n \geq 2$. Let T be an invertible linear transformation $T : \mathbb{C}^n \to \mathbb{C}^n$ with characteristic polynomial f. Then T induces $t : \mathbb{C}P^{n-1} \to \mathbb{C}P^{n-1}$. Note that $t \in \mathrm{PGL}_n(\mathbb{C}) = \mathrm{GL}_n(\mathbb{C})/Z$, where Z is the center of $\mathrm{GL}_n(\mathbb{C})$ (i. e., the group of scalar multiples of the identity matrix), and this group is connected. So there is a path from t to id; that is, $t \simeq$ id. Thus, the Lefschetz number is $\tau(t) = \tau(\mathrm{id}) = \chi(\mathbb{C}P^{n-1}) = n$. By Theorem 2.6.32, t has a fixed point; that is, there is a 1-dimensional subspace $\ell \subset \mathbb{C}^n$ with $T(\ell) = \ell$. The corresponding eigenvalue λ is a root of f. $\square$

Exercises

Exercise 2.6.7. Let M_g be the orientable surface of genus g. Compute $\chi(M_g; \mathbb{Z})$ and $\chi(M_g; \mathbb{Z}_2)$.

Exercise 2.6.8. Let K be the Klein bottle. Compute $\chi(K; \mathbb{Z})$ and $\chi(K; \mathbb{Z}_2)$.

Exercise 2.6.9. Let X and Y be finite cell complexes. Show that $\chi(X \times Y) = \chi(X)\chi(Y)$.

Exercise 2.6.10. Suppose X is a finite cell complex that is the union of two subcomplexes A and B. Show that $\chi(X) = \chi(A) + \chi(B) - \chi(A \cap B)$.

Exercise 2.6.11. Prove that $\mathrm{PGL}_n(\mathbb{C})$ is connected.

Exercise 2.6.12. Corollary 2.6.33 seems like cheating. This result is really a consequence of the completeness of the real numbers and the Intermediate Value Theorem. Where is the completeness of $\mathbb{R}$ hiding in the proof?

Exercise 2.6.13. Verify Theorem 2.6.32 in the following example. Let $f : T \to T$ be the map of the torus obtained by rotating it $180°$ about the axis through the hole in the center. This map has no fixed points. Show that $\tau(f) = 0$.

Exercise 2.6.14. Show that a map $S^n \to S^n$ has a fixed point unless its degree is equal to the degree of the antipodal map.

Exercise 2.6.15. Let X be a finite simplicial complex and let $f : X \to X$ be a simplicial homeomorphism. Show that the Lefschetz number $\tau(f)$ equals the Euler characteristic of the set of fixed points of f. In particular, $\tau(f)$ is the number of fixed points if the fixed points are isolated. (Hint: subdivide X to make the fixed point set a subcomplex.)

2.7 Persistent homology and topological data analysis

All the preceding homology theory depends on having a topology on our set of points: a simplicial structure, a cell structure, a manifold structure, etc. However, many problems of interest in mathematics and data analysis begin with a discrete set of data points in some Euclidean space. As a topological space, this set is not interesting–it is discrete and therefore has homology only in degree zero. But what if the points were sampled from some topological space? Can we deduce something about the underlying space from the data? Is there some way to compute homology using the set of data?

This is the motivating question in topological data analysis (TDA). There are many useful statistical techniques for analyzing data, but algebraic topology is another powerful tool with well-developed computational methods. The yoga of TDA is to begin with a data set, construct a simplicial complex from it in some way, compute the homology of this complex, and then make inferences about the (assumed) manifold underlying the data. This immediately raises many questions: how do we know the data come from a manifold? (we do not); how do we build a complex out of the data? (there are many approaches to this); how do we know the homology we compute is that of the underlying manifold? (we have to prove something about this).

One important property of homology is that it is functorial; this turns out to be the crucial insight in TDA. There are many ways one might build a simplicial complex from data, but the simplest one to visualize is the following. Begin with a set X in euclidean space (or more generally in some metric space). For $r > 0$, take an r-ball around each point in X. For small r, this is just a disjoint collection of balls, but if we allow r to grow, the balls will begin to intersect. When two such balls meet, we can join the corresponding data points with an edge. When three meet, we fill in the corresponding triangle, and so on. This gives us a *nested family* of complexes $V(r)$: if $r < r'$ then $V(r) \subseteq V(r')$. We then get an induced map between the homology groups of these spaces and we can track homology classes as the parameter varies. The mantra is that classes that persist over long ranges of r values are significant and correspond to genuine topological features in the underlying manifold. Those classes that arise and then get filled in quickly are less significant (perhaps they are just noise).

This is the idea behind persistent homology. In this section, we will discuss the basics of this theory. We will not attempt to present the full state of the art generality here, choosing instead to focus on the simpler discrete filtration case.

2.7.1 Filtrations, barcodes, and persistence diagrams

Throughout this section, homology will be taken with coefficients in a fixed field F. Integral homology could be used, but there are technical problems with it that will become evident as we proceed.

We begin with a simple, but important, example. Let K be a simplicial complex and let $f : K \to \mathbb{R}$ be a function that is monotone in the sense that if $\sigma < \tau$ then $f(\sigma) \leq f(\tau)$ (here f assigns a number to each *simplex* of K, not each point). If $a \in \mathbb{R}$, set $K(a) = f^{-1}(-\infty, a]$. Because f is monotone, $K(a)$ is a subcomplex of K. Let m be the number of simplices in K. Then there are $n + 1 \leq m + 1$ subcomplexes

$$\emptyset = K_0 \subseteq K_1 \subseteq \cdots \subseteq K_n = K.$$

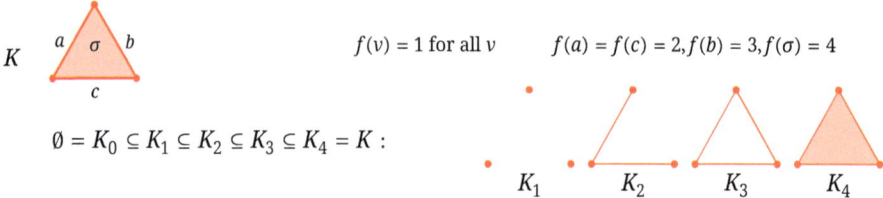

$$K$$

$$f(v) = 1 \text{ for all } v \qquad f(a) = f(c) = 2, f(b) = 3, f(\sigma) = 4$$

$$\emptyset = K_0 \subseteq K_1 \subseteq K_2 \subseteq K_3 \subseteq K_4 = K :$$

$$K_1 \qquad K_2 \qquad K_3 \qquad K_4$$

Figure 2.10: The filtration of a function on a 2-simplex.

That is, if $a_1 < a_2 < \cdots < a_n$ are the distinct function values of the simplices and $a_0 = -\infty$, then $K_i = K(a_i)$. We call this sequence of subcomplexes the *filtration* of f. We think of this as adding a collection of simplices at each stage. See Figure 2.10 for an example.

More generally, a *filtration* of a simplicial complex K is a sequence of subcomplexes

$$\emptyset = K_0 \subseteq K_1 \subseteq \cdots \subseteq K_n = K.$$

Now, if $i \le j$, we have an inclusion map $K_i \hookrightarrow K_j$ and an induced map

$$f_p^{i,j} : H_p(K_i) \to H_p(K_j).$$

(Recall that homology is with coefficients in a field F, which we will omit from the notation.) When we pass from K_i to K_{i+1}, we might get new homology classes, some might die, and some might merge with other classes.

Definition 2.7.1. The *pth persistent homology groups* are the images of the homomorphisms induced by the inclusions $K_i \hookrightarrow K_j$, $i \le j$:

$$H_p^{i,j} = \operatorname{im} f_p^{i,j} \qquad 0 \le i \le j \le n.$$

The *pth persistent Betti numbers* are $\beta_p^{i,j} = \dim H_p^{i,j}$.

Note that $H_p^{i,i} = H_p(K_i)$. The persistent classes consist of classes in K_i that are still alive in K_j; that is,

$$H_p^{i,j} = Z_p(K_i)/(B_p(K_j) \cap Z_p(K_i)).$$

Definition 2.7.2. A class $\gamma \in H_p(K_i)$ is *born* at K_i if $\gamma \notin H_p^{i-1,i}$. If γ is born at K_i, then γ dies at K_j if it merges with an older class as we go from K_{j-1} to K_j:

$$f_p^{i,j-1}(\gamma) \notin H_p^{i-1,j-1}, \quad \text{but } f_p^{i,j}(\gamma) \in H_p^{i-1,j}.$$

Definition 2.7.3. Suppose γ is born at K_i and dies at K_j. If the filtration $\{K_i\}$ arises from a function (i. e., $K_i = K(a_i)$ for $f : K \to \mathbb{R}$), we call the difference in function values the *persistence* of γ: pers(γ) = $a_j - a_i$. We also have the *index persistence*, still denoted pers(γ) = $j - i$. If γ is born at K_i but never dies, we say that γ *lives to infinity* and set index persistence to ∞.

Figure 2.11 shows an example of a hollow tetrahedron (homeomorphic to S^2) with a filtration. The dots under each subcomplex represent generators of the relevant homology group and the persistence of each class is shown.

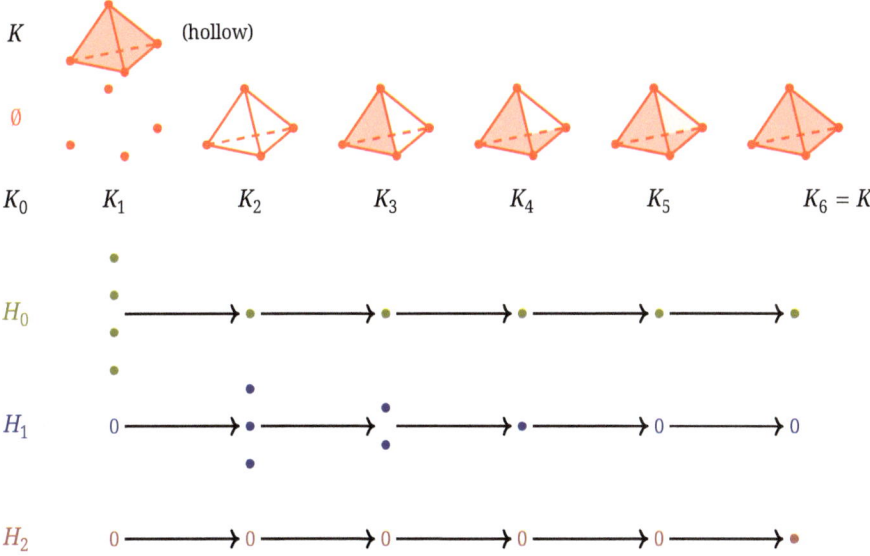

Figure 2.11: A filtered tetrahedron and its persistence.

There are two primary devices to visualize persistence. They are equivalent, but there are advantages to each. We begin with the *barcode*. Given a filtered complex, and a fixed homological degree p, the persistences of the various homology classes are represented by intervals $[a_i, a_j]$ or $[a_i, \infty)$ (these may be interval persistence values). We then draw these as a collection of bars lying above an axis parametrizing the interval values. For the filtered tetrahedron in Figure 2.11, the barcodes for $p = 0, 1, 2$ are shown in Figure 2.12.

This is a convenient graphical representation, and one can see quickly how many homology classes there are and how long they persist (longer bars mean greater persistence). However, we will want to compare these objects, and barcodes are less appealing for that purpose. The better choice for that is the *persistence diagram*, which encodes the same data but in a different way. Again, suppose we have a filtration. Fix a homological degree p and let $\mu_p^{i,j}$ be the number of independent p-dimensional classes born at K_i and dying at K_j. Note that we have

$$\mu_p^{i,j} = \underbrace{(\beta_p^{i,j-1} - \beta_p^{i,j})}_{\substack{\text{classes born at or before } K_i, \\ \text{dying at } K_j}} - \underbrace{(\beta_p^{i-1,j-1} - \beta_p^{i-1,j})}_{\substack{\text{classes born at or before } K_{i-1}, \\ \text{dying at } K_j}}$$

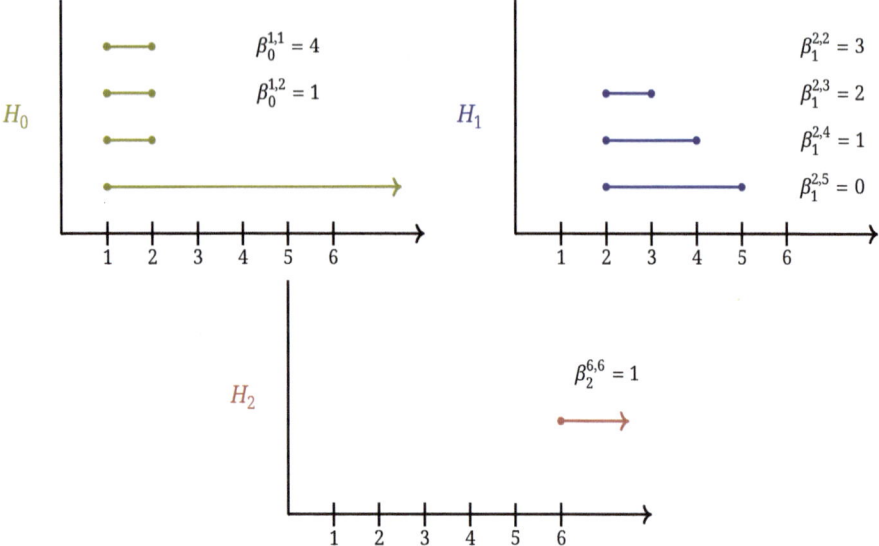

Figure 2.12: The barcodes for a filtered tetrahedron.

To construct the *pth persistence diagram*, $\mathrm{Dgm}_p(f)$, draw each point (a_i, a_j) with multiplicity $\mu_p^{i,j}$. This gives a multiset in the extended plane $\overline{\mathbb{R}}^2 = (\mathbb{R} \cup \{\pm\infty\})^2$. Note that since classes must die after they are born, these points all live above the diagonal $y = x$ in the plane. For technical reasons which we will explain in the next section, we also include the diagonal in the diagram, each point with infinite multiplicity. The persistence diagrams for the example in Figure 2.11 are shown in Figure 2.13. Barcodes and persistence diagrams are equivalent, as it is clear how to go from one to the other.

The persistence algorithm

Computing homology is expensive, especially when the complex contains a large number of simplices. Moreover, in the persistent homology arena, in principle we must compute the homology of each subcomplex in the filtration, as well as the induced maps in homology. This requires finding bases for these vector spaces, and there is no guarantee that we will make consistent choices of generating cycles. Luckily, if we are simply interested in finding the birth and death points for various homology classes (and therefore the corresponding persistence diagram), we can avoid a lot of this work. We now present a practical algorithm for computing persistence diagrams. We assume the coefficient field is $F = \mathbb{Z}_2$ for this purpose.

As a first step, suppose that $N \subset M$ are simplicial complexes that differ by a single simplex: $M - N = \{\sigma\}$. Say $\dim \sigma = p$. We have the long exact sequence of the pair (M, N)

$$0 \longrightarrow \tilde{H}_p(N) \xrightarrow{i_*} \tilde{H}_p(M) \xrightarrow{D} H_p(M, N) \longrightarrow \tilde{H}_{p-1}(N) \xrightarrow{i_*} \tilde{H}_{p-1}(M) \longrightarrow 0.$$

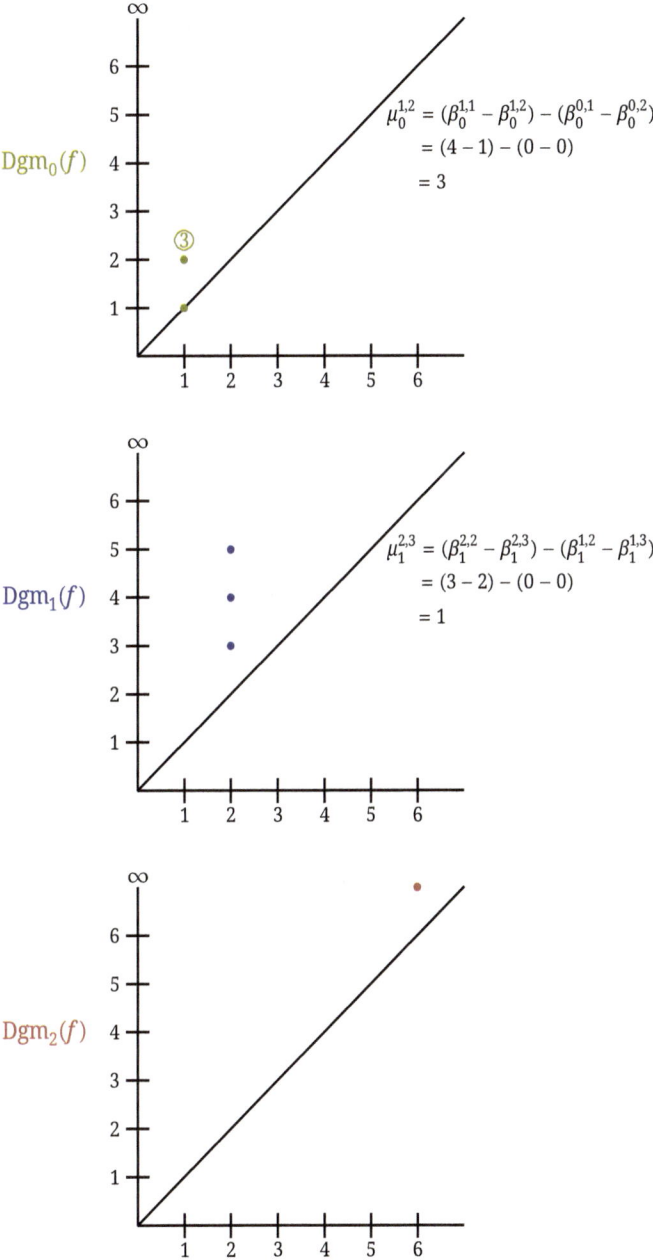

$$\mu_0^{1,2} = (\beta_0^{1,1} - \beta_0^{1,2}) - (\beta_0^{0,1} - \beta_0^{0,2})$$
$$= (4 - 1) - (0 - 0)$$
$$= 3$$

$$\mu_1^{2,3} = (\beta_1^{2,2} - \beta_1^{2,3}) - (\beta_1^{1,2} - \beta_1^{1,3})$$
$$= (3 - 2) - (0 - 0)$$
$$= 1$$

Figure 2.13: The persistence diagrams for a hollow tetrahedron.

The vector space $H_q(M, N)$ is trivial for $q \neq p$ and has dimension 1 for $q = p$. It follows that M and N have isomorphic homology groups, except possibly in degrees p and $p - 1$. There are two cases:

(1) D is surjective. Then $\tilde{\beta}_p(M) = \tilde{\beta}_p(N) + 1$.

(2) $D \equiv 0$. Then $\tilde{\beta}_{p-1}(M) = \tilde{\beta}_{p-1}(N) - 1$.

In the first case, we call σ *positive* because its addition has created a new p-dimensional homology class. In the second case, we call σ *negative* because its addition has killed a $(p - 1)$-dimensional class.

Given a simplicial complex K, fix an ordering $\sigma_1, \sigma_2, \ldots, \sigma_m$ of the simplices of K such that each $K_i = \{\sigma_1, \ldots, \sigma_i\}$ is a subcomplex. Set $K_0 = \emptyset$. Then each subcomplex is obtained from the previous by attaching a single simplex, and we can label it as positive or negative. We then obtain an incremental algorithm for the Betti numbers:

– Initialize each $\tilde{\beta}_p = 0$.

– For $i = 1, \ldots, m$, if σ_i is positive, then $\tilde{\beta}_p = \tilde{\beta}_p + 1$; if σ_i is negative, then $\tilde{\beta}_{p-1} = \tilde{\beta}_{p-1} - 1$.

So, we need only categorize each simplex as positive or negative. We now present an algorithm to do this. If $f : K \to \mathbb{R}$ is a monotonic function on the simplices of K, then a *compatible ordering* of the simplices is an ordering $\sigma_1, \ldots, \sigma_m$ such that $f(\sigma_i) < f(\sigma_j)$ implies that $i < j$ or σ_i is a face of σ_j. Since f is monotonic, such orderings exist.

Given such an ordering, every initial segment forms a subcomplex. Define an $m \times m$ matrix ∂ over $\mathbb{Z}_2$ by

$$\partial_{ij} = \begin{cases} 1 & \text{if } \sigma_i \text{ is a codimension-1 face of } \sigma_j \\ 0 & \text{otherwise} \end{cases}$$

This stores all the individual boundary matrices $\partial_k : C_k(K; \mathbb{Z}_2) \to C_{k-1}(K; \mathbb{Z}_2)$ at once.

Definition 2.7.4. Let low(j) be the row index of the lowest 1 in column j of a matrix over $\mathbb{Z}_2$. If column j is 0, then leave low(j) undefined. Call a 0–1 matrix R *reduced* if low(j) $\neq$ low(k) for j and k corresponding to nonzero columns.

Algorithm 1 is the *persistence algorithm*. Note that since the algorithm adds columns to the right, at each stage we have $R = \partial V$, where V is upper triangular. Since ∂ is upper triangular, it follows that R is as well. Note further that the number of zero columns of R that correspond to p-simplices is the rank of Z_p (the group of p-cycles) and the number of nonzero columns corresponding to $(p + 1)$-simplices gives the rank of B_p. This allows us to compute $\beta_p(K)$. But there's more: the matrix R actually gives us information about the persistence pairing.

Proposition 2.7.5. *The lowest 1's in a reduced matrix are unique, even though R is not (since we could possibly reduce R further).*

Algorithm 1: Persistence reduction.

$R = \partial$
   for $j = 1, \ldots, m$ **do**
      while $\exists j_0 < j$ with $\mathrm{low}(j_0) = \mathrm{low}(j)$ **do**
         add column j_0 to column j
      end while
   end for

Proof. Consider the lower left submatrix R_i^j obtained by removing the first $i - 1$ rows and the last $m - j$ columns of R. Since left-to-right column operations preserve the rank of every such submatrix, the rank of R_i^j is the same as that of ∂_i^j. Set

$$r_R(i,j) = \mathrm{rank}\,R_i^j - \mathrm{rank}\,R_{i+1}^j + \mathrm{rank}\,R_{i+1}^{j-1} - \mathrm{rank}\,R_i^{j-1}.$$

Note that $r_R(i,j) = r_\partial(i,j)$. A linear combination of any collection of nonzero columns of R_i^j is nonzero (since R is reduced) and so the rank of R_i^j is just the number of nonzero columns. If R_{ij} is a lowest 1, then R_i^j has one more nonzero column than the other 3 matrices above, so that $r_R(i,j) = 1$. If R_{ij} is not a lowest 1 then there are two cases:

(1) If none of the columns from 1 to $j - 1$ has a lowest 1 in row i, then R_i^j and R_{i+1}^j have the same number of nonzero columns, and so do R_i^{j-1} and R_{i+1}^{j-1}.

(2) If one of these columns has a lowest 1 in row i, then R_i^j has one more nonzero column than R_{i+1}^j and R_i^{j-1} and one more nonzero column than R_{i+1}^{j-1}.

In either case, $r_R(i,j) = 0$. Since the ranks of the lower left submatrices of R are the same as those of ∂, we have a characterization of the lowest 1's that does not depend on the reduction process. □

This immediately gives the following result.

Lemma 2.7.6 (Pairing Lemma). $i = \mathrm{low}(j) \Leftrightarrow r_\partial(i,j) = 1$. *In particular, the pairing between rows and columns defined by the lowest 1's in the reduced matrix does not depend on R.*

But what do these pairings mean? Note that column j reaches its final form at the end of the jth iteration of the outer loop of Algorithm 1. At this time, we have the reduced matrix for the subcomplex consisting of the first j simplices in the ordering. There are two cases:

(1) Column j is 0. Then σ_j corresponds to a new cycle. Call σ_j *positive* (a new homology class has been born).

(2) Column j is nonzero. Then it stores the boundary of the chain accumulated in column j of the matrix V. Call σ_j *negative*, since its addition kills a homology class.

The class that dies in Case 2 above is represented by column j. We need to verify that it is born at the time of the simplex of its lowest 1, σ_i with $i = \text{low}(j)$, is added. But this is clear: the cycle in column j just died and all other cycles that die with it have 1's below row i; otherwise we could further reduce the matrix to obtain $\text{low}(j) < i$, a contradiction. So we have proved the following.

Proposition 2.7.7. *The lowest 1's in the reduced matrix R correspond to points in the persistence diagram:*

$$(a_i, a_j) \in \text{Dgm}_p(f) \Leftrightarrow i = \text{low}(j) \quad and \quad \dim \sigma_i = p.$$

In this case, σ_j has dimension $p + 1$. We have $(a_i, \infty) \in \text{Dgm}_p(f)$ if and only if column i is 0, since in this case row i has no lowest 1; that is, σ_i is positive and it does not get paired with a negative simplex.

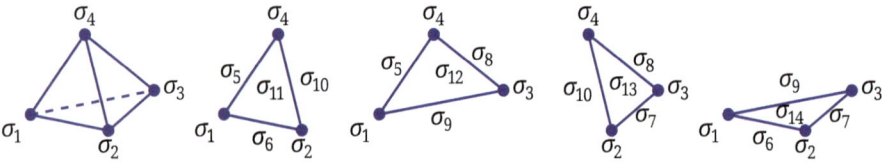

Figure 2.14: An ordering of the simplices in a hollow tetrahedron.

Example 2.7.8. Let us perform the algorithm on a filtered hollow tetrahedron. The simplices are ordered as in Figure 2.14. The matrix ∂ is then a 14×14 matrix over $\mathbb{Z}_2$:

		σ_1	σ_2	σ_3	σ_4	σ_5	σ_6	σ_7	σ_8	σ_9	σ_{10}	σ_{11}	σ_{12}	σ_{13}	σ_{14}
	$+ \sigma_1$	0	0	0	0	1	1	0	0	1	0	0	0	0	0
	$+ \sigma_2$	0	0	0	0	0	1	1	0	0	1	0	0	0	0
	$+ \sigma_3$	0	0	0	0	0	0	1	1	1	0	0	0	0	0
	$+ \sigma_4$	0	0	0	0	1	0	0	1	0	1	0	0	0	0
	$- \sigma_5$	0	0	0	0	0	0	0	0	0	0	1	1	0	0
	$- \sigma_6$	0	0	0	0	0	0	0	0	0	0	1	0	0	1
$\partial =$	$- \sigma_7$	0	0	0	0	0	0	0	0	0	0	0	0	1	1
	$+ \sigma_8$	0	0	0	0	0	0	0	0	0	0	0	1	1	0
	$+ \sigma_9$	0	0	0	0	0	0	0	0	0	0	0	1	0	1
	$+ \sigma_{10}$	0	0	0	0	0	0	0	0	0	0	1	0	1	0
	$- \sigma_{11}$	0	0	0	0	0	0	0	0	0	0	0	0	0	0
	$- \sigma_{12}$	0	0	0	0	0	0	0	0	0	0	0	0	0	0
	$- \sigma_{13}$	0	0	0	0	0	0	0	0	0	0	0	0	0	0
	$+ \sigma_{14}$	0	0	0	0	0	0	0	0	0	0	0	0	0	0

The result of the algorithm is the following matrix. The circled entries are the lowest 1's in each column.

$$
R =
\begin{array}{c}
 \\
+\,\sigma_1 \\
+\,\sigma_2 \\
+\,\sigma_3 \\
+\,\sigma_4 \\
-\,\sigma_5 \\
-\,\sigma_6 \\
-\,\sigma_7 \\
+\,\sigma_8 \\
+\,\sigma_9 \\
+\,\sigma_{10} \\
-\,\sigma_{11} \\
-\,\sigma_{12} \\
-\,\sigma_{13} \\
+\,\sigma_{14}
\end{array}
\begin{array}{ccccccccccccccc}
\sigma_1 & \sigma_2 & \sigma_3 & \sigma_4 & \sigma_5 & \sigma_6 & \sigma_7 & \sigma_8 & \sigma_9 & \sigma_{10} & \sigma_{11} & \sigma_{12} & \sigma_{13} & \sigma_{14} \\
0 & 0 & 0 & 0 & 1 & 1 & 0 & 0 & 1 & 0 & 0 & 0 & 0 & 0 \\
0 & 0 & 0 & 0 & 0 & ① & 1 & 0 & 0 & 0 & 0 & 0 & 0 & 0 \\
0 & 0 & 0 & 0 & 0 & 0 & ① & 0 & 0 & 0 & 0 & 0 & 0 & 0 \\
0 & 0 & 0 & 0 & ① & 0 & 0 & 0 & 0 & 0 & 0 & 0 & 0 & 0 \\
0 & 0 & 0 & 0 & 0 & 0 & 0 & 0 & 0 & 0 & 1 & 1 & 1 & 0 \\
0 & 0 & 0 & 0 & 0 & 0 & 0 & 0 & 0 & 0 & 1 & 0 & 1 & 0 \\
0 & 0 & 0 & 0 & 0 & 0 & 0 & 0 & 0 & 0 & 0 & 0 & 1 & 0 \\
0 & 0 & 0 & 0 & 0 & 0 & 0 & 0 & 0 & 0 & 0 & 1 & ① & 0 \\
0 & 0 & 0 & 0 & 0 & 0 & 0 & 0 & 0 & 0 & 0 & ① & 0 & 0 \\
0 & 0 & 0 & 0 & 0 & 0 & 0 & 0 & 0 & 0 & ① & 0 & 0 & 0 \\
0 & 0 & 0 & 0 & 0 & 0 & 0 & 0 & 0 & 0 & 0 & 0 & 0 & 0 \\
0 & 0 & 0 & 0 & 0 & 0 & 0 & 0 & 0 & 0 & 0 & 0 & 0 & 0 \\
0 & 0 & 0 & 0 & 0 & 0 & 0 & 0 & 0 & 0 & 0 & 0 & 0 & 0 \\
0 & 0 & 0 & 0 & 0 & 0 & 0 & 0 & 0 & 0 & 0 & 0 & 0 & 0
\end{array}
$$

The persistence pairing is then

$$(\sigma_4, \sigma_5), (\sigma_3, \sigma_7), (\sigma_2, \sigma_6), (\sigma_{10}, \sigma_{11}), (\sigma_9, \sigma_{12}), (\sigma_8, \sigma_{13}),$$

with σ_1 and σ_{14} being unpaired. The latter two correspond to homology classes in K in degrees 0 and 2, respectively. The corresponding persistence diagrams are shown in Figure 2.15.

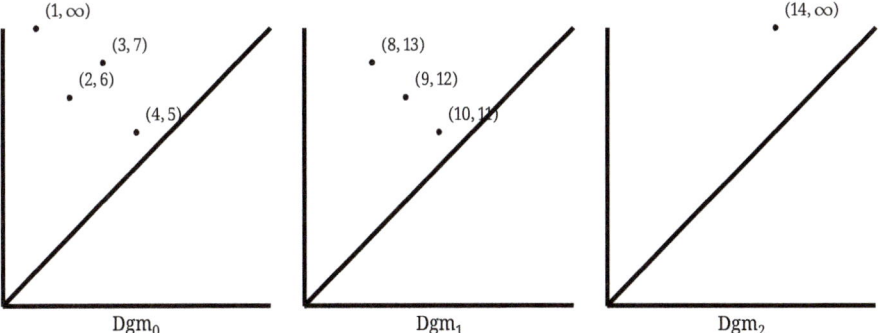

Figure 2.15: The persistence diagrams for Example 2.7.8.

Exercises

Exercise 2.7.1. Prove that, given a persistence diagram, one can read off the persistent Betti numbers as follows: $\beta_p^{k,\ell}$ is the number of points in the upper left quadrant of the persistence diagram with lower right corner (a_k, a_ℓ).

Exercise 2.7.2. Alter the persistence algorithm by adding each jth column to columns on its right rather than adding columns on its left to it. Show that this version generates the same lowest 1's as the standard algorithm and give an example for which this and the standard algorithm produce different reduced matrices.

2.7.2 Stability

In the previous section, we discussed persistence in the context of a filtered simplicial complex, but there are other natural scenarios where the same concept arises. Suppose M is a triangulable space (e. g., a smooth manifold, possibly with boundary) and that $f : M \rightarrow \mathbb{R}$ is a continuous function. If $a \in \mathbb{R}$, consider the *sublevel set* $M_a = f^{-1}(-\infty, a]$. If M is a smooth manifold, then M_a is a closed submanifold with boundary. Note that if $a \leq b$, then $M_a \subseteq M_b$, and so we obtain an $\mathbb{R}$-valued *sublevel set filtration* $\{M_a\}_{a \in \mathbb{R}}$ of the space M. For a fixed homological degree, we then have induced maps, for $a \leq b$:

$$f_p^{a,b} : H_p(M_a; F) \rightarrow H_p(M_b; F)$$

and the corresponding *persistent homology groups*

$$H_p^{a,b} = \mathrm{im}\, f_p^{a,b}.$$

In general, this could be rather wild, but in practice there are only finitely many $a \in \mathbb{R}$ where the topology of the sublevel sets changes, so that we really only have finitely many such persistent homology groups. More on this in Section 2.7.3 below.

Let us consider the following vague question: suppose two filtrations of a simplicial complex K are not that different. Should we then expect that the corresponding persistence diagrams are not that different? There are some undefined terms here, but in essence we want to somehow define a distance between these objects and then show that passage from filtrations to diagrams is continuous with respect to these distances.

Here is a simple example. Consider the two functions $f, g : [0,1] \rightarrow \mathbb{R}$ shown in Figure 2.16. These functions are close in the sense that $\|f - g\|_\infty = \sup_{x \in [0,1]} |f(x) - g(x)|$ is small. We have the sublevel set filtrations $\{f^{-1}(-\infty, a]\}$ and $\{g^{-1}(-\infty, a]\}$ of $[0, 1]$. Since $[0, 1]$ is 1-dimensional, the only interesting persistence diagrams are Dgm_0; these are shown in Figure 2.17.

Note that the two diagrams are similar away from the diagonal, but that $\mathrm{Dgm}_0(g)$ has four extra points near the diagonal. These arise from the small perturbations in the

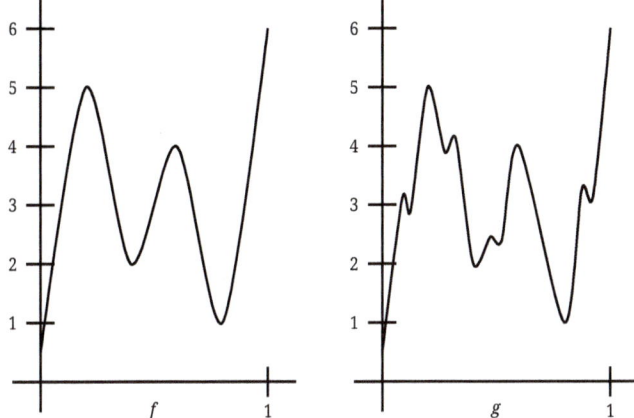

Figure 2.16: Two functions on $[0, 1]$ that are close to each other.

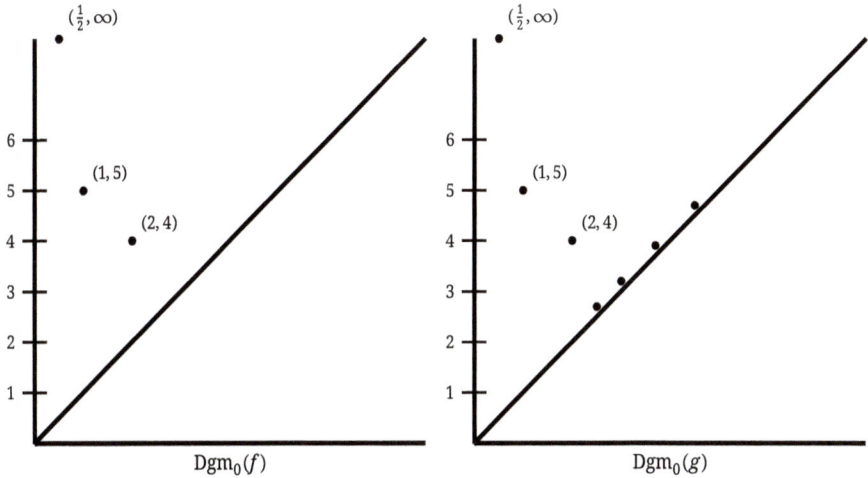

Figure 2.17: The 0-th persistence diagrams for the functions in Figure 2.16.

graph of g. Our feeling here is that f and g are very similar, and so we want to think of them as having roughly the same persistence diagrams. To make that precise, we need to put a metric on the set of persistence diagrams. There are many choices, but we will focus on the bottleneck distance.

Recall that a persistence diagram is a finite multiset of points in the extended plane $\overline{\mathbb{R}}^2$, along with all the points on the diagonal, counted with infinite multiplicity. If $x = (x_1, x_2)$ and $y = (y_1, y_2)$ are points in a diagram, set

$$\|x - y\|_\infty = \max\{|x_1 - y_1|, |x_2 - y_2|\}.$$

Definition 2.7.9. Let X and Y be two persistence diagrams and let $\eta : X \to Y$ be a bijection (which exists because we have included the diagonal). The *bottleneck distance* between X and Y is

$$W_\infty = \inf_{\eta:X\to Y} \sup_{x\in X} \|x - \eta(x)\|_\infty.$$

The definition of W_∞ is illustrated in Figure 2.18. Even though the two diagrams have different numbers of off-diagonal points, we may still construct bijections between them by matching points to points on the diagonal. Then the idea is to find the smallest square centered on the +'s that catch all the dots. The bottleneck distance is then half the length of the side of this square. The function W_∞ satisfies the following properties:
(1) $W_\infty(X,Y) = 0 \Leftrightarrow X = Y$
(2) $W_\infty(X,Y) = W_\infty(Y,X)$
(3) $W_\infty(X,Z) \le W_\infty(X,Y) + W_\infty(Y,Z)$

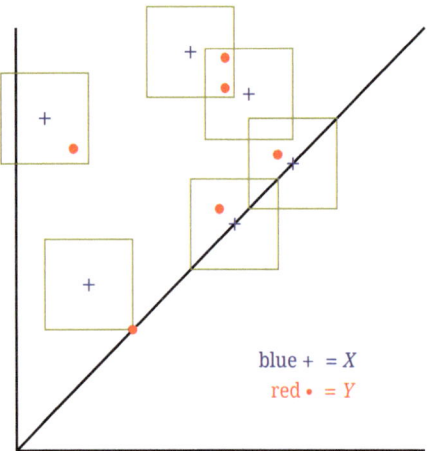

blue + = X
red • = Y

Figure 2.18: A bijection between two persistence diagrams.

Thus, W_∞ is an extended metric on the set of persistence diagrams (we can have $W_\infty(X,Y) = \infty$). The most difficult of these properties to prove is the triangle inequality. Moreover, we can get around the possibility of infinite distances as follows. The only way we get a point at infinity in a persistence diagram is if there is a positive simplex that never gets killed by a negative one. This does not happen in reduced homology if and only if K is homologically trivial. But we can assume this, without loss of generality, by adding additional simplices at the end of the filtration that do not alter the evolution along the filtration (e. g., we could take the cone on K, which is contractible). Thus, we may assume that there is a bijection between the lowest 1's in a reduced matrix and the off-diagonal entries in the persistence diagram.

We are now ready to address *stability*.

Theorem 2.7.10 (Stability Theorem). *Suppose K is a simplicial complex and that f, g : $K \to \mathbb{R}$ are monotonic functions. For $p \geq 0$, let $X_p = \mathrm{Dgm}_p(f)$ and $Y_p = \mathrm{Dgm}_p(g)$. Then*

$$W_\infty(X_p, Y_p) \leq \|f - g\|_\infty = \sup_{\sigma \in K} |f(\sigma) - g(\sigma)|.$$

Proof. As noted above, we may assume there are no points at infinity in any diagram. Since f and g are assumed monotonic, each element of the straight-line homotopy F : $K \times [0, 1] \to \mathbb{R}$ defined by

$$F(\sigma, t) = (1 - t)f(\sigma) + tg(\sigma)$$

is also monotonic. Denote the map $F(-, t)$ by f_t. Fix a homological degree p. Each f_t has a persistence diagram in dimension p; drawing t along a third coordinate axis yields a 3-dimensional representation of how the persistence diagrams evolve from $f_0 = f$ to $f_1 = g$. Each off-diagonal point of $X_t = \mathrm{Dgm}_p(f_t)$ is of the form $x(t) = (f_t(\sigma), f_t(\tau), t)$, with σ, τ simplices in K that form a persistence pair. Note that there are only finitely many values at which the pairings change; denote these by $0 = t_0 < t_1 < \cdots < t_n < t_{n+1} = 1$.

Within each interval (t_i, t_{i+1}) the pairing is constant and each pair σ, τ gives rise to a line segment of points $x(t)$ joining points in the planes $t = t_i$ and $t = t_{i+1}$. If the endpoint is an off-diagonal point at t_{i+1}, then there is some other unique line segment that begins at that point (perhaps corresponding to the same simplex pair, but perhaps not if the pairings switch). It is also possible that the line segment hits the diagonal at t_{i+1}, in which case there is no continuation. No matter what, the line segments form polygonal paths that monotonically increase in t. Each path begins at an off-diagonal point in $X = X_0$ or at a diagonal point in some X_{t_i}, and ends at an off-diagonal point in $Y = X_1$ or at a diagonal point in some X_{t_j}. Such a path is called a *vine* and the collection of all of them is a *vineyard*.

Differentiate $x(t) = (1 - t)(f(\sigma), f(\tau), 0) + t(g(\sigma), g(\tau), 1)$ to get

$$\frac{\partial x}{\partial t}(t) = (g(\sigma) - f(\sigma), g(\tau) - f(\tau), 1).$$

Projecting the endpoints of a line segment in a vine back into $\mathbb{R}^2$, we get two points whose ∞-distance is $t_{i+1} - t_i$ times the larger of the differences between f and g and the two simplices. Let a be the simplex in K that maximizes this difference; we therefore get $\|f - g\|_\infty = |f(a) - g(a)|$. This is an upper bound on the slope of any line segment in the vineyard, and so is an upper bound on the ∞-distance between the projected endpoints of any vine. □

This handles stability for filtrations, but what about the sorts of functions like those in Figure 2.16? We still get such a theorem, but we need some mild hypotheses on the functions. Suppose M is triangulable and $f : M \to \mathbb{R}$ is continuous. We have the sublevel set filtration $\{M_a\}$ and the associated persistence diagrams for each $p \geq 0$.

Definition 2.7.11. A real number a is a *homological critical value* for f if there is no $\varepsilon > 0$ such that the map

$$f_p^{a-\varepsilon,a+\varepsilon} : H_p(M_{a-\varepsilon}) \to H_p(M_{a+\varepsilon})$$

is an isomorphism for each dimension p. The function f is *tame* if it has only finitely many homological critical values.

Now observe the following. If f is tame, let $a_1 < a_2 < \cdots < a_n$ be the homological critical values and choose real numbers b_0 to b_n with $b_{i-1} < a_i < b_{i+1}$ for all i. Let $b_{-1} = a_0 = -\infty$ and $a_{n+1} = b_{n+1} = \infty$. Then we have the sequence of homology groups

$$0 = H_p(M_{b_{-1}}) \to H_p(M_{b_0}) \to \cdots \to H_p(M_{b_n}) \to H_p(M_{b_{n+1}}) = H_p(M)$$

The multiplicity of the pair (a_i, a_j) is then defined as

$$\mu_p^{a_i,a_j} = (\beta_p^{b_i,b_{j-1}} - \beta_p^{b_i,b_j}) - (\beta_p^{b_{i-1},b_{j-1}} - \beta_p^{b_{i-1},b_j}).$$

It is now an easy exercise to see that the persistence diagram of f in dimension p consists of the points (a_i, a_j) with multiplicity $\mu_p^{a_i,a_j}$. The proof of the following theorem is similar to that of Theorem 2.7.10.

Theorem 2.7.12. *Let M be a triangulable space and let $f, g : M \to \mathbb{R}$ be tame functions. For each dimension p, let $X_p = \mathrm{Dgm}_p(f)$ and $Y_p = \mathrm{Dgm}(g)$. Then for all $p \geq 0$,*

$$W_\infty(X_p, Y_p) \leq \|f - g\|_\infty.$$

Exercises

Exercise 2.7.3. Let T be a torus embedded in $\mathbb{R}^3$ and let $f : T \to \mathbb{R}$ be the height function with minimum value 0, as in Figure 2.19. Compute the persistence diagrams for this function. Do the same for the surface M_g of genus g.

Exercise 2.7.4. Prove the triangle inequality for W_∞. You may use the fact that $\| \cdot \|_\infty$ satisfies the triangle inequality.

Exercise 2.7.5. Let q be a positive real number and let X and Y be persistence diagrams. Define the *q-Wasserstein distance* between X and Y to be

$$W_q(X, Y) = \left[\inf_{\eta:X \to Y} \sum_{x \in X} \|x - \eta(x)\|_\infty^q \right]^{1/q}.$$

Show that W_q is a metric. Does it satisfy a stability theorem?

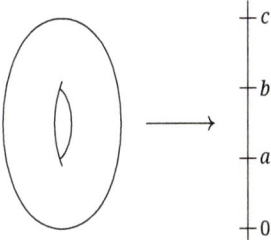

Figure 2.19: The height function on the torus.

Exercise 2.7.6. Prove that the persistence diagram of a tame function f in dimension p consists of the points (a_i, a_j) with multiplicity $\mu_p^{a_i,a_j}$.

2.7.3 Persistence modules

We have defined persistent homology using field coefficients. Of course, one could define this over the integers, but there are several issues with this. One is that computing integral homology takes more time–it amounts to computing the Smith normal forms of various matrices. Since we are usually more interested in the evolution of Betti numbers through a filtration, and these can be computed via standard theorems of linear algebra (dimension = rank + nullity), field coefficients offer a definite advantage.

There is one more important fact: if F is a field, then the polynomial ring $F[x]$ is a (graded) principal ideal domain. This is not true for $\mathbb{Z}[x]$. For example, the ideal $(2, x)$ is not principal.

Why does this matter? By a *graded $F[x]$-module*, we mean an $F[x]$-module M with a direct sum decomposition

$$M = \bigoplus_{n \geq 0} M_n,$$

where the action of x is given by a map $x : M_n \to M_{n+1}$. Scalars act as scalar multiplication. For example, the rank-1 free module $F[x]$ may be written

$$F[x] = \bigoplus_{n \geq 0} x^n \cdot F,$$

with x acting as multiplication in the polynomial ring.

Definition 2.7.13. Let K be a simplicial complex with a filtration

$$\emptyset = K_{-1} \subset K_0 \subset K_1 \subset \cdots \subset K_r = K.$$

Fix a homological degree i and consider

$$M = \bigoplus_{j=0}^{r} H_i(K_j; F).$$

This is a graded $F[x]$-module where the map $x : H_i(K_j; F) \rightarrow H_i(K_{j+1}; F)$ is the map induced by the inclusion $K_j \hookrightarrow K_{j+1}$. We call M a *persistence module*.

Theorem 2.7.14 (Classification theorem for graded $F[x]$-modules). *If N is a graded $F[x]$-module, then*

$$N \cong \underbrace{\bigoplus_{i=1}^{m} x^{s_i} F[x]}_{\text{shifted free modules}} \oplus \underbrace{\bigoplus_{j=1}^{n} x^{b_j} F[x]/(x^{d_j})}_{\text{torsion modules}}.$$

In the case of a persistence module M arising from a filtered simplicial complex K, we get

$$M \cong \underbrace{\bigoplus_{i=1}^{m} x^{s_i} F[x]}_{\substack{\text{classes born at filtration} \\ \text{level } s_i \text{ that live forever}}} \oplus \underbrace{\bigoplus_{j=1}^{n} x^{b_j} F[x]/(x^{d_j})}_{\substack{\text{classes born at level } b_j, \\ \text{dying at level } d_j}}.$$

Thus, over a field barcodes/persistence diagrams encode everything about persistence; that is, they are a *complete invariant*. There is no analogue of Theorem 2.7.14 over $\mathbb{Z}$, so we cannot expect the integral barcode to be a complete invariant.

Example 2.7.15. Consider the filtered tetrahedron K of Figure 2.11. Note that the indexing begins with $K_0 = \emptyset$. The persistence modules are

$$H_0 \cong xF[x] \oplus \left(xF[x]/(x^2)\right)^3$$
$$H_1 \cong x^2 F[x]/(x^3) \oplus x^2 F[x]/(x^4) \oplus x^2 F[x]/(x^5)$$
$$H_2 \cong x^6 F[x]$$

An obvious question to ask is what we should do about sublevel set filtrations. A priori, these are not indexed by the natural numbers. However, if the function is tame, then in any given homological degree there are only finitely many homological critical values, and so we can define an integer based filtration of the space M by setting $M_i = M_{a_i}$, where a_i is a homological critical value. Then we can apply the decomposition arising from Theorem 2.7.14.

An alternate approach is to consider the $\mathbb{R}$-graded module

$$M = \bigoplus_{r \in \mathbb{R}} H_p(M_r; F).$$

It turns out that there is a decomposition theorem analogous to Theorem 2.7.14 for these modules. Assume that each $H_p(M_r; F)$ is finite-dimensional. Then there is an isomorphism

$$M \cong \bigoplus_I F_I,$$

where I ranges over a set of intervals in $\mathbb{R}$ and F_I is the *interval module*

$$F_I = \bigoplus_{r \in I} F.$$

Even more generally, we might consider a partially ordered set P and a functor $M : P \to \underline{\text{Vect}}_F$ from P to the category of finite-dimensional vector spaces over F. Call such an object a *persistence module*. So far we have considered the cases $P = (\mathbb{N}, \leq)$ and $P = (\mathbb{R}, \leq)$. Another interesting case is $P = (\mathbb{N}^n, \preceq)$, where $(a_1, \ldots, a_n) \preceq (b_1, \ldots, b_n)$ if $a_i \leq b_i$ for $1 \leq i \leq n$. This gives rise to *multiparameter persistence*, which is very useful in topological data analysis. The unfortunate fact, however, is that the representation theory of these modules is usually wild, in the sense that there is no simple set of indecomposables like the interval modules. Still, much of the preceding discussion can be extended into this setting, including defining distances between persistence modules (e. g., the *interleaving distance*) and showing these are stable with respect to small perturbations. This is further afield than we wish to go here, however.

Exercises

Exercise 2.7.7. Suppose you are told that a persistence module has three generators in degrees $0, 0, 2$ and that one class dies in degree 4. Does this determine the module uniquely? If not, exhibit two nonisomorphic modules with these characteristics. Are there others?

2.7.4 Vietoris–Rips complexes

In the beginning of this section, we described a method for building a space out of a finite set of points in some metric space. Namely, we imagine small ε-balls around each point. At first, these balls do not intersect, but if we allow ε to increase, intersections will occur. We could then join the corresponding centers by edges, fill in triangles, etc. In this section, we will formalize this construction.

Suppose we have a finite metric space (X, d). The example to keep in mind is a set of points in some Euclidean space $\mathbb{R}^n$ with the standard metric. Let $\varepsilon \geq 0$ be a real number.

Definition 2.7.16. The *Vietoris–Rips complex* at scale ε is the simplicial complex $\mathrm{VR}(X,\varepsilon)$ with vertex set X in which a set $\{x_0, x_1, \ldots, x_k\}$ forms a k-simplex precisely when all the pairwise distances $d(x_i, x_j) \leq 2\varepsilon$ for all i, j.

Note that if $\varepsilon \leq \varepsilon'$, then $\mathrm{VR}(X,\varepsilon) \subseteq \mathrm{VR}(X,\varepsilon')$. Thus, we have an increasing family of complexes, indexed by ε. Since the space X is finite, this is actually a finite filtration, and for $\varepsilon \geq (1/2) \max\{d(x_i, x_j)\}$, $\mathrm{VR}(X,\varepsilon)$ is an n-simplex, where X consists of $n+1$ points.

The Vietoris–Rips complex is easy to describe, but it is a bit coarse in the sense that once all the edges joining all possible pairs in a particular set of points exist, the entire simplex on that set is in the complex. So if we think of this complex as somehow modeling the union of balls around the points in X, there are scales at which we are adding too many simplices, and therefore possibly losing topological information. It seems plausible that there is a certain range of ε values where things are ok, but there is no real way to know a priori what those values are.

An alternative to the Vietoris–Rips construction is the *Čech complex.*

Definition 2.7.17. Let X be a finite subset of a metric space (Z, d) and let $\varepsilon \geq 0$ be a real number. The *Čech complex* at scale ε is the simplicial complex $C(X,\varepsilon)$ with vertex set X in which a set $\{x_0, x_1, \ldots, x_k\}$ forms a k-simplex precisely when there is some $z \in Z$ satisfying $d(z, x_i) \leq \varepsilon$ for all i.

We usually have $Z = \mathbb{R}^n$ with the standard metric here; in that case, note that $C(X,\varepsilon)$ does not necessarily embed in $\mathbb{R}^n$. Still it is a simplicial complex, and if $\varepsilon \leq \varepsilon'$, then $C(X,\varepsilon) \subseteq C(X,\varepsilon')$. Again, since X is finite, there are only finitely many ε where these complexes change, so we get a finite collection of nested complexes.

And, in the case of $Z = \mathbb{R}^n$, this is really modeling the topology of the union of the ε-balls around the points in X. Indeed, if we consider the ball $B_\varepsilon(x)$ for some $x \in X$, then

$$\bigcap_{i=0}^{k} B_\varepsilon(x_i) \neq \emptyset \Leftrightarrow \text{there exists } z \in \mathbb{R}^n \text{ with } x_i \in B_\varepsilon(z), \quad i = 0, \ldots, n.$$

Definition 2.7.18. Suppose $\mathcal{F}$ is a finite collection of sets, all subsets of some larger set. The *nerve* of $\mathcal{F}$ is the simplicial complex whose simplices are all $X \subset \mathcal{F}$ such that $\bigcap X \neq \emptyset$.

Theorem 2.7.19 (Nerve Theorem). *Let $\mathcal{F}$ be a finite collection of closed, convex sets in Euclidean space. Then the nerve of $\mathcal{F}$ and the union of the sets in $\mathcal{F}$ have the same homotopy type.*

One of the projects at the end of this chapter provides a proof of this theorem. In the case of a collection $\mathcal{F}$ of closed balls around a finite set $X \subset \mathbb{R}^n$, Theorem 2.7.19 tells us that the Čech complex $C(X,\varepsilon)$ has the homotopy type of the union of the balls in $\mathbb{R}^n$. So if we think of the set X as being sampled from some underlying manifold M, then it is plausible that there is some ε where $C(X,\varepsilon)$ has the homotopy type of M, and therefore that we can compute the homology of M via the complex $C(X,\varepsilon)$. Of course, there are many parameters that might influence the veracity of this claim, such as the density of the samples X, the geometry of the manifold (local curvature, for example), etc. A theorem of Niyogi–Smale–Weinberger asserts that if the samples are sufficiently dense with respect to a parameter related to the size of the largest tubular neighborhood of M that embeds in $\mathbb{R}^n$, then the union of ε-balls around the sample points deformation retracts to M. Thus, if we compute persistent homology using the Čech filtration, then there is a range of parameters where we are capturing the homology of the underlying manifold.

The drawback of the Čech filtration, however, is that it is difficult to compute. That is, deciding if a collection of points forms a simplex requires finding another point in the ambient space Z satisfying certain distance relationships to the points. Another way to think of this is that we are looking for the smallest ball in Z that contains a given set of points; once that radius r is reached, we get a simplex in $C(X, r)$.

Note that for a given ε, $C(X, \varepsilon) \subseteq \mathrm{VR}(X, \varepsilon)$. Indeed, if $\{x_0, x_1, \ldots, x_k\}$ form a k-simplex in $C(X, \varepsilon)$, then the triangle inequality implies that $d(x_i, x_j) \leq 2\varepsilon$ for all i, j and so this is also a simplex in $\mathrm{VR}(X, \varepsilon)$. On the other hand, we also have the following.

Proposition 2.7.20. *For each $\varepsilon \geq 0$, $\mathrm{VR}(X, \varepsilon) \subseteq C(X, \varepsilon \sqrt{2})$.*

Proof. To begin, consider the standard k-simplex Δ^k embedded in $\mathbb{R}^{k+1}$ as the convex hull of the endpoints of the standard basis vectors. Each edge then has length $\sqrt{2}$. The distance from the origin to Δ^k is just the distance to the barycenter $z = (1/(k+1), 1/(k+1), \ldots, 1/(k+1))$. Note that z is also the center of the smallest ball containing all the vertices. If we call the radius of that ball r_k, then $r_k^2 = 1 - \|z\|^2 = k/(k+1)$. As k goes to infinity, this approaches 1 from below.

Now, any set X of $k+1$ or fewer points for which the same k-ball of radius r_k is the minimal ball containing all of the points has a pair of points at distance $\sqrt{2}$ or larger. It follows that every simplex of diameter $\sqrt{2}$ or less belongs to $C(X, r_k)$. Multiplying distances by $\varepsilon \sqrt{2}$, we get $\mathrm{VR}(X, \varepsilon) \subseteq C(X, \varepsilon r_k \sqrt{2})$. Since $r_k \leq 1$ for all k, the latter is a subcomplex of $C(X, \varepsilon \sqrt{2})$. □

Proposition 2.7.20 suggests that using the Vietoris–Rips filtration is reasonable since there is a range of ε-values where the Čech complexes capture the topology of the underlying manifold, and the Vietoris–Rips complexes are sandwiched between. Again, there is no guarantee that $\mathrm{VR}(X, \varepsilon)$ has the same homotopy type as the underlying manifold, but in practice it is unlikely to be significantly different.

We therefore have the following pipeline for homology inference. Given a finite set X in a metric space Z (usually euclidean space), we compute the Vietoris–Rips filtration $\{\mathrm{VR}(X, \varepsilon)\}_{\varepsilon \geq 0}$. This is a finite filtration. We then fix a field F and compute the persistent homology of this filtration. The associated barcodes provide information about which homological features persist for long ε-intervals, and are therefore likely to be actual features of the space M underlying X.

Exercises

Exercise 2.7.8. Prove that $VR(X, \varepsilon)$ is, in fact, a simplicial complex.

Exercise 2.7.9. Prove that $C(X, \varepsilon)$ is a simplicial complex.

Project: the Jordan–Brouwer Separation Theorem

In this project, we outline a proof of the Jordan–Brouwer Separation Theorem, a special case of which is the classic Jordan Curve Theorem in the plane.

Consider the following result: Let e^r be a closed cell of dimension r in S^n. Then $\tilde{H}_k(S^n - e^r) = 0$ for all $k \geq 0$. Prove this by induction on r as follows:

(1) The base case $r = 0$ is easy (why?).

(2) Suppose $r > 0$ and that the theorem is true for $r - 1$. Let z be a k-cycle in $S^n - e^r$ and let $\varphi : I^r \to e^r$ be a homeomorphism. For $t \in I$, let $e^{r-1}(t) = \varphi(t \times I^{r-1})$; this is a closed $(r-1)$-cell. Use the inductive hypothesis to deduce that $z = \partial w_t$ for some $(k+1)$-chain w_t in $S^n - e^{r-1}(t)$.

(3) Note that the support $|w_t|$ of w_t does not meet the compact set $e^{r-1}(t)$. Let ε_t be the (positive) distance between these two sets. Since φ is uniformly continuous, there is a $\delta_t > 0$ so that if $x, y \in I^r$ are less than δ_t apart, their images are less than ε_t apart. Let I_t be an open interval centered at t of width $< \delta_t$ and let $e^r(t) = \varphi(I_t \times I^{r-1})$; this is an open r-cell. Deduce that $z = \partial w_t$ in $S^n - e^r(t)$.

(4) The collection of I_t cover I. By the Lebesgue number lemma, there is a $\rho > 0$ such that every closed interval of length $< \rho$ lies in some I_t. Choose m with $1/m < \rho$ and consider the intervals $I_0 = [0, 1/m], I_1 = [1/m, 2/m], \ldots, I_{m-1} = [(m-1)/m, 1]$. Let e^r_j be the image of $I_j \times I^{r-1}$. Deduce there is a chain w_j in $S^n - e^r_j$ with $z = \partial w_j$.

(5) By induction on j, we are reduced to proving the following: Let J_1, J_2 be closed subintervals of I with $J_1 \cap J_2 = \{t\}$. Let $e' = \varphi(J_1 \times I^{r-1})$ and $e'' = \varphi(J_2 \times I^{r-1})$. Suppose there are $(k+1)$-chains w', w'' in $S^n - e'$ and $S^n - e''$, respectively, such that $\partial w' = z = \partial w''$. Then there is a $(k+1)$-chain w in $S^n - (e' \cup e'')$ with $z = \partial w$. Prove this using excision on the space $S^n - e^{r-1}(t)$, with open subsets $S^n - e'$ and $S^n - e''$.

Deduce the following: S^n cannot be disconnected by removing a closed cell.

Now prove the following result: Let s^r be a subspace of S^n, homeomorphic to S^r. Then $r \leq n$. If $r = n$, then $s^r = S^n$, and if $r < n$, we have

$$\tilde{H}_k(S^n - s^r) \cong \begin{cases} R & k = n - r - 1 \\ 0 & \text{otherwise.} \end{cases}$$

(Hint: if $\varphi : S^r \to S^n$ is the map with image s^r, consider the images of the closed northern and southern hemispheres of S^r. These give closed cells in S^n, and the previous theorem is then applicable.)

The case $r = 1$ yields a *knot* in S^n. The case of most interest is that of a knot in S^3, where s^1 is the image of S^1. Compute the homology of the knot complement $S^3 - s^1$. Note that it does not depend on φ. However, $\pi_1(S^3 - s^1)$ definitely depends on the embedding φ.

Now prove the Jordan–Brouwer Separation Theorem: For any s^{n-1} inside S^n, $S^n - s^{n-1}$ consists of two connected components, each having s^{n-1} as a common frontier. The first statement follows from the preceding work, so it remains to show the frontier assertion. That the frontier of each component is contained in s^{n-1} is straightforward (prove it!), so you need only show that if $x \in s^{n-1}$, then every neighborhood U of x intersects both the complementary components. To do this, find a set A with $x \in A \subset U \cap s^{n-1}$ and $s^{n-1} - A$ is a closed $(n-1)$-cell e^{n-1}. Use the results above to deduce that $S^n - e^{n-1}$ is connected and then conclude that U meets both components.

Now regard $\mathbb{R}^n$ as S^n minus a point to prove the following: if $n \geq 2$ and $s^{n-1} \subset \mathbb{R}^n$ is homeomorphic to S^{n-1}, then $\mathbb{R}^n - s^{n-1}$ has two components, both having s^{n-1} as frontier.

When $n = 1$, this is the Jordan Curve Theorem. Look up a classical proof of this theorem to compare with this approach. What are the main differences?

Project: the Nerve Theorem

In this project, we will outline a proof of Theorem 2.7.19. Let K be a simplicial complex and let X be a topological space. A *carrier* C for K in X is an assignment of a subset $C(\sigma) \subset X$ to every simplex σ of K such that $C(\sigma) \subset C(\tau)$ whenever σ is a face of τ. (In other words, C is a functor from the partially ordered set of simplices of K to the partially ordered set of subsets of X.) We say that C *carries* a continuous map $f : |K| \to X$ if for each simplex σ, we have $f(|\sigma|) \subset C(\sigma)$. Similarly, we way that C carries a homotopy $F : |K| \times [0,1] \to X$ if for each t the map $F_t(x) = F(x,t)$ is carried by C.

Carrier Lemma. *Let C be a carrier for K in X. If the subset $C(\sigma) \subset X$ is contractible for each simplex σ, then there exists a continuous map $f : |K| \to X$ carried by C. Moreover, any two continuous maps $f, g : |K| \to X$ are homotopic, and we can choose a homotopy $F : |K| \times [0,1] \to X$ that is also carried by C.*

Prove the Carrier Lemma as follows. Order the simplices of K as $\{\sigma_1, \ldots, \sigma_m\}$ so that for each i the set

$$K_i = \bigcup_{j \leq i} \sigma_j$$

is a subcomplex. Then we have a filtration $\{K_i\}_{1 \leq i \leq m}$ with $K_m = K$. We will prove the homotopy statement first. Proceed by induction on i.

(1) In the base case, $i = 1$, we have σ_1 must be a vertex (why?). The maps f and g send σ_1 to points x_0 and x_1 in X. These lie in the contractible set $C(\sigma_1)$. Deduce that there is a path in $C(\sigma_1)$ from x_0 to x_1 and that we have a homotopy carried by C of the restrictions of f and g to K_1.

(2) For the inductive step, suppose for some $i > 1$ the restrictions of f and g to K_{i-1} admit a homotopy $F : |K_{i-1}| \times [0,1] \to X$ carried by C. To extend this to $|K_i| \times [0,1]$ we must define it on $|\sigma_i| \times [0,1]$. Let $B \subset |K_{i-1}|$ be the union of the realizations of all faces $\tau \le \sigma_i$ other than σ_i. Show that $F(B \times [0,1]) \subset C(\sigma_i)$.

(3) Let $d = \dim \sigma_i$. The product $|\sigma_i| \times [0,1]$ is homeomorphic to $|\Delta^d| \times [0,1]$ and this is homeomorphic to $|\Delta^{d+1}|$. Deduce that the boundary of $|\sigma_i| \times [0,1]$ is homeomorphic to $|\partial \Delta^{d+1}| \cong (|\partial \Delta^d| \times [0,1]) \cup (|\Delta^d| \times \{0,1\})$.

(4) The first piece of the union is homeomorphic to $B \times [0,1]$ and the second is homeomorphic to two disjoint copies of $|\sigma_i|$. Use the first part of the inductive step above, along with the fact that $f(|\sigma_i|)$ and $g(|\sigma_i|)$ are both contained in $C(\sigma_i)$ to deduce the result.

(5) Prove the first statement (the existence of f) in a similar manner.

Now if $f : K \to L$ is a simplicial map, and if τ is a simplex in L, the *fiber of f under τ* is the following collection of simplices in K:

$$\tau/f = \{\sigma \in K \mid f(\sigma) \le \tau\}.$$

Each fiber is a subcomplex of K, and if $\tau \le \tau'$, we have $\tau/f \subseteq \tau'/f$.

Quillen's Fiber Theorem. *Let $f : K \to L$ be a simplicial map. If the fiber τ/f is contractible for every simplex τ in L, then the induced continuous map $|f| : |K| \to |L|$ admits a homotopy inverse $G : |L| \to |K|$. In particular, K and L are homotopy equivalent.*

Prove this theorem as follows.

(1) For each simplex τ of L, let $C(\tau) \subset |K|$ be the geometric realization of τ/f. Use the Carrier Lemma to produce a map $G : |L| \to |K|$. What can we say about this map?

(2) Prove that $|f| \circ G$ is homotopic to the identity on L by noting the containment $|f| \circ G(|\tau|) \subset |\tau|$ yields a carrier for L in $|L|$. This carries two maps: $|f| \circ G$ and the identity on L. Use the Carrier Lemma to deduce the result.

(3) Prove that $G \circ |f|$ is homotopic to the identity on K: the construction of G gives the containment

$$G(|f(\sigma)|) \subset C(f(\sigma)) = |f(\sigma)/f|.$$

Use this to define a carrier C_K for K in $|K|$, and show that it carries both $G \circ |f|$ and the identity on K. Use the Carrier Lemma again.

Finally, prove the following, more general version of Theorem 2.7.19.

Nerve Theorem. *Let $\mathcal{U} = \{U_a\}_{a \in A}$ be a finite open cover of a topological space X. Let N be the nerve of this cover. If each simplex σ in N has contractible support $\mathrm{Supp}(\sigma) \subset X$, then $|N|$ is homotopy equivalent to X.*

(1) Let $X(\mathcal{U})$ be the subset of the product $X \times |N|$ containing all pairs (x, u) such that there is a simplex σ in N satisfying both $s \in \mathrm{Supp}(\sigma)$ and $u \in |\sigma|$. There are natural projection maps from $X(\mathcal{U})$ to both X and $|N|$. Prove that the fibers of these maps are contractible.

(2) There is a stronger version of Quillen's Fiber Theorem that applies to sufficiently well-behaved continuous maps between metric spaces (such as the ones above), which implies that these maps are homotopy equivalences. Deduce the Nerve Theorem.

Project: example persistent homology calculations

This is an open-ended project whose goal is to get students to compute persistent homology of various datasets. There is no preferred platform, but we suggest the following software.

(1) **JavaPlex.** Available by visiting https://appliedtopology.github.io/javaplex/. This library is used in Matlab and it is quite flexible. It allows for the computation of Vietoris–Rips complexes, witness complexes (which we did not mention in the text, but which are a convenient tool for topological approximation), and the associated barcodes. There is an excellent tutorial available at the site.

(2) **Ripser.** Available at https://github.com/Ripser/ripser. Ripser is written in C++ and computes only Vietoris–Rips barcodes, but it is very good at it. Users who do not want to download and install the software can try it out in a browser by visiting https://live.ripser.org/.

(3) **Eirene.** Available at https://github.com/Eetion/Eirene.jl. Another powerful program to compute barcodes from data, written in Julia.

(4) **Dionysus 2.** Available at https://mrzv.org/software/dionysus2/. Computes persistence, vineyards, bottleneck distance, omni-field persistence (multiple finite field coefficients at once), and more.

(5) **Perseus.** Available by visiting https://people.maths.ox.ac.uk/nanda/perseus/index.html. Makes use of discrete Morse theory to speed persistence computations.

Now for some data. The author generated some datasets on the manifolds $\mathbb{R}P^2$ (embedded in both $\mathbb{R}^4$ and $\mathbb{R}^5$, the latter being an isometric embedding), $\mathrm{SO}(3) = \mathbb{R}P^3$, and the Grassmann manifold $G(2, 4)$. These are available at https://github.com/niveknosdunk/grassmann. The Matlab files for generating the sets are also at this site, and users may modify them to generate different sets, larger sets, etc.

For example, barcodes associated to 1000 points sampled from $\mathbb{R}P^2$ embedded in $\mathbb{R}^4$ (a non-isometric embedding) are shown in Figure 2.20. These were generated in Eirene. Note that there is one long bar in each homological degree, which is what we expect.

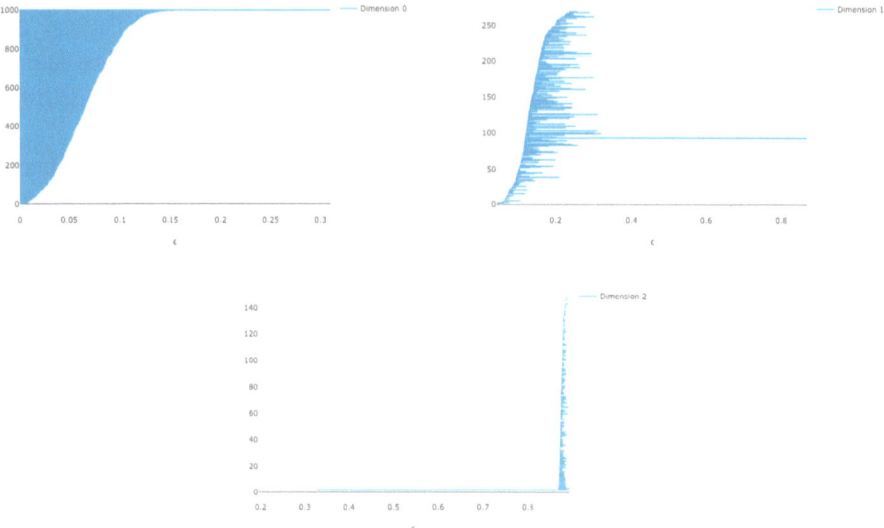

Figure 2.20: Barcodes for $\mathbb{R}P^2$ in $\mathbb{R}^4$, H_0 (top left), H_1 (top right), H_2 (bottom).

Some suggested explorations:
(1) Use the datasets https://github.com/niveknosdunk/grassmann to generate barcodes. Note that for $G(2, 4)$, a significant amount of memory is required for the calculations, as this is a 4-manifold.
(2) Generate random collections of points in the unit square in $\mathbb{R}^2$, or the unit cube in $\mathbb{R}^d$. Try 100, 500, 1000. What do the barcodes look like? Is this what you would expect? How large can d be before you run out of memory (or the calculation takes a long time to complete)?
(3) Generate points on the torus embedded in $\mathbb{R}^3$ (this is a surface of revolution, so you can parametrize it easily). Compute the associated barcodes.

Bibliographic notes
The presentation in this chapter is influenced by a number of sources: Hatcher [2], Greenberg & Harper, [1], Munkres [5], Griffiths & Morgan [3], etc., and some of the exercises are borrowed from those as well. The discussion of persistent homology owes much to the book of Edelsbrunner and Harer [12]. The Nerve Theorem and acyclic carrier material is assembled from various places, including Nanda's notes [13].

3 Cohomology and duality

3.1 Cohomology of spaces

3.1.1 Motivation from calculus

You were actually introduced to cohomology during your first course in multivariable calculus, you just did not know it. Let us begin with a simple question. Suppose $U \subseteq \mathbb{R}^2$ is an open set and that $f : U \to \mathbb{R}^2$ is a smooth function. Write $f = (f_1, f_2)$. Is there a smooth function $F : U \to \mathbb{R}$ such that

$$\frac{\partial F}{\partial x_1} = f_1 \quad \text{and} \quad \frac{\partial F}{\partial x_2} = f_2?$$

Note that since such an F has equal mixed partials,

$$\frac{\partial^2 F}{\partial x_1 \partial x_2} = \frac{\partial^2 F}{\partial x_2 \partial x_1},$$

a necessary condition for the existence of F is that

$$\frac{\partial f_1}{\partial x_2} = \frac{\partial f_2}{\partial x_1}. \tag{3.1}$$

Is this condition also sufficient?

The answer is no. Consider the map $f : \mathbb{R}^2 - \{0\} \to \mathbb{R}^2$ defined by

$$f(x_1, x_2) = \left(\frac{-x_2}{x_1^2 + x_2^2}, \frac{x_1}{x_1^2 + x_2^2} \right).$$

It is easy to check the condition for the partials of the coordinate functions, but there is no smooth $F : \mathbb{R}^2 - \{0\} \to \mathbb{R}^2$ with $\partial F/\partial x_1 = f_1$ and $\partial F/\partial x_2 = f_2$.

There is something we can say, however. We say that a subset $U \subseteq \mathbb{R}^d$ is *star-shaped* if there is a point $x_0 \in U$ such that for any $x \in U$, the line segment joining x_0 and x lies entirely in U. The simplest example is a convex set, where any $x_0 \in U$ will work, but there are many other examples.

Theorem 3.1.1. *Let $U \subseteq \mathbb{R}^2$ be a star-shaped open set and suppose $f : U \to \mathbb{R}^2$ satisfies (3.1). Then there is a smooth $F : U \to \mathbb{R}$ such that*

$$\frac{\partial F}{\partial x_1} = f_1 \quad \text{and} \quad \frac{\partial F}{\partial x_2} = f_2.$$

Proof. Assume, without loss of generality, that $x_0 = 0 \in U$ is a point such that the line segment from x_0 to x lies in U for every $x \in U$. Define $F : U \to \mathbb{R}$ by

https://doi.org/10.1515/9783111014852-003

$$F(x_1, x_2) = \int_0^1 \left[x_1 f_1(tx_1, tx_2) + x_2 f_2(tx_1, tx_2) \right] dt.$$

Then we have the following

$$\frac{\partial F}{\partial x_1}(x_1, x_2) = \int_0^1 \left[f_1(tx_1, tx_2) + tx_1 \frac{\partial f_1}{\partial x_1}(tx_1, tx_2) + tx_2 \frac{\partial f_2}{\partial x_1}(tx_1, tx_2) \right] dt$$

$$\frac{d}{dt} tf_1(tx_1, tx_2) = f_1(tx_1, tx_2) + tx_1 \frac{\partial f_1}{\partial x_1}(tx_1, tx_2) + tx_2 \frac{\partial f_1}{\partial x_2}(tx_1, tx_2).$$

Putting these together, we have

$$\frac{\partial F}{\partial x_1}(x_1, x_2) = \int_0^1 \left[\frac{d}{dt} tf_1(tx_1, tx_2) + tx_2 \underbrace{\left(\frac{\partial f_2}{\partial x_1}(tx_1, tx_2) - \frac{\partial f_1}{\partial x_2}(tx_1, tx_2) \right)}_{=0} \right] dt$$

$$= \int_0^1 \left[\frac{d}{dt} tf_1(tx_1, tx_2) \right] dt$$

$$= tf_1(tx_1, tx_2)\big|_{t=0}^{t=1}$$

$$= f_1(x_1, x_2).$$

A similar calculation shows that $\partial F / \partial x_2 = f_2$. □

What makes this work? The topology of U ensures that the integral is defined, so it makes sense to try to understand conditions on U that guarantee solutions to these types of questions. Note that in the example above of the function defined on $U = \mathbb{R}^2 - \{0\}$, the domain is not star-shaped.

Definition 3.1.2. If $U \subseteq \mathbb{R}^2$ is open, $C^\infty(U, \mathbb{R}^k)$ is the set of smooth functions $f : U \to \mathbb{R}^k$. Note that this is actually a real vector space, where addition is defined pointwise. Define two operations as follows:

$$\text{grad} : C^\infty(U, \mathbb{R}) \to C^\infty(U, \mathbb{R}^2) \quad \text{grad}(\varphi) = \left(\frac{\partial \varphi}{\partial x_1}, \frac{\partial \varphi}{\partial x_2} \right)$$

and

$$\text{curl} : C^\infty(U, \mathbb{R}^2) \to C^\infty(U, \mathbb{R}) \quad \text{curl}(\varphi_1, \varphi_2) = \frac{\partial \varphi_1}{\partial x_2} - \frac{\partial \varphi_2}{\partial x_1}.$$

Note that $\text{curl} \circ \text{grad} = 0$. We then define

$$H^1(U) = \ker(\text{curl})/\text{im}(\text{grad}).$$

Note that $H^1(U) = 0$ if U is star-shaped, by Theorem 3.1.1. However, $H^1(\mathbb{R}^2 - \{0\}) \neq 0$, as the first example shows.

Definition 3.1.3. Suppose $U \subseteq \mathbb{R}^k$ is open and $f : U \to \mathbb{R}$ is smooth. Then we set

$$\mathrm{grad}(f) = \left(\frac{\partial f}{\partial x_1}, \frac{\partial f}{\partial x_2}, \ldots, \frac{\partial f}{\partial x_k} \right),$$

and define

$$H^0(U) = \ker(\mathrm{grad}).$$

Theorem 3.1.4. *An open set $U \subseteq \mathbb{R}^k$ is connected if and only if $H^0(U) = \mathbb{R}$.*

Proof. Suppose $\mathrm{grad}(f) = 0$. Then f is locally constant: each $x_0 \in U$ has a neighborhood $V(x_0)$ with $f(x) = f(x_0)$ for all $x \in V(x_0)$. If U is connected, then every locally constant function is constant, since if $x_0 \in U$, the set

$$f^{-1}(f(x_0)) = \{x \in U \mid f(x) = f(x_0)\}$$

is closed (because f is continuous), and open since f is locally constant. Thus, $H^0(U) = \mathbb{R}$, the space of constant functions on U.

Conversely, if U is not connected, then there is a smooth surjection $f : U \to \{0,1\}$. This is locally constant and so $\mathrm{grad}(f) = 0$. Thus $\dim H^0(U) > 1$. □

Now let us extend this to three variables. Let $U \subseteq \mathbb{R}^3$ be open. We now have three operations:

$$\mathrm{grad} : C^\infty(U, \mathbb{R}) \to C^\infty(U, \mathbb{R}^3) \quad \mathrm{grad}(f) = \left(\frac{\partial f}{\partial x_1}, \frac{\partial f}{\partial x_2}, \frac{\partial f}{\partial x_3} \right)$$

$$\mathrm{curl} : C^\infty(U, \mathbb{R}^3) \to C^\infty(U, \mathbb{R}^3) \quad \mathrm{curl}(f_1, f_2, f_3) = \left(\frac{\partial f_3}{\partial x_2} - \frac{\partial f_2}{\partial x_3}, \frac{\partial f_1}{\partial x_3} - \frac{\partial f_3}{\partial x_1}, \frac{\partial f_2}{\partial x_1} - \frac{\partial f_1}{\partial x_2} \right)$$

$$\mathrm{div} : C^\infty(U, \mathbb{R}^3) \to C^\infty(U, \mathbb{R}) \quad \mathrm{div}(f_1, f_2, f_3) = \frac{\partial f_1}{\partial x_1} + \frac{\partial f_2}{\partial x_2} + \frac{\partial f_3}{\partial x_3}$$

Note that $\mathrm{curl} \circ \mathrm{grad} = 0$ and $\mathrm{div} \circ \mathrm{curl} = 0$, so that we have a complex

$$0 \to C^\infty(U, \mathbb{R}) \xrightarrow{\mathrm{grad}} C^\infty(U, \mathbb{R}^3) \xrightarrow{\mathrm{curl}} C^\infty(U, \mathbb{R}^3) \xrightarrow{\mathrm{div}} C^\infty(U, \mathbb{R}) \to 0.$$

Define $H^0(U)$ and $H^1(U)$ as before and set

$$H^2(U) = \ker(\mathrm{div})/\mathrm{im}(\mathrm{curl}).$$

Theorem 3.1.5. *If $U \subseteq \mathbb{R}^3$ is star-shaped, then*

$$H^1(U) = H^2(U) = 0.$$

Proof. The proof that $H^1(U) = 0$ for star-shaped subsets of $\mathbb{R}^2$ works in this context, with the obvious modification. Suppose that $F : U \to \mathbb{R}^3$ has div$(F) = 0$. Define $G : U \to \mathbb{R}^3$ by

$$G(\vec{x}) = \int_0^1 (F(t\vec{x}) \times t\vec{x}) \, dt,$$

where $\times$ is the cross-product in $\mathbb{R}^3$. It is a straightforward calculation to show that

$$\text{curl}(F(t\vec{x}) \times t\vec{x}) = \frac{d}{dt}(t^2 F(t\vec{x})).$$

It then follows that

$$\text{curl}(G(\vec{x})) = \int_0^1 \frac{d}{dt}(t^2 F(t\vec{x})) \, dt = F(\vec{x}). \qquad \square$$

Example 3.1.6. If $U \subseteq \mathbb{R}^3$ is not star-shaped, it is possible to have $H^1(U) \neq 0$ and $H^2(U) \neq 0$. Let $S = \{(x_1, x_2, x_3) \in \mathbb{R}^3 \mid x_1^2 + x_2^2 = 1 \text{ and } x_3 = 0\}$ (i.e., S is the unit circle in the $x_1 x_2$-plane) and let $U = \mathbb{R}^3 - S$. Note that U is not star-shaped. Define $f : U \to \mathbb{R}^3$ by

$$f(x_1, x_2, x_3) = \left(\frac{-2x_1 x_3}{x_3^2 + (x_1^2 + x_2^2 - 1)^2}, \frac{-2x_2 x_3}{x_3^2 + (x_1^2 + x_2^2 - 1)^2}, \frac{x_1^2 + x_2^2 - 1}{x_3^2 + (x_1^2 + x_2^2 - 1)^2} \right).$$

A direct calculation shows that curl$(f) = 0$, so that f determines a class $[f] \in H^1(U)$. Now consider the curve $y(t) = (\sqrt{1 + \cos t}, 0, \sin t)$, $-\pi \le t \le \pi$; note that im$(y) \subset U$. Suppose grad$(F) = f$ on U. Then

$$\int_{-\pi+\varepsilon}^{\pi-\varepsilon} \frac{d}{dt} F(y(t)) \, dt = F(y(\pi - \varepsilon)) - F(y(-\pi + \varepsilon)) \to 0 \qquad \text{as } \varepsilon \to 0.$$

However, the Chain Rule gives the following:

$$\frac{d}{dt} F(y(t)) = f_1(y(t)) y_1'(t) + f_2(y(t)) y_2'(t) + f_3(y(t)) y_3'(t)$$

$$= \sin^2 t + 0 + \cos^2 t$$

$$= 1,$$

so that

$$\int_{-\pi+\varepsilon}^{\pi-\varepsilon} \frac{d}{dt} F(y(t)) \, dt \to 2\pi \quad \text{as } \varepsilon \to 0.$$

This contradiction shows that $[f] \neq 0$ in $H^1(U)$.

We can extend these ideas further to subsets $U \subseteq \mathbb{R}^n$ as follows. We define real vector spaces $\Omega^p(U)$ by

$$\Omega^p(U) = \{\text{differentiable } p\text{-forms on } U\} = \left\{\text{smooth maps } \omega : U \to \textstyle\bigwedge^p \mathbb{R}^n\right\},$$

where $\bigwedge^p \mathbb{R}^n$ is the pth exterior power of $\mathbb{R}^n$. There is an operator, called the *exterior derivative*, $d : \Omega^p(U) \to \Omega^{p+1}(U)$, which agrees with div, grad, curl when it makes sense. We have $d^2 = 0$ and therefore a complex

$$0 \longrightarrow \Omega^0(U) \xrightarrow{d} \Omega^1(U) \xrightarrow{d} \cdots \xrightarrow{d} \Omega^{n-1}(U) \xrightarrow{d} \Omega^n(U) \longrightarrow 0.$$

Here, $\Omega^0(U)$ is simply the vector space of smooth real-valued functions on U. We then set

$$H^p(U) = \frac{\ker d : \Omega^p(U) \to \Omega^{p+1}(U)}{\operatorname{im} d : \Omega^{p-1}(U) \to \Omega^p(U)} = \frac{\text{closed } p\text{-forms}}{\text{exact } p\text{-forms}}.$$

These are higher-dimensional versions of the problem of finding antiderivatives: $H^p(U)$ measures how many closed p-forms are not exact.

Lemma 3.1.7 (Poincaré Lemma). *If U is a star-shaped subset of $\mathbb{R}^n$, then $H^0(U) = \mathbb{R}$ and $H^p(U) = 0$ for $p > 0$.*

The proof of this is similar to the low-dimensional cases we proved above, and is left as an exercise using the definition of d presented below.

Since p-forms are built from local information (open sets in $\mathbb{R}^n$), we can assemble them together to build a corresponding theory on a manifold M. So this whole idea is essentially getting at the question of whether we can solve PDEs on manifolds. A closed form $\omega \in \Omega^p(M)$ satisfies $d\omega = 0$, which is basically a collection of homogeneous differential equations. Finding η with $d\eta = \omega$ gives (local) solutions for this system. This is an old question, certainly predating homology theory.

All of this works only on manifolds, though, and we are interested in other spaces. So the question is: can we build a theory of functions on cell complexes, say, that is analogous to this local obstruction problem? The answer is yes, and we will develop these ideas in this chapter.

Exercises

Exercise 3.1.1. Prove that curl $\circ$ grad $= 0$.

Exercise 3.1.2. Prove that div $\circ$ curl $= 0$.

Exercise 3.1.3. Prove that $\operatorname{curl}(F(t\vec{x}) \times t\vec{x}) = \frac{d}{dt}(t^2 F(t\vec{x}))$.

Exercise 3.1.4. Prove that curl(f) = 0 for the function f in Example 3.1.6.

Exercise 3.1.5. Prove that if the open set $U \subset \mathbb{R}^n$ has k connected components, then $H^0(U) \cong \mathbb{R}^k$.

Exercise 3.1.6. Suppose U is an open set in $\mathbb{R}^n$. A smooth real-valued function f on U is a 0-form, i.e., an element of $\Omega^0(U)$. Let $x_1, \ldots, x_n$ be a coordinate system in U. The 1-forms $dx_1, dx_2, \ldots, dx_n$ form a basis of $\Omega^1(U) = \{$smooth $\omega : U \to \mathbb{R}^n\}$ (here x_i is the ith coordinate function $U \to \mathbb{R}$). Given a collection $i_1, \ldots, i_p$ of indices between 1 and n, we have the corresponding element of $\Omega^p(U)$ given by $dx_{i_1} \wedge dx_{i_2} \wedge \cdots \wedge dx_{i_p}$. Note that we may multiply any such element by a smooth real-valued function to get another p-form. We are now ready to define d. Suppose $\omega = g \, dx_{i_1} \wedge \cdots \wedge dx_{i_p}$. Then we set

$$d\omega = \sum_j \frac{\partial g}{\partial x_j} dx_j \wedge dx_{i_1} \wedge \cdots \wedge dx_{i_p}.$$

Note that if j is one of the indices appearing in ω, then $dx_j \wedge dx_{i_1} \wedge \cdots \wedge dx_{i_p} = 0$ by the properties of the exterior product. Prove that $d^2 = 0$ and that d agrees with div, grad, curl, when it makes sense. Also, prove the Poincaré Lemma 3.1.7.

3.1.2 Cochains

We begin with some formalities. Suppose we have a chain complex of free abelian groups, $(C_\bullet, \partial)$:

$$\cdots \to C_{n+1} \xrightarrow{\partial} C_n \xrightarrow{\partial} C_{n-1} \xrightarrow{\partial} \cdots$$

Let G be any abelian group, and apply the functor $\mathrm{Hom}_{\mathbb{Z}}(-, G)$ degreewise to $C_\bullet$:

$$\cdots \to \mathrm{Hom}_{\mathbb{Z}}(C_{n-1}, G) \xrightarrow{\delta} \mathrm{Hom}_{\mathbb{Z}}(C_n, G) \xrightarrow{\delta} \mathrm{Hom}_{\mathbb{Z}}(C_{n+1}, G) \xrightarrow{\delta} \cdots$$
$$\| \qquad\qquad\qquad \| \qquad\qquad\qquad \|$$
$$C^{n-1} \qquad\qquad\qquad C^n \qquad\qquad\qquad C^{n+1}$$

Note that since $\mathrm{Hom}_{\mathbb{Z}}(-, G)$ is contravariant, the arrows are reversed. The map δ is the dual map to ∂ and is called the *coboundary map*. It is defined as follows: if $f \in \mathrm{Hom}_{\mathbb{Z}}(C_n, G)$ and $\alpha \in C_{n+1}$, then

$$\delta(f)(\alpha) = f(\partial\alpha).$$

Since $\partial^2 = 0$, we have $\delta^2 = 0$, so that $(C^\bullet, \delta)$ is a complex called the *cochain complex* and we can define the *cohomology groups*

$$H^n(C_\bullet; G) = \frac{\ker \delta : C^n \to C^{n+1}}{\mathrm{im}\, \delta : C^{n-1} \to C^n} = \frac{n\text{-cocycles}}{n\text{-coboundaries}}.$$

Example 3.1.8. Consider the chain complex $C_\bullet$.

$$0 \to \mathbb{Z} \xrightarrow{0} \mathbb{Z} \to \mathbb{Z} \oplus \mathbb{Z} \xrightarrow{0} \mathbb{Z} \to 0$$

$$1 \longmapsto (0,2)$$

$$C_3 \quad C_2 \quad C_1 \quad C_0$$

The homology groups of $C_\bullet$ are

$$H_0(C_\bullet) = \mathbb{Z}$$
$$H_1(C_\bullet) = \frac{\mathbb{Z} \oplus \mathbb{Z}}{(0,2)} \cong \mathbb{Z} \oplus \mathbb{Z}_2$$
$$H_2(C_\bullet) = 0$$
$$H_3(C_\bullet) = \mathbb{Z}$$

Let us consider $G = \mathbb{Z}$. Then $\mathrm{Hom}_{\mathbb{Z}}(C_\bullet, \mathbb{Z})$ is the cochain complex

$$0 \to \mathbb{Z} \xrightarrow{0} \mathbb{Z} \oplus \mathbb{Z} \xrightarrow{0} \mathbb{Z} \xrightarrow{\delta} \mathbb{Z} \to 0$$

$$C^0 \quad C^1 \quad C^2 \quad C^3$$

The canonical generators of $C^1 = \mathrm{Hom}(\mathbb{Z} \oplus \mathbb{Z}, \mathbb{Z}) \cong \mathbb{Z} \oplus \mathbb{Z}$ are functions f_1, f_2, defined by $f_1(a,b) = a$ and $f_2(a,b) = b$. We then compute δ as

$$\delta(f_1)(n) = f_1(\partial n) = f_1(0, 2n) = 0$$
$$\delta(f_2)(n) = f_2(\partial n) = f_2(0, 2n) = 2n$$

It follows that $\delta : \mathbb{Z} \oplus \mathbb{Z} \to \mathbb{Z}$ is the map $(a,b) \mapsto 2b$ so that $\ker \delta = \{(a,0)\} \cong \mathbb{Z}$ and $\mathrm{im}\, \delta = 2\mathbb{Z} \subset \mathbb{Z} = C^2$. We therefore find that

$$H^0(C_\bullet; \mathbb{Z}) = \mathbb{Z}$$
$$H^1(C_\bullet; \mathbb{Z}) = \mathbb{Z}$$
$$H^2(C_\bullet; \mathbb{Z}) = \mathbb{Z}/\mathrm{im}\,\delta \cong \mathbb{Z}_2$$
$$H^3(C_\bullet; \mathbb{Z}) = \mathbb{Z}$$

If we take $G = \mathbb{Z}_2$, then the calculation of δ above shows that it is the zero map ($2b = 0$ in $\mathbb{Z}_2$). We therefore deduce that

$$H^0(C_\bullet; \mathbb{Z}_2) = \mathbb{Z}_2$$
$$H^1(C_\bullet; \mathbb{Z}_2) = \mathbb{Z}_2 \oplus \mathbb{Z}_2$$
$$H^2(C_\bullet; \mathbb{Z}_2) = \mathbb{Z}_2$$
$$H^3(C_\bullet; \mathbb{Z}_2) = \mathbb{Z}_2$$

One obvious question arises here: since the cochain complex is obtained from the chain complex by applying a functor componentwise, is it possible to determine the cohomology groups $H^\bullet(C_\bullet; G)$ in terms of $H_\bullet(C_\bullet)$ and G? The answer is yes, but it is slightly complicated.

First observe that there is a natural map

$$h : H^n(C_\bullet; G) \to \operatorname{Hom}_{\mathbb{Z}}(H_n(C_\bullet), G)$$

defined as follows. If $f : C_n \to G$ is a cocycle ($\delta f = 0$), then $f\partial = 0$ and so f vanishes on $B_n \subseteq C_n$. Thus, f induces a map $f_0 : Z_n/B_n \to G$; that is, an element of $\operatorname{Hom}_{\mathbb{Z}}(H_n(C_\bullet), G)$. If f lies in im δ, then $f = \delta g = g\partial$ for some g and therefore f is 0 on Z_n; that is, $f_0 \equiv 0$. So we can define h by the formula

$$h([f]) = [f_0] \in \operatorname{Hom}_{\mathbb{Z}}(H_n(C_\bullet), G).$$

Lemma 3.1.9. *The map h is surjective.*

Proof. Consider the short exact sequence

$$0 \to Z_n \to C_n \to B_{n-1} \to 0.$$

This splits (though not canonically), since B_{n-1} is free and so there is a projection $p :$ $C_n \to Z_n$ restricting to the identity on Z_n. So, if we have a map $\varphi_0 : Z_n \to G$, we have $\varphi = \varphi_0 \circ p : C_n \to G$ extending φ_0. In particular, this extends homomorphisms $Z_n \to G$ vanishing on B_n to maps $C_n \to G$ that still vanish on B_n; that is, it extends maps $H_n(C_\bullet) \to G$ to elements of ker δ. Passing to quotients, we get a map $\operatorname{Hom}_{\mathbb{Z}}(H_n(C_\bullet), G) \to H^n(C_\bullet; G)$. If we follow this by h we get the identity on $\operatorname{Hom}_{\mathbb{Z}}(H_n(C_\bullet), G)$ and hence h is surjective and we get a *split* short exact sequence

$$0 \to \ker h \to H^n(C_\bullet; G) \overset{h}{\underset{}{\rightleftarrows}} \operatorname{Hom}_{\mathbb{Z}}(H_n(C_\bullet), G) \to 0. \qquad \square$$

What is the kernel of the map h? To describe it, we need to take a short detour.

Lemma 3.1.10. *Let R be a commutative ring with identity and suppose we have an exact sequence of R-modules*

$$A' \to A \to A'' \to 0.$$

If B is any R-module, then there is an exact sequence

$$0 \to \operatorname{Hom}_R(A'', B) \to \operatorname{Hom}_R(A, B) \to \operatorname{Hom}_R(A', B).$$

The proof of this is a straightforward exercise. Now, if $A' \to A$ happens to be injective, one would hope that the corresponding map $\mathrm{Hom}_R(A, B) \to \mathrm{Hom}_R(A', B)$ is surjective. That is not true, however (see the exercises). However, we have the following result.

Lemma 3.1.11. *Suppose we have a split short exact sequence of R-modules*

$$0 \to A' \to A \to A'' \to 0.$$

Then for any R-module B, the sequence

$$0 \to \mathrm{Hom}_R(A'', B) \to \mathrm{Hom}_R(A, B) \to \mathrm{Hom}_R(A', B) \to 0$$

is a split short exact sequence.

Proof. In light of Lemma 3.1.10, we need only show the surjectivity of the map $\mathrm{Hom}_R(A, B) \to \mathrm{Hom}_R(A', B)$ and that the sequence splits. But the proof of this is entirely similar to the proof of Lemma 3.1.9. Indeed, since the sequence splits, we have a projection $A \to A'$, restricting to the identity on the image of A' in A. We now mimic the construction of the splitting of the map h to get a splitting of the map $\mathrm{Hom}_R(A, B) \to \mathrm{Hom}_R(A', B)$, which completes the proof. □

Now, if we have a split short exact sequence of chain complexes

$$0 \to C'_\bullet \to C_\bullet \to C''_\bullet \to 0,$$

then by Lemma 3.1.11 we get a short exact sequence of cochain complexes

$$0 \to \mathrm{Hom}_R(C''_\bullet, G) \to \mathrm{Hom}_R(C_\bullet, G) \to \mathrm{Hom}_R(C'_\bullet, G) \to 0$$

for any R-module G. The Snake Lemma 2.3.10 now applies to show that we can construct a long exact sequence in cohomology.

Theorem 3.1.12. *Given a split short exact sequence of chain complexes*

$$0 \to C'_\bullet \to C_\bullet \to C''_\bullet \to 0,$$

and a module G, there is a functorial long exact sequence

$$\cdots \to H^q(C''_\bullet; G) \to H^q(C_\bullet; G) \to H^q(C'_\bullet; G) \xrightarrow{\delta} H^{q+1}(C''_\bullet; G) \to \cdots$$

Ext modules

By a *free resolution* of an R-module M, we mean an exact sequence of free R-modules

$$\cdots \to C_n \to C_{n-1} \to \cdots \to C_2 \to C_1 \to C_0 \to M \to 0.$$

Observe that this is a chain complex. Such objects are unique up to chain homotopy, and there is a canonical way to produce one. Take a free module C_0 surjecting onto M (take the free module generated by the elements of M, for example), and then take C_1 to be a free module surjecting onto the kernel of the projection $C_0 \to M$, etc. This process may not be the most efficient, in the sense that the modules C_i may be infinitely generated, but it does work. If N is any R-module, we have the cochain complex $\mathrm{Hom}_R(C_\bullet, N)$, and the cohomology of this complex depends only on M and N, up to isomorphism.

Definition 3.1.13. The nth Ext-module is the nth cohomology of this chain complex:

$$\mathrm{Ext}^n(M, N) = H^n(\mathrm{Hom}_R(C_\bullet, N)).$$

The functor $\mathrm{Ext}^n(M, N)$ is contravariant in M and covariant in N. Note that $\mathrm{Ext}^0(M, N)$ is just $\mathrm{Hom}_R(M, N)$.

Note that if R happens to be a principal ideal domain, then any module has a free resolution of the form

$$0 \to C_1 \to C_0 \to M \to 0,$$

since submodules of free modules are free. It follows that $\mathrm{Ext}^n(M, N) = 0$ for $n > 1$. We will drop the superscript and simply refer to $\mathrm{Ext}(M, N)$ in this case. Note that we have an exact sequence

$$0 \to \mathrm{Hom}(M, N) \to \mathrm{Hom}(C_0, N) \to \mathrm{Hom}(C_1, N) \to \mathrm{Ext}(M, N) \to 0.$$

We note the following computations.

Example 3.1.14. If M is a free module, then it has a free presentation

$$0 \to 0 \to M \to M \to 0.$$

It follows that $\mathrm{Ext}(M, N) = 0$ for any module N.

Example 3.1.15. If r is a nonzero element of R, then there is a short exact sequence

$$0 \to R \xrightarrow{a} R \to R/rR \to 0,$$

where $a(x) = rx$ for $x \in R$. This is a free presentation of R/rR. For any R-module N, we have $\mathrm{Hom}(R, N) \cong N$ and the homomorphism $a^* : \mathrm{Hom}(R, N) \to \mathrm{Hom}(R, N)$ corresponds to the map $a^* : N \to N$ given by $a^*(n) = rn$. Thus, we have an isomorphism

$$\mathrm{Ext}(R/rR, N) \cong \mathrm{coker}(a^*) = N/rN.$$

Since Hom commutes with direct sums, the same is true for Ext. This allows us to compute the Ext-modules for any pair M and N over a PID by using the structure theorem for such modules.

We are now ready to identify the group ker h in Lemma 3.1.9.

Theorem 3.1.16. *Suppose $(C_\bullet, \partial)$ is a chain complex of free R-modules, where R is a PID. Let G be an R-module. Then for each $q \geq 0$ there is a functorial short exact sequence*

$$0 \to \mathrm{Ext}(H_{q-1}(C_\bullet), G) \to H^q(C_\bullet; G) \overset{h}{\to} \mathrm{Hom}(H_q(C_\bullet), G) \to 0,$$

and this sequence is split (though not naturally).

Proof. The proof of Lemma 3.1.9 is the starting point for this. We have a short exact sequence of chain complexes

$$0 \to Z_\bullet \to C_\bullet \to B_\bullet \to 0,$$

and this splits because $B_\bullet$ is a complex of free modules. Note that the qth module in $B_\bullet$ is $B_{q-1} \subset C_{q-1}$. We then get a long exact sequence in cohomology from Theorem 3.1.12:

$$\cdots \to H^q(B_\bullet; G) \to H^q(C_\bullet; G) \to H^q(Z_\bullet; G) \overset{\delta}{\to} H^{q+1}(B_\bullet; G) \to \cdots$$

Since $Z_\bullet$ and $B_\bullet$ have trivial boundary maps, we have

$$H^q(Z_\bullet; G) = \mathrm{Hom}(Z_q, G)$$
$$H^q(B_\bullet; G) = \mathrm{Hom}(B_{q-1}, G)$$

Also, the map $\delta : H^q(Z_\bullet; G) \to H^{q+1}(B_\bullet; G)$ is just the map $\mathrm{Hom}(Z_q, G) \to \mathrm{Hom}(B_q, G)$ induced by the inclusion $B_q \subset Z_q$; call this map γ_q. We therefore have short exact sequences

$$0 \to \mathrm{coker}\,\gamma_{q-1} \to H^q(C_\bullet; G) \to \ker\gamma_q \to 0$$

and it remains to identify the terms on each end. Note that we have exact sequences

$$0 \to B_q \overset{\gamma_q}{\to} Z_q \to H_q(C_\bullet) \to 0$$

and this is a free presentation of $H_q(C_\bullet)$. We then get an exact sequence

$$0 \to \mathrm{Hom}(H_q(C_\bullet), G) \to \mathrm{Hom}(Z_q, G) \overset{\gamma_q}{\to} \mathrm{Hom}(B_q, G) \to \mathrm{Ext}(H_q(C_\bullet), G) \to 0.$$

It follows that $\ker\gamma_q \cong \mathrm{Hom}(H_q(C_\bullet), G)$ and $\mathrm{coker}\,\gamma_q \cong \mathrm{Ext}(H_q(C_\bullet), G)$. This completes the proof. $\qquad\square$

Example 3.1.17. Consider the chain complex $C_\bullet$ of Example 3.1.8

$$0 \to \mathbb{Z} \xrightarrow{0} \mathbb{Z} \to \mathbb{Z} \oplus \mathbb{Z} \xrightarrow{0} \mathbb{Z} \to 0$$

$$1 \longmapsto (0,2)$$

Let us use Theorem 3.1.16 to compute $H^\bullet(C_\bullet; \mathbb{Z}_2)$. For each q, we have a split short exact sequence

$$0 \to \mathrm{Ext}(H_{q-1}(C_\bullet), \mathbb{Z}_2) \to H^q(C_\bullet; \mathbb{Z}_2) \to \mathrm{Hom}(H_q(C_\bullet), \mathbb{Z}_2) \to 0.$$

For $q = 0$, we simply have $H^0(C_\bullet; \mathbb{Z}_2) = \mathrm{Hom}(H_0(C_\bullet), \mathbb{Z}_2) = \mathrm{Hom}(\mathbb{Z}, \mathbb{Z}_2) = \mathbb{Z}_2$. For $q = 1$, since $H_0(C_\bullet) = \mathbb{Z}$, the Ext-group in the kernel vanishes and we have

$$H^1(C_\bullet; \mathbb{Z}_2) \cong \mathrm{Hom}(H_1(C_\bullet), \mathbb{Z}_2) = \mathrm{Hom}(\mathbb{Z} \oplus \mathbb{Z}_2, \mathbb{Z}_2) \cong \mathbb{Z}_2 \oplus \mathbb{Z}_2.$$

For $q = 2$, we have

$$0 \to \mathrm{Ext}(\mathbb{Z} \oplus \mathbb{Z}_2, \mathbb{Z}_2) \to H^2(C_\bullet; \mathbb{Z}_2) \to \mathrm{Hom}(0, \mathbb{Z}_2) \to 0,$$

so that

$$H^2(C_\bullet; \mathbb{Z}_2) = \mathrm{Ext}(\mathbb{Z} \oplus \mathbb{Z}_2, \mathbb{Z}_2) = \mathrm{Ext}(\mathbb{Z}, \mathbb{Z}_2) \oplus \mathrm{Ext}(\mathbb{Z}_2, \mathbb{Z}_2) = \mathrm{Ext}(\mathbb{Z}_2, \mathbb{Z}_2) \cong \mathbb{Z}_2.$$

Finally, since $H_2(C_\bullet) = 0$, we have $H^3(C_\bullet; \mathbb{Z}_2) \cong \mathrm{Hom}(H_3(C_\bullet), \mathbb{Z}_2) = \mathrm{Hom}(\mathbb{Z}, \mathbb{Z}_2) \cong \mathbb{Z}_2$. Note that this agrees with the calculation in Example 3.1.8.

Exercises

Exercise 3.1.7. Prove Lemma 3.1.10.

Exercise 3.1.8. Consider the short exact sequence of $\mathbb{Z}$-modules

$$0 \to \mathbb{Z}_2 \to \mathbb{Z}_4 \to \mathbb{Z}_2 \to 0$$

and let $B = \mathbb{Z}_2$. Show that the induced map

$$\mathrm{Hom}(\mathbb{Z}_4, \mathbb{Z}_2) \to \mathrm{Hom}(\mathbb{Z}_2, \mathbb{Z}_2)$$

is not surjective.

Exercise 3.1.9. The notation $\mathrm{Ext}(M, N)$ is suggestive: "Ext" is short for "extensions" and this exercise aims to make this precise. An *extension* of N by M is a short exact sequence

$$0 \to N \to E \to M \to 0$$

(some authors call this an extension of M by N). If we have another extension $0 \to N \to E' \to M \to 0$, we say the two are *equivalent* if there is a commutative diagram

$$0 \to N \to E \to M \to 0$$
$$\downarrow\text{id} \qquad \downarrow\varphi \qquad \downarrow\text{id}$$
$$0 \to N \to E' \to M \to 0.$$

Prove that φ is an isomorphism and that this is an equivalence relation on the collection of all extensions. Denote the set of equivalence classes by $E(M, N)$. Note that there is a distinguished element of this set called the *trivial extension*:

$$0 \to N \to M \oplus N \to M \to 0.$$

Say that an extension *splits* if there is a map $M \to E$ with the composite $M \to E \to M$ equal to the identity. Prove that an extension splits if and only if it is equivalent to the trivial extension. Now given a free resolution of M: $0 \to C_1 \to C_0 \to M \to 0$, this determines an element of $E(M, N)$. Conversely, given an extension $0 \to N \to E \to M \to 0$, there is a commutative diagram

$$0 \to C_1 \to C_0 \to M \to 0$$
$$\downarrow\psi_1 \qquad \downarrow\psi_2 \qquad \downarrow\text{id}$$
$$0 \to N \to E \to M \to 0$$

for any free resolution. Prove that this correspondence induces an isomorphism $E(M, N) \cong \text{Ext}(M, N)$.

3.1.3 Calculations

With these formalities in place, we now turn our attention to the cohomology of a topological space. We already have a few chain complexes at our disposal: simplicial chains, cellular chains, and singular chains on X. We begin with singular cohomology. Recall that the singular n-chains on X are defined to be the free abelian group on the continuous maps $\sigma : \Delta^n \to X$. Applying the construction of the previous section, we have the *singular cochain complex*

$$S^n(X; G) = \text{Hom}_{\mathbb{Z}}(S_n(X), G),$$

with coboundary operator δ dual to the boundary operator ∂. So, a cochain $\varphi \in S^n(X; G)$ is a map assigning to any $\sigma : \Delta^n \to G$ a value $\varphi(\sigma) \in G$. The coboundary map $\delta : S^n(X; G) \to S^{n+1}(X; G)$ is then defined as

$$\delta\varphi(\sigma) = \varphi(\partial\sigma) = \sum_{i=0}^{n+1}(-1)^i\varphi(\sigma|_{\langle v_0,\dots,\hat{v}_i,\dots,v_{n+1}\rangle}),$$

where $\sigma : \Delta^{n+1} \to X$ is an $(n+1)$-simplex. Since $\partial^2 = 0$, we have $\delta^2 = 0$ and we define the n-th singular cohomology group of X with coefficients in G to be

$$H^n(X;G) = \frac{\ker \delta : S^n(X;G) \to S^{n+1}(X;G)}{\operatorname{im} \delta : S^{n-1}(X;G) \to S^n(X;G)} = \frac{Z^n(X;G)}{B^n(X;G)} = \frac{n\text{-cocycles}}{n\text{-coboundaries}}.$$

Note that by Theorem 3.1.16, we have split short exact sequences

$$0 \to \operatorname{Ext}(H_{q-1}(X);G) \to H^q(X;G) \to \operatorname{Hom}(H_q(X),G) \to 0.$$

Example 3.1.18. Consider $X = S^n$. Using the short exact sequence above, we see that

$$H^i(S^n;\mathbb{Z}) = \begin{cases} \mathbb{Z} & i = 0, n \\ 0 & \text{otherwise} \end{cases}$$

Example 3.1.19. Suppose we take a field F as the coefficient group. The calculations we performed to prove Theorem 3.1.16 are quite general. In particular, if we consider a chain complex $C_\bullet$ of F-modules (i. e., a complex of F-vector spaces), then the homology groups $H^n(C_\bullet)$ are also F-vector spaces. In particular, they are free modules over F. It follows that the Ext-modules $\operatorname{Ext}(H_n(C_\bullet), F)$ all vanish. Thus, $H^n(C_\bullet; F) \cong \operatorname{Hom}_F(H_n(C_\bullet), F)$ for all n. Thus, if we consider the singular chains on X with F-coefficients, we have

$$H^n(X;F) \cong \operatorname{Hom}_F(H_n(X;F), F);$$

that is, over a field, homology and cohomology are dual vector spaces.

The obvious question that arises is: if we can compute cohomology purely algebraically via Theorem 3.1.16, why bother? The answer is that cohomology actually has a much richer structure, which we will explore in Section 3.2 below. For now, let us forge ahead.

Example 3.1.20. Let us compute $H^0(X;G)$. Note that $S_0(X) = \mathbb{Z}\{\text{points of } X\}$. Thus, a cochain $\varphi \in S^0(X;G)$ is an arbitrary function $\varphi : X \to G$ (not necessarily continuous). Suppose $\delta\varphi = 0$. Then if $\sigma : \langle v_0, v_1 \rangle \to X$ is a 1-simplex, we have $\varphi(\sigma(v_1)) - \varphi(\sigma(v_0)) = 0$; that is, $\varphi(\sigma(v_1)) = \varphi(\sigma(v_0))$. This implies that φ is constant on each path component of X. Thus,

$$H^0(X;G) = \ker \delta$$
$$= \{\text{functions from path components of } X \text{ to } G\}$$
$$= \operatorname{Hom}_{\mathbb{Z}}(H_0(X), G).$$

In particular, the rank of $H^0(X;\mathbb{Z})$ is the number of path components of X.

We also have *relative cohomology*. Recall that for a pair (X, A), we have an exact sequence of chain complexes

$$0 \to C_\bullet(A) \xrightarrow{i} C_\bullet(X) \xrightarrow{j} C_\bullet(X,A) \to 0.$$

Dualizing this, we obtain a sequence of cochain complexes

$$0 \to S^\bullet(X,A;G) \xrightarrow{j^*} S^\bullet(X;G) \xrightarrow{i^*} S^\bullet(A;G) \to 0.$$

Lemma 3.1.21. *This sequence is exact.*

Proof. The map i^* restricts a cochain on X to a cochain on A. If $f : S_n(A) \to G$ is a cochain, then f can be extended to all of $S_n(X)$ by assigning the value 0 to all simplices not in A; it follows that i^* is surjective. The kernel of i^* consists of all cochains taking the value 0 on $S_n(A)$. These correspond to maps $S_n(X,A) = S_n(X)/S_n(A) \to G$ and so $\ker i^* = \mathrm{Hom}(S_n(X,A), G) = \mathrm{im}\, j^*$. The map j^* is clearly injective. $\square$

The relative coboundary $\delta : S^n(X,A;G) \to S^{n+1}(X,A;G)$ is just the restriction of the absolute coboundary on X, and so the groups $H^n(X,A;G)$ are defined. The maps i^* and j^* commute with δ, since i and j commute with ∂. We therefore have a long exact cohomology sequence

$$\cdots \to H^n(X,A;G) \xrightarrow{j^*} H^n(X;G) \xrightarrow{i^*} H^n(A;G) \xrightarrow{\delta} H^{n+1}(X,A;G) \to \cdots$$

Moreover, we have a commutative diagram

$$
\begin{array}{ccc}
H^n(A;G) & \xrightarrow{\;\;\delta\;\;} & H^{n+1}(X,A;G) \\
\downarrow h & & \downarrow h \\
\mathrm{Hom}(H_n(A),G) & \xrightarrow{\;\;\partial\;\;} & \mathrm{Hom}(H_{n+1}(X,A),G)
\end{array}
$$

Example 3.1.22. Consider the pair (D^n, S^{n-1}). The long exact sequence has the form

$$\cdots \to H^{n-1}(D^n) \to H^{n-1}(S^{n-1}) \xrightarrow[\cong]{\delta} H^n(D^n, S^{n-1}) \to H^n(D^n) \to \cdots,$$
$$\qquad\qquad \| \qquad\qquad\qquad\qquad\qquad\qquad\qquad\qquad \|$$
$$\qquad\qquad 0 \qquad\qquad\qquad\qquad\qquad\qquad\qquad\qquad\;\; 0$$

where the groups $H^k(D^n) = 0$ for $k > 0$ by Theorem 3.1.16.

Variance
Cohomology is a *contravariant* functor. If $f : (X,A) \to (Y,B)$ is a continuous map, we get a map $f^\sharp : S^n(Y,B;G) \to S^n(X,A;G)$, defined as the dual of the map $f_\sharp : S_n(X,A) \to S_n(Y,B)$. Explicitly, if $\sigma : \Delta^n \to X$ is an n-simplex and if $\varphi \in S^n(Y;G)$, then

$$f^\sharp(\varphi)(\sigma) = \varphi(f \circ \sigma),$$

and we restrict this to the relative submodules. Since $\partial f_{\#} = f_{\#}\partial$, we have $f^{\#}\delta = \delta f^{\#}$, and so we obtain maps

$$f^* : H^n(Y,B;G) \to H^n(X,A;G).$$

These maps commute with the exact sequences from Theorem 3.1.16:

$$
\begin{array}{ccccccccc}
0 & \to & \mathrm{Ext}(H_{n-1}(X,A),G) & \to & H^n(X,A;G) & \xrightarrow{h} & \mathrm{Hom}(H_n(X,A),G) & \to & 0 \\
& & {\scriptstyle (f_*)^*}\uparrow & & {\scriptstyle f^*}\uparrow & & {\scriptstyle (f_*)^*}\uparrow & & \\
0 & \to & \mathrm{Ext}(H_{n-1}(Y,B),G) & \to & H^n(Y,B;G) & \to & \mathrm{Hom}(H_n(Y,B),G) & \to & 0
\end{array}
$$

Theorem 3.1.23. *If $f,g : (X,A) \to (Y,B)$ are homotopic maps then*

$$f^* = g^* : H^\bullet(Y,B;G) \to H^\bullet(X,A;G).$$

Proof. Dualize the homology proof. The details are left as an exercise. □

Excision

Suppose $Z \subset A \subset X$ with $\overline{Z} \subset \mathrm{int}(A)$.

Theorem 3.1.24. *The inclusion map $i : (X - Z, A - Z) \to (X,A)$ induces isomorphisms*

$$i^* : H^n(X,A;G) \to H^n(X - Z, A - Z;G).$$

Proof. One could dualize the homology proof (exercise). Here is another approach. Consider the commutative diagram

$$
\begin{array}{ccccccccc}
0 & \to & \mathrm{Ext}(H_{n-1}(X,A),G) & \to & H^n(X,A;G) & \to & \mathrm{Hom}(H_n(X,A),G) & \to & 0 \\
& & {\scriptstyle (i_*)^*}\downarrow & & {\scriptstyle i^*}\downarrow & & {\scriptstyle (i_*)^*}\downarrow & & \\
0 & \to & \mathrm{Ext}(H_{n-1}(X-Z,A-Z),G) & \to & H^n(X-Z,A-Z;G) & \to & \mathrm{Hom}(H_n(X-Z,A-Z),G) & \to & 0
\end{array}
$$

Since excision holds for singular homology, the maps $(i_*)^*$ on the left and the right are isomorphisms. By the Five Lemma, the map i^* is an isomorphism. □

Mayer–Vietoris

If $X = A \cup B$, then we may dualize the exact sequence

$$0 \to S_n(A \cap B) \to S_n(A) \oplus S_n(B) \to S_n^{\mathcal{U}}(X) \to 0,$$

where $\mathcal{U} = \{A, B\}$. Since the complex $S^{\mathcal{U}}_{\bullet}(X)$ is quasi-isomorphic to $S_{\bullet}(X)$, the same is true when we dualize and pass to cohomology. We therefore obtain the Mayer–Vietoris exact sequence

$$\cdots \to H^n(X; G) \to H^n(A; G) \oplus H^n(B; G) \to H^n(A \cap B; G) \to H^{n+1}(X; G) \to \cdots$$

Simplicial cohomology

If X is a simplicial complex, then we may dualize the simplicial chain complex $C_{\bullet}(X)$ to get the simplicial cochain complex $C^{\bullet}(X; G)$. We also have the relative complexes $C^{\bullet}(X, A; G)$. The corresponding cohomology groups agree with the singular groups, so we denote them $H^n(X, A; G)$.

Cellular cohomology

If X is a cell complex, then we may dualize the cellular chain complex to obtain the cellular cochains. Again, the corresponding cohomology groups agree with the singular groups.

The cohomology groups of the spaces from the previous chapter are now easily computable from the corresponding cellular chain complexes. We leave most of these as exercises, but we present one calculation here.

Example 3.1.25. Consider the real projective space $\mathbb{R}P^n$. Recall that the cellular chain complex is

$$0 \to \mathbb{Z} \to \mathbb{Z} \to \cdots \xrightarrow{0} \mathbb{Z} \xrightarrow{\times 2} \mathbb{Z} \xrightarrow{0} \mathbb{Z} \to 0,$$

where the reader may fill in whether the initial map is 0 or multiplication by 2 depending on the parity of n. Dualizing the multiplication by 2 map, we obtain the same map, and the corresponding cochain complex is

$$0 \to \mathbb{Z} \xrightarrow{0} \mathbb{Z} \xrightarrow{\times 2} \cdots \to \mathbb{Z} \to \mathbb{Z} \to 0,$$

where, again, the last map is either 0 or multiplication by 2 depending on the parity of n. It follows that if n is even, we have

$$H^k(\mathbb{R}P^n; \mathbb{Z}) \cong \begin{cases} \mathbb{Z} & k = 0 \\ \mathbb{Z}_2 & 0 < k \leq n, k \text{ even} \\ 0 & \text{otherwise.} \end{cases}$$

If n is odd, then we have

$$H^k(\mathbb{R}P^n;\mathbb{Z}) \cong \begin{cases} \mathbb{Z} & k = 0, n \\ \mathbb{Z}_2 & 0 < k < n, k \text{ even} \\ 0 & \text{otherwise.} \end{cases}$$

If we take $\mathbb{Z}_2$ coefficients, then all the maps in the complex $C^{\bullet}(\mathbb{R}P^n;\mathbb{Z}_2)$ are 0 and we obtain $H^k(\mathbb{R}P^n;\mathbb{Z}_2) \cong \mathbb{Z}_2, 0 \le k \le n$. Or we could use Theorem 3.1.16:

$$0 \longrightarrow \text{Ext}(H_{k-1}(\mathbb{R}P^n),\mathbb{Z}_2) \longrightarrow H^k(\mathbb{R}P^n;\mathbb{Z}_2) \longrightarrow \text{Hom}(H_k(\mathbb{R}P^n),\mathbb{Z}_2) \longrightarrow 0$$

If k is odd, then we have

$$H^k(\mathbb{R}P^n;\mathbb{Z}_2) \cong \text{Hom}(H_k(\mathbb{R}P^n),\mathbb{Z}_2) \cong \mathbb{Z}_2$$

for odd $k \le n$. If k is even, then

$$H^k(\mathbb{R}P^n;\mathbb{Z}_2) \cong \text{Ext}(H_{k-1}(\mathbb{R}P^n),\mathbb{Z}_2) \cong \mathbb{Z}_2$$

for even $k \le n$. We therefore recover that $H^k(\mathbb{R}P^n;\mathbb{Z}_2) \cong \mathbb{Z}_2$ for $0 \le k \le n$.

Exercises

Exercise 3.1.10. We also have a notion of *reduced cohomology*. Recall that we have the augmentation map $\varepsilon : S_0(X) \rightarrow \mathbb{Z}$, taking a 0-simplex v to $1 \in \mathbb{Z}$. If we dualize the augmented chain complex, we obtain a cochain complex

$$0 \rightarrow \text{Hom}(\mathbb{Z}, G) \xrightarrow{\varepsilon^*} S^0(X;G) \rightarrow S^1(X;G) \rightarrow \cdots$$

The reduced cohomology groups of X are the cohomology groups of this complex. As in the case of homology, $\tilde{H}^n(X;G) = H^n(X;G)$ for $n > 0$. Show that $\tilde{H}^0(X;G) \cong H^0(X;G)/\{\text{constant functions}\}$.

Exercise 3.1.11. Prove that the diagram relating the connecting homomorphisms in relative cohomology with the projections to the Hom-modules (following the proof of Lemma 3.1.21) commutes.

Exercise 3.1.12. Prove Theorem 3.1.23.

Exercise 3.1.13. Prove Theorem 3.1.24 by dualizing the proof of excision in singular homology.

Exercise 3.1.14. Prove that the simplicial and cellular cohomology groups agree with the singular groups.

Exercise 3.1.15. Let $[X, S^1]$ denote the set of pointed homotopy classes of maps $f : X \to S^1$. Since S^1 is an abelian group, we can multiply such functions pointwise to obtain another function: $f \cdot g(x) = f(x)g(x)$.

- Show that $[X, S^1]$ is an abelian group with respect to the operation of pointwise multiplication. To do this, you essentially must prove that if $f \simeq f'$ and $g \simeq g'$ then $f \cdot g \simeq f' \cdot g'$.

- Define a map $\varphi : [X, S^1] \to H^1(X; \mathbb{Z})$ by $\varphi([f]) = f^*(\alpha)$, where α is a generator of $H^1(S^1; \mathbb{Z})$. Prove that this is a group homomorphism.

- Prove that φ is in fact an isomorphism.

Exercise 3.1.16. Compute $H^\bullet(X; \mathbb{Z})$ and $H^\bullet(X; \mathbb{Z}_2)$ for $X = \mathbb{C}P^n, K, M_g$, where K is the Klein bottle and M_g is the orientable surface of genus g.

Exercise 3.1.17. Compute the cohomology of the lens space $L_m(\ell_1, \dots, \ell_n)$ with $\mathbb{Z}$ and $\mathbb{Z}_m$ coefficients.

Exercise 3.1.18. Let U be an open set in the complex plane and let $\varphi : U \to \mathbb{C}$ be an analytic function. Define a singular 1-cochain $c(\varphi)$ by

$$c(\varphi)(\sigma) = \int_\sigma \varphi \, dz,$$

where σ is a singular 1-simplex in U. Prove that $c(\varphi)$ is a cocycle.

Exercise 3.1.19. Define a map $f : \mathbb{R}P^2 \to S^2$ by collapsing the 1-cell to a point. Compute the induced map $f^* : H^\bullet(S^2; R) \to H^\bullet(\mathbb{R}P^2; R)$ for $R = \mathbb{Z}$ and $\mathbb{Z}_2$.

Exercise 3.1.20. Prove that if $f : S^n \to S^n$ has degree d, then $f^* : H^n(S^n; R) \to H^n(S^n; R)$ is multiplication by d for any coefficient ring R.

3.2 Cup products and the cohomology ring

We deliberately did not compute many examples of cohomology groups in the previous section. Indeed, in most cases, we can compute them algebraically via Theorem 3.1.16 using the calculation of the homology groups. The real utility of cohomology is that it comes with a ring structure that is also a topological invariant. For example, the complex projective plane $\mathbb{C}P^2$ and the space $S^2 \vee S^4$ have the same integral homology and cohomology groups. However, their cohomology rings differ, which allows us to deduce that these spaces cannot be homotopy equivalent. We begin by defining a product in cohomology, called the *cup product*.

3.2.1 Definition of the cup product

Let R be a ring (usually $\mathbb{Z}$, $\mathbb{Z}_n$, or $\mathbb{Q}$). Let $\varphi \in S^n(X; R)$ and $\psi \in S^m(X; R)$.

Definition 3.2.1. The *cup product* of φ and ψ is the cochain $\varphi \smile \psi \in S^{n+m}(X; R)$, whose value on a singular simplex $\sigma : \Delta^{n+m} \to X$ is

$$(\varphi \smile \psi)(\sigma) = \varphi(\ \underbrace{\sigma|_{\langle v_0,\dots,v_n\rangle}}_{\text{front } n\text{-face of } \sigma}\ )\psi(\ \underbrace{\sigma|_{\langle v_n,\dots,v_{n+m}\rangle}}_{\text{back } m\text{-face of } \sigma}\ )$$

For this to be useful, it must descend to cohomology.

Lemma 3.2.2. *If $\varphi \in S^n(X; R)$ and $\psi \in S^m(X; R)$, then*

$$\delta(\varphi \smile \psi) = (\delta\varphi) \smile \psi + (-1)^n \varphi \smile (\delta\psi).$$

Proof. Let $\sigma : \Delta^{n+m+1} \to X$ be a singular $(n + m + 1)$-simplex. Then

$$(\delta\varphi \smile \psi)(\sigma) = \sum_{i=0}^{n+1}(-1)^i \varphi(\sigma|_{\langle v_0,\dots,\hat{v}_i,\dots,v_{n+1}\rangle})\psi(\sigma|_{\langle v_{n+1},\dots,v_{n+m+1}\rangle})$$

and

$$(-1)^n(\varphi \smile \delta\psi)(\sigma) = \sum_{i=n}^{n+m+1}(-1)^i \varphi(\sigma|_{\langle v_0,\dots,v_n\rangle})\psi(\sigma|_{\langle v_n,\dots,\hat{v}_i,\dots,v_{n+m+1}\rangle}).$$

Adding these, the last term of the first sum cancels the first term of the second. The remainder is exactly $\delta(\varphi \smile \psi)(\sigma) = (\varphi \smile \psi)(\partial\sigma)$. $\qquad\square$

Corollary 3.2.3. *The cup product of two cocycles is a cocycle. The cup product of a cocycle and a coboundary is a coboundary.*

Proof. If $\delta\varphi = 0 = \delta\psi$, then $\partial(\varphi \smile \psi) = \delta\varphi \smile \psi \pm \varphi \smile \delta\psi = 0$, since the cup product of any cochain with the zero cochain is zero. If $\delta\varphi = 0$ and $\psi = \delta\eta$, then

$$\delta(\varphi \smile \eta) = \delta\varphi \smile \eta \pm \varphi \smile \delta\eta = \pm\varphi \smile \psi$$

(and similarly if δ is a coboundary and ψ is a cocycle). $\qquad\square$

It follows that we have an induced cup product

$$\smile: H^n(X; R) \times H^m(X; R) \to H^{n+m}(X; R),$$

and in fact, since $\smile$ is bilinear, we can replace $\times$ by the tensor product $\otimes_R$. This product is associative and distributive, since it is on the chain level. If R has an identity 1_R, then the class $1 \in H^0(X; R)$ defined by $1(\sigma) = 1_R$ for all $\sigma \in C_0(X; R)$ is the identity for $\smile$. We therefore define the *cohomology ring*

$$H^\bullet(X;R) = \bigoplus_{i \geq 0} H^i(X;R),$$

with the product being $\smile$. This is a *graded ring*. Moreover, we can extend these constructions to relative cohomology

$$H^n(X;R) \times H^m(X,A;R) \xrightarrow{\ \smile\ } H^{n+m}(X,A;R)$$

$$H^n(X,A;R) \times H^m(X;R) \xrightarrow{\ \smile\ } H^{n+m}(X,A;R)$$

$$H^n(X,A;R) \times H^m(X,A;R) \xrightarrow{\ \smile\ } H^{n+m}(X,A;R)$$

Proposition 3.2.4. *If $f : X \to Y$ is continuous, then*

$$f^* : H^\bullet(Y;R) \to H^\bullet(X;R)$$

is a ring homomorphism.

Proof. Let $\varphi \in S^n(Y;R)$, $\psi \in S^m(Y;R)$, and $\sigma : \Delta^{n+m} \to X$. Then

$$
\begin{aligned}
(f^\sharp\varphi \smile f^\sharp\psi)(\sigma) &= f^\sharp\varphi(\sigma|_{\langle v_0,\dots,v_n \rangle}) f^\sharp\psi(\sigma|_{\langle v_n,\dots,v_{n+m} \rangle}) \\
&= \varphi(f\sigma|_{\langle v_0,\dots,v_n \rangle}) \psi(f\sigma|_{\langle v_n,\dots,v_{n+m} \rangle}) \\
&= (\varphi \smile \psi)(f\sigma) \\
&= f^\sharp(\varphi \smile \psi)(\sigma). \qquad \square
\end{aligned}
$$

Theorem 3.2.5. *Suppose R is commutative and consider classes $\alpha \in H^p(X;R)$, $\beta \in H^q(X;R)$. Then*

$$\alpha \smile \beta = (-1)^{pq}\beta \smile \alpha.$$

In particular, if p is odd, then $2(\alpha \smile \alpha) = 0 \in H^{2p}(X;R)$. Thus, if $H^{2p}(X;R)$ has no 2-torsion, then $\alpha \smile \alpha = 0$.

We therefore say that the cup product is *graded commutative.*

Proof. This seemingly simple statement is actually very complicated to prove. We will have to introduce some new ideas, and the casual reader may choose to skip them. However, if one has done the project from the previous chapter on the Nerve Theorem, then one idea from that makes an appearance here in a more algebraic form.

One fundamental issue is this: given an $(p + q)$-simplex σ, the product $\alpha \smile \beta$ is defined using the front p-face and back q-face of σ, while the product $\beta \smile \alpha$ uses the front q-face and the back p-face. So it would seem that we need some mechanism to "shuffle" the faces of a simplex. Of course, this requires some sort of chain homotopy to get a statement about cohomology classes. So, let us get started.

Let π be a permutation of the set $\{0, 1, \ldots, p\}$. We can think of this as giving an affine map $\pi : \Delta^p \to \Delta^p$ sending the vertex v_i to $v_{\pi(i)}$. So given a singular p-simplex in X, we get a new singular p-simplex by precomposing with π. Extending linearly gives a map $\pi : S_p(X) \to S_p(X)$. Now, if $i_0, \ldots, i_q$ are $q+1$ integers between 0 and p, let $a_{i_0,\ldots,i_q} : \Delta^q \to \Delta^p$ be the affine map sending v_j to v_{i_j} for all j. Then if σ is a singular p-simplex in X, the map $\sigma \circ a_{i_0,\ldots,i_q}$ is a singular q-simplex. Note that since the i_j are not required to be distinct, there is a special case: the sequence $0, \ldots, j-1, j, j, j+1, \ldots p$ is called the jth degeneracy operator of degree p. Denote the submodule of $S_q(X)$ generated by all simplices of the form $\sigma \circ a_{i_0,\ldots,i_q}$ by $C(\sigma)_q$. Since the boundary operator takes $C(\sigma)_q$ into $C(\sigma)_{q-1}$, we have a subcomplex $C_\bullet(\sigma)$.

We claim the complex $C_\bullet(\sigma)$ is acyclic, that is, its homology vanishes in degrees $q > 0$. To see this, define an operator $D : C(\sigma)_q \to C(\sigma)_{q+1}$ by

$$D(\sigma \circ a_{i_0,\ldots,i_q}) = \sigma \circ a_{0,i_0,\ldots,i_q}.$$

This is simply σ composed with the join of v_0 and $\langle i_0, \ldots, i_q \rangle$. One checks easily that for $q > 0$ and any $z \in C(\sigma)_q$ we have

$$\partial(Dz) = z - D(\partial z),$$

so that if z is a cycle, $z = \partial(Dz)$.

Now, for any p, let ρ_p be the permutation reversing the order: $\rho_p(i) = p - i$. Define a map on $S_\bullet(X)$ by

$$\rho(z) = \epsilon_p \rho_p(z)$$

for $z \in S_p(X)$, where $\epsilon_p = (-1)^{p(p+1)/2}$. We claim that ρ is a chain map. To see this, let σ be a singular p-simplex and note that

$$\partial\rho(\sigma) = \epsilon_p \partial(\sigma \circ \rho_p) = \epsilon_p \sum_{i=0}^{p} (-1)^{p-i} \sigma \circ \rho_{p,\hat{i}},$$

where $\rho_{p,i}$ is the permutation reversing the set $\{0, \ldots, \hat{i}, \ldots, p\}$. We also have

$$\rho(\partial\sigma) = \epsilon_{p-1} \sum_{i=0}^{p} (-1)^i \sigma \circ \rho_{p,\hat{i}}.$$

But since

$$\epsilon_p(-1)^{p-i} = (-1)^{p(p+1)/2}(-1)^{p-i} = (-1)^{(p^2+3p-2i)/2},$$

while

$$\epsilon_{p-1}(-1)^i = (-1)^{(p-1)p/2}(-1)^i = (-1)^{(p^2-p+2i)/2},$$

we see that these two expressions are equal; thus, $\partial\rho = \rho\partial$.

It follows that ρ induces a map on homology, which we claim is the identity. That is, we claim there is a chain homotopy $T : S_n(X) \to S_{n+1}(X)$ such that $\mathrm{id} - \rho = \partial T + T\partial$. It is possible to write out a map T explicitly, but we defined the complexes $C_*(\sigma)$ for a reason. Note that for any p-simplex σ, both σ and $\rho(\sigma)$ lie in $C(\sigma)_p$; in the language of the previous chapter, this means that C carries both id and ρ. Since each $C_*(\sigma)$ is acyclic, we call this an *acyclic carrier*. Note also that ρ is the identity map on $S_0(X)$; thus id $- \rho$ is the zero map on $S_0(X)$.

We now need the following result. Suppose $\phi : S_*(X) \to S_*(Y)$ is a chain map that is 0 on $S_0(X)$. Then if ϕ has an acyclic carrier, ϕ is chain homotopic to the 0 map. To see this, we construct $T : S_p(X) \to S_{p+1}(Y)$ inductively, with the base case being $T \equiv 0$ on $S_0(X)$. Assume T is defined in dimensions less than p so that $\phi = T\partial + \partial T$ and such that if τ is a simplex of dimension less than p, $T(\tau) \in C(\tau)$. Let σ be a p-simplex and denote its ith face by $\sigma^{(i)}$. Note that $T(\partial\sigma) \subseteq \cup_i C(\sigma^{(i)}) \subseteq C(\sigma)$. Moreover, since C carries ϕ, we also have $\phi(\sigma) \in C(\sigma)$. It follows that

$$\phi(\sigma) - T\partial(\sigma) \in C(\sigma).$$

Note that this chain is a cycle:

$$\partial(\phi(\sigma) - T\partial(\sigma)) = \phi(\partial(\sigma)) - \partial T\partial(\sigma) = (\phi - \partial T)\partial\sigma,$$

and since $\phi = T\partial + \partial T$ in dimension $p - 1$,

$$(\phi - \partial T)\partial\sigma = T(\partial\partial\sigma) = 0.$$

Since $p > 0$ and $C(\sigma)$ is acyclic, there exists a chain $z \in C(\sigma)_{p+1}$ with $\partial z = \phi(\sigma) - T\partial(\sigma)$. We define $T(\sigma) = z$. This completes the inductive step. Thus, taking $\phi = \mathrm{id} - \rho$, we see that ρ induces the identity on $H_*(X)$.

Finally, let us get back to the main problem at hand. The transpose ρ^* of ρ is the map on $S^*(X)$ given by

$$\rho^*(f)(z) = f(\rho(z)).$$

Since ρ^* commutes with the coboundary map, it induces a map in cohomology, which must be the identity (transpose the operator T in the above). Now, the front p-face of a simplex Δ^{p+q} may be realized as a map $\varphi_p : \Delta^p \to \Delta^{p+q}$, and the back q-face is realized as a map $\beta_q : \Delta^q \to \Delta^{p+q}$. One checks easily that

$$\rho_{p+q} \circ \varphi_p = \beta_p \circ \rho_p$$
$$\rho_{p+q} \circ \beta_q = \varphi_q \circ \rho_q$$

Then if we have $c \in S^p(X)$ and $d \in S^q(X)$, we have

$$\rho^*(c \smile d)(\sigma) = (c \smile d)(\rho(\sigma))$$
$$= \epsilon_{p+q}(c \smile d)(\rho_{p+q}(\sigma))$$
$$= \epsilon_{p+q}c(\rho_{p+q}(\sigma \circ \varphi_p))d(\rho_{p+q}(\sigma \circ \beta_q))$$
$$= \epsilon_{p+q}c(\rho_p(\sigma \circ \beta_p))d(\rho_q(\sigma \circ \varphi_q))$$
$$= \epsilon_{p+q}\epsilon_p\epsilon_q\rho^*(c)(\sigma \circ \beta_p)\rho^*(d)(\sigma \circ \varphi_q)$$
$$= \epsilon_{p+q}\epsilon_p\epsilon_q(\rho^*(d) \smile \rho^*(c))(\sigma).$$

At the last step, we have used the fact that R is commutative. Now $\epsilon_{p+q}\epsilon_p\epsilon_q = (-1)^{pq}$, and so we see that

$$\rho^*(c \smile d) = (-1)^{pq}\rho^*(d) \smile \rho^*(c).$$

Since ρ^* induces the identity on cohomology, the cohomology classes α and β of c and d satisfy

$$\alpha \smile \beta = (-1)^{pq}\beta \smile \alpha. \qquad \square$$

Exercises

Exercise 3.2.1. Prove that $\smile$ is associative and distributive at the chain level.

Exercise 3.2.2. Prove that the operator D in the proof of Theorem 3.2.5 satisfies $\partial D = \text{id} - D\partial$.

Exercise 3.2.3. Check that in the proof of Theorem 3.2.5 we have

$$\rho_{p+q} \circ \varphi_p = \beta_p \circ \rho_p$$
$$\rho_{p+q} \circ \beta_q = \varphi_q \circ \rho_q$$

Exercise 3.2.4. Let X be a one-point space. Then as an R-module, $H^0(X; R) = R$ with generator the cohomology class of the cocycle 1. Prove that $H^\bullet(X; R)$ is isomorphic to R as a ring.

3.2.2 Calculations of cohomology rings

We now compute several examples of cohomology rings. We begin with the easiest non-trivial space.

Example 3.2.6. The cohomology groups of S^n are R in degrees 0 and n, and 0 otherwise. If we denote the generator of $H^n(S^n; R)$ by α, then we necessarily have $\alpha \smile \alpha = 0$, since $H^{2n}(S^n; R) = 0$. As this was the only possible nonzero cup product, we see that the cohomology ring $H^\bullet(S^n; R)$ is $R[\alpha]/(\alpha^2)$, where α has degree n.

Exterior algebras

This is as good a time as any to remind the reader about *exterior algebras*. Given a commutative ring R, the kth exterior power of the free module with basis $a_1, \ldots, a_n$ is the R-module with basis all products $a_{i_1} a_{i_2} \cdots a_{i_k}$, with $i_1 < i_2 < \cdots < i_k$, subject to the relations $a_i a_j = -a_j a_i$ and $a_i^2 = 0$ for all i, j. This module is in fact free. We denote this by $\bigwedge_R^k[a_1, \ldots, a_n]$. Note that the rank of this module is $\binom{n}{k}$. The exterior algebra is

$$\bigwedge_R = \bigoplus_{k \geq 0} \bigwedge_R^k[a_1, \ldots, a_n].$$

With this notation, we may then write the following

$$H^\bullet(S^n; \mathbb{Z}) \cong \bigwedge_{\mathbb{Z}}[a], a \in H^n(S^n; \mathbb{Z}).$$

We typically only write this expression for $H^\bullet(S^n; \mathbb{Z})$ when n is odd, preferring the truncated polynomial algebra description for even-dimensional spheres.

Example 3.2.7. Consider the torus T. We know the integral homology groups (Example 2.3.23), and since they are all torsion-free, the cohomology groups are simply the Hom duals of the homology. The generating cycles are shown in Figure 2.9; denote them by α and β. The corresponding dual basis of $H^1(T)$ is then α^* and β^*, where

$$\alpha^*(\alpha) = 1 = \beta^*(\beta)$$

and

$$\alpha^*(\beta) = 0 = \beta^*(\alpha).$$

In Examples 2.3.23 and 2.3.25, we computed the homology of T in three different ways, but none of them really suffice here. We need an explicit 2-cycle representing the generator of $H_2(T)$, and we never constructed one. We have a cellular chain, given by the interior of the square, but that does not align well with the generators we have chosen here. To that end, let us consider the square shown in Figure 3.1.

Consider the chain $z = \langle v_0, v_1, u_1 \rangle - \langle v_0, u_0, u_1 \rangle$. One checks easily that under the quotient map $\pi : I^2 \to T$, $\zeta = \pi_*(z)$ is a cycle on T generating $H_2(T)$. Moreover, α is the class of the loop $\pi_*(\langle u_0, u_1 \rangle)$ and β is the class of $\pi_*(\langle v_0, u_0 \rangle)$. Note that the front 1-face of ζ consists of α from the first 2-simplex and β from the second, with the back 1-face being β from the first 2-simplex and α from the second. We now compute

$$(\alpha^* \smile \beta^*)(\zeta) = \alpha^*(\alpha)\beta^*(\beta) - \alpha^*(\beta)\beta^*(\alpha)$$
$$= 1 \cdot 1 - 0 \cdot 0$$
$$= 1$$

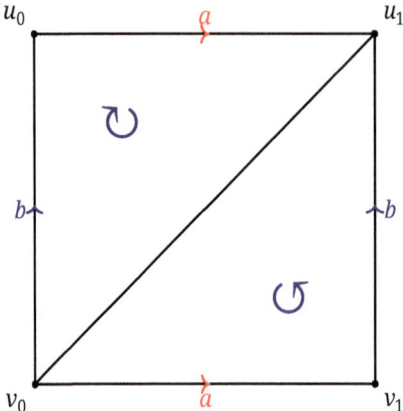

u_0 a u_1

b b

v_0 a v_1

Figure 3.1: A generating 2-cycle on the torus.

It follows that $a^* \smile \beta^*$ generates $H^2(T)$, and is in fact cohomologous to ζ^*, as the above calculation shows. It follows that the cohomology algebra is

$$H^\bullet(T;\mathbb{Z}) = \mathbb{Z}[a^*,\beta^*]/((a^*)^2, (\beta^*)^2, a^*\beta^* + \beta^*a^*).$$

The last relation comes from the anti-commutativity of the cup product. We may also write this as follows:

$$H^\bullet(T;\mathbb{Z}) \cong \bigwedge_{\mathbb{Z}}[\alpha,\beta], \alpha,\beta \in H^1(T;\mathbb{Z}).$$

Example 3.2.8. Let K be the Klein bottle. We will compute $H^\bullet(K;\mathbb{Z}_2)$, leaving the integral case as an exercise. The cohomology *groups* of K are the same as those of the torus, but we need to check the cup product structure. Let's do this very explicitly from the cochain level. Consider the square shown in Figure 3.2. The images of a, b, and c under the projection $I^2 \to K$ are all 1-cycles; similarly, the projection of $U + L$ is a 2-cycle generating $H_2(K;\mathbb{Z}_2)$. If we compute the coboundary of the duals of a, b, and c, we get the following

$$\delta(a^*)(U) = a^*(\partial U) = a^*(a + b + c) = 1$$
$$\delta(a^*)(L) = a^*(\partial L) = a^*(a + b + c) = 1$$
$$\delta(b^*)(U) = b^*(\partial U) = b^*(a + b + c) = 1$$
$$\delta(b^*)(L) = b^*(\partial L) = b^*(a + b + c) = 1$$
$$\delta(c^*)(U) = c^*(\partial U) = c^*(a + b + c) = 1$$
$$\delta(c^*)(L) = c^*(\partial L) = c^*(a + b + c) = 1$$

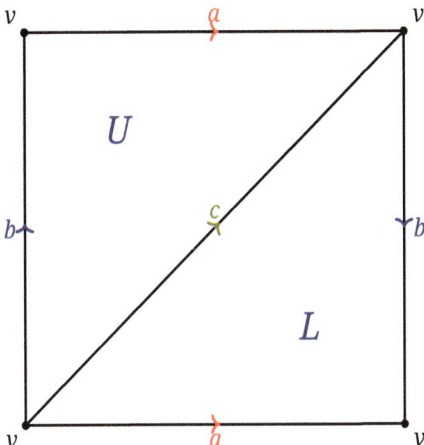

Figure 3.2: A square projecting to the Klein bottle.

It follows that ker δ^1 is generated by $a^* + b^*$ and $b^* + c^*$. The image of δ^1 is spanned by $U^* + L^*$, and the kernel of δ^2 is all of $C^2(K; \mathbb{Z}_2)$. Taking U^*, $U^* + L^*$ as a basis, we have that $H^2(K; \mathbb{Z}_2)$ is generated by the class of $\mu = U^*$ and $H^1(K; \mathbb{Z}_2)$ is generated by the classes of $\alpha = a^* + b^*$ and $\beta = b^* + c^*$. We now compute the cup products

$$(\alpha \smile \alpha)(U) = \alpha(a)\alpha(b) = 1 \cdot 1 = 1$$
$$(\alpha \smile \alpha)(L) = \alpha(c)\alpha(a) = 0 \cdot 1 = 0$$
$$(\beta \smile \beta)(U) = \beta(a)\beta(b) = 0 \cdot 1 = 0$$
$$(\beta \smile \beta)(L) = \beta(c)\beta(a) = 1 \cdot 0 = 0$$
$$(\alpha \smile \beta)(U) = \alpha(a)\beta(b) = 1 \cdot 1 = 1$$
$$(\alpha \smile \beta)(L) = \alpha(c)\beta(a) = 0 \cdot 1 = 0$$

This shows that $\alpha \smile \alpha = \mu$, $\alpha \smile \beta = \mu$ and $\beta \smile \beta = 0$. Thus, we have an isomorphism of rings

$$H^{\bullet}(K; \mathbb{Z}_2) \cong \mathbb{Z}_2[\alpha, \beta]/(\alpha^3, \beta^2, \alpha^2\beta).$$

Note that this is not the same *ring* as $H^{\bullet}(T; \mathbb{Z}_2)$, so that the cup product distinguishes between spaces with the same cohomology groups.

Example 3.2.9. Since $H^1(\mathbb{R}P^2; \mathbb{Z}) \cong \mathrm{Hom}(H_1(\mathbb{R}P^2), \mathbb{Z}) \cong \mathrm{Hom}(\mathbb{Z}_2, \mathbb{Z}) = 0$, there are no interesting cup products in $H^{\bullet}(\mathbb{R}P^2; \mathbb{Z})$. However, with $\mathbb{Z}_2$ coefficients, we have $H^i(\mathbb{R}P^2; \mathbb{Z}_2) \cong \mathbb{Z}_2$ for $0 \le i \le 2$, so there is perhaps a nontrivial product

$$\smile : H^1(\mathbb{R}P^2; \mathbb{Z}_2) \times H^1(\mathbb{R}P^2; \mathbb{Z}_2) \to H^2(\mathbb{R}P^2; \mathbb{Z}_2).$$

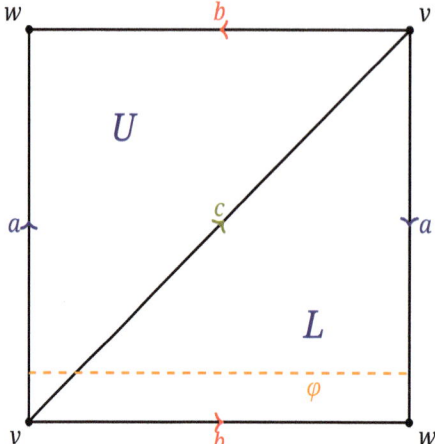

Figure 3.3: A cell decomposition of $\mathbb{R}P^2$.

Consider the cellular decomposition of $\mathbb{R}P^2$ shown in Figure 3.3. This gives a chain complex over $\mathbb{Z}_2$ computing the homology:

$$0 \to \mathbb{Z}_2\{U, L\} \to \mathbb{Z}_2\{a, b, c\} \to \mathbb{Z}_2\{u, v\} \to 0$$

$$U \longmapsto a + b + c$$

$$V \longmapsto a + b + c$$

$$a \longmapsto w + v$$

$$b \longmapsto w + v$$

$$c \longmapsto v + v$$

It follows that $H_2(\mathbb{R}P^2; \mathbb{Z}_2)$ is generated by the cycle $U + L$. Moreover, the kernel Z_1 of $\partial : C_1 \to C_0$ is generated by $a + b$ and c. The module $B_1 = \text{im}\, \partial_2$ is generated by $a + b + c$, so that, using the generating set $a + b$ and $a + b + c$ for Z_1, we see that $H_1(\mathbb{R}P^2; \mathbb{Z}_2)$ is generated by $a + b$. Define a cochain $\varphi : C_1 \to \mathbb{Z}_2$ by setting $\varphi(a) = 1$, $\varphi(b) = 0$ and $\varphi(c) = 1$. Then φ is a cocycle and it is dual to the homology generator $a + b$. Its cohomology class therefore generates $H^1(\mathbb{R}P^2; \mathbb{Z}_2)$. We now compute

$$(\varphi \smile \varphi)(U) = \varphi(b)\varphi(c) = 0$$
$$(\varphi \smile \varphi)(L) = \varphi(a)\varphi(c) = 1$$

We conclude that $\varphi \smile \varphi$ is a cocycle whose cohomology class is dual to the class of $U + L$ and is therefore a generator of $H^2(\mathbb{R}P^2; \mathbb{Z}_2)$. Thus,

$$H^2(\mathbb{RP}^2; \mathbb{Z}_2) \cong \mathbb{Z}_2[a]/(a^3),$$

where a is a generator in degree 1.

To extend this further, we will need to work a bit.

Theorem 3.2.10. *There are ring isomorphisms*

$$H^\bullet(\mathbb{RP}^n; \mathbb{Z}_2) = \mathbb{Z}_2[a]/(a^{n+1}) \quad and \quad H^\bullet(\mathbb{RP}^\infty; \mathbb{Z}_2) = \mathbb{Z}_2[a], \quad \deg a = 1$$

$$H^\bullet(\mathbb{CP}^n; \mathbb{Z}) = \mathbb{Z}[a]/(a^{n+1}) \quad and \quad H^\bullet(\mathbb{CP}^\infty; \mathbb{Z}) = \mathbb{Z}[a], \quad \deg a = 2$$

Proof. We begin with the real case. Note that this actually says that a^k generates $H^k(\mathbb{RP}^n; \mathbb{Z}_2)$, where a is a generator of $H^1(\mathbb{RP}^n; \mathbb{Z}_2)$. This is true trivially for $\mathbb{RP}^1 = S^1$ and direct calculation in Example 3.2.9 shows that it is true for $\mathbb{RP}^2$. Assume inductively that $H^\bullet(\mathbb{RP}^i; \mathbb{Z}_2) \cong \mathbb{Z}_2[a]/(a^{i+1})$ for $i < n$. Note that the inclusion $u : \mathbb{RP}^k \hookrightarrow \mathbb{RP}^n$ induces an isomorphism in cohomology $u^* : H^\ell(\mathbb{RP}^n; \mathbb{Z}_2) \to H^\ell(\mathbb{RP}^k; \mathbb{Z}_2)$ for $\ell \le k$. It therefore suffices to show that if $a_i \in H^i(\mathbb{RP}^n; \mathbb{Z}_2)$ and $a_j \in H^j(\mathbb{RP}^n; \mathbb{Z}_2)$ are generators with $i + j = n$, then $a_i \smile a_j \ne 0$ in $H^n(\mathbb{RP}^n; \mathbb{Z}_2)$.

Recall that $\mathbb{RP}^n = S^n/(\mathbb{Z}_2)$, where $\mathbb{Z}_2$ acts via the antipodal map. Let $S^i = \{(x_0, \ldots, x_i, 0, \ldots, 0) \mid \sum x_j^2 = 1\} \subset S^n$ and $S^j = \{(0, \ldots, 0, x_{n-j}, \ldots, x_n) : \sum x_k^2 = 1\} \subset S^n$. Since $i + j = n$, we have $i = n - j$ so that $S^i \cap S^j = \{(0, \ldots, \pm 1, \ldots, 0)\}$. Note that these spheres are invariant under the $\mathbb{Z}_2$ action, so that passing to quotients we obtain $\mathbb{RP}^i \subset \mathbb{RP}^n$ and $\mathbb{RP}^j \subset \mathbb{RP}^n$, with $\mathbb{RP}^i \cap \mathbb{RP}^j = \{p\} = \{[0 : \cdots : 0 : \underset{i}{\underline{1}} : 0 : \cdots : 0]\}$. Note that since $\mathbb{RP}^n = \mathbb{RP}^{n-1} \cup D^n$, $\mathbb{RP}^n - \{p\}$ deformation retracts to $\mathbb{RP}^{n-1}$. We also have $D^n = D^i \times D^j$, and passing to quotients, we get $D^i = \{[x_0 : \cdots : x_i : 0 : \cdots : 0]\}$ and $D^j = \{[0 : \cdots : 0 : x_{n-j} : \cdots : x_n]\}$, so that, in homogeneous coordinates, $D^n = \{[x_0 : \cdots : x_{i-1} : 1 : x_{i+1} : \cdots : x_n]\}$.

Consider the commutative diagram (all cohomology with $\mathbb{Z}_2$-coefficients):

$$
\begin{array}{ccc}
H^i(\mathbb{RP}^n) \times H^j(\mathbb{RP}^n) & \xrightarrow{\;\smile\;} & H^n(\mathbb{RP}^n) \\
\uparrow & & \uparrow \\
H^i(\mathbb{RP}^n, \mathbb{RP}^n - \mathbb{RP}^j) \times H^j(\mathbb{RP}^n, \mathbb{RP}^n - \mathbb{RP}^i) & \xrightarrow{\;\smile\;} & H^n(\mathbb{RP}^n, \mathbb{RP}^n - \{p\}) \\
\downarrow & & \downarrow \\
H^i(D^n, D^n - D^j) \times H^j(D^n, D^n - D^i) & \xrightarrow{\;\smile\;} & H^n(D^n, D^n - \{0\})
\end{array}
$$

The goal is to show that the top horizontal map is surjective.

We claim that all the vertical maps are isomorphisms. To see this, note that the bottom right map

$$H^n(\mathbb{RP}^n, \mathbb{RP}^n - \{p\}) \to H^n(D^n, D^n - \{0\})$$

is an isomorphism by excision. The top right map comes from the long exact sequence of the pair $(\mathbb{R}P^n, \mathbb{R}P^n - \{p\}) \simeq (\mathbb{R}P^n, \mathbb{R}P^{n-1})$:

$$\cdots \to H^{n-1}(\mathbb{R}P^{n-1}) \overset{\cong}{\twoheadrightarrow} H^{n-1}(\mathbb{R}P^n) \overset{0}{\twoheadrightarrow} H^n(\mathbb{R}P^n, \mathbb{R}P^{n-1}) \overset{\cong}{\twoheadrightarrow} H^n(\mathbb{R}P^n) \to 0$$

(the last group is $H^n(\mathbb{R}P^{n-1}) = 0$). The first map is an isomorphism by the inductive hypothesis, which makes the next map 0, which makes the map in question an isomorphism.

For the left vertical arrows, consider the commutative diagram

$$H^i(\mathbb{R}P^n) \underset{\text{LES}}{\overset{\cong}{\longleftarrow}} H^i(\mathbb{R}P^n, \mathbb{R}P^i) \overset{(*)}{\longleftarrow} H^i(\mathbb{R}P^n, \mathbb{R}P^n - \mathbb{R}P^j) \overset{(***)}{\longrightarrow} H^i(D^n, D^n - D^j)$$

$$\text{ind.}\downarrow\cong \qquad \cong\downarrow \qquad\qquad \cong\downarrow \qquad\qquad\qquad \downarrow(**)$$

$$H^i(\mathbb{R}P^i) \underset{\text{LES}}{\overset{\cong}{\longleftarrow}} H^i(\mathbb{R}P^i, \mathbb{R}P^{i-1}) \overset{\cong}{\longleftarrow} H^i(\mathbb{R}P^i, \mathbb{R}P^i - \{p\}) \underset{\text{excision}}{\overset{\cong}{\longrightarrow}} H^i(D^i, D^i - \{0\})$$

We aim to show that $(* * *)$ is an isomorphism. First note that the bottom arrow in the middle is an isomorphism, since $\mathbb{R}P^i - \{p\}$ deformation retracts to $\mathbb{R}P^{i-1}$. The left vertical arrow is an isomorphism by the inductive hypothesis. This then makes the second vertical arrow an isomorphism. Let us consider $(*)$. We claim that $\mathbb{R}P^n - \mathbb{R}P^j$ deformation retracts to $\mathbb{R}P^{i-1}$. Indeed, if $v \in \mathbb{R}P^n - \mathbb{R}P^j$, then at least one of the first i coordinates of v is nonzero; that is, there is an ℓ, $0 \le \ell \le i - 1$ with $x_\ell \ne 0$. Let $f_t(v) = f_t[x_0 : \cdots : x_n] = [x_0 : \cdots : x_{i-1} : tx_i : \cdots : tx_n]$. Then $f_1(v) = v$ and $f_0(v) \in \mathbb{R}P^{i-1}$. This implies that $(*)$ is an isomorphism, which, in turn, implies that the third vertical map is an isomorphism.

Now, $(**)$ is an isomorphism: we have $D^n = D^i \times D^j$, which implies $D^n - D^j = (D^i - \{0\}) \times D^j$, and therefore

$$H^i(D^n, D^n - D^j) \cong H^i(D^n, (D^i - \{0\}) \times D^j)$$
$$\cong H^i(D^i, D^i - \{0\}),$$

where the last isomorphism follows from the homotopy equivalence

$$(D^i \times D^j, (D^i - \{0\}) \times D^j) \simeq (D^i, D^i - \{0\}).$$

Finally, all this implies that $(* * *)$ is an isomorphism, because every other map in the diagram is. Repeating this with the roles of i and j interchanged, shows that both left vertical arrows in the original diagram are isomorphisms.

It remains to show that the bottom map in the diagram is surjective. We just showed that

$$H^i(D^n, D^n - D^j) \cong H^i(D^i, D^i - \{0\}) \cong H^i(D^i, S^{i-1})$$

and this is generated by the dual of the homology class $[D^i]$. The same is true for $H^j(D^n, D^n - D^i)$, generated by the dual of $[D^j]$. Clearly

$$[D^i]^* \smile [D^j]^* = [D^n]^* \in H^n(D^n, S^{n-1}).$$

Since the bottom map is surjective, so are the other two, and the proof is complete.

For $\mathbb{C}P^n$, the argument is identical, except that H^{2k} replaces H^k everywhere. We obtain the results for $\mathbb{R}P^\infty$ and $\mathbb{C}P^\infty$ by passing to the limit. □

Once we prove Poincaré Duality, we will be able to provide a shorter proof of this result, although that is like using a sledgehammer to drive a nail.

Proposition 3.2.11. *Suppose $\{X_\alpha\}_{\alpha \in A}$ is a collection of spaces. Fix a base point $x_\alpha \in X_\alpha$ for each α and assume x_α has a contractible neighborhood in X_α. Then we have ring isomorphisms*

(1) $H^\bullet(\coprod_\alpha X_\alpha; R) \cong \prod_\alpha H^\bullet(X_\alpha; R)$
(2) $\tilde{H}^\bullet(\bigvee_\alpha X_\alpha; R) \cong \prod_\alpha \tilde{H}^\bullet(X_\alpha; R)$

Proof. The first statement follows from the fact that if M_α is a collection of R-modules, then

$$\text{Hom}\left(\bigoplus_\alpha M_\alpha, R\right) \cong \prod_\alpha \text{Hom}(M_\alpha, R).$$

For the second statement, it suffices to consider the case of two spaces X and Y. The Mayer–Vietoris sequence for the space $X \vee Y$, covered by the union of $X \vee U_y$ and $Y \vee U_x$, where U_x and U_y are the contractible neighborhoods of the respective base points, yields an isomorphism

$$H^n(X \vee Y) \to H^n(X) \times H^n(Y)$$

for $n > 0$, and in degree 0 an exact sequence

$$0 \to H^0(X \vee Y) \to H^0(X) \times H^0(Y) \to H^0(U_x \vee U_y) \to 0.$$

The kernel of the latter map is all pairs of functions (f, g) with $f(x) = g(y)$. But with reduced cohomology this latter group is trivial, since $U_x \vee U_y$ is contractible. □

Example 3.2.12. Consider the spaces $\mathbb{C}P^2$ and $S^2 \vee S^4$. The integral homology groups of these spaces are all $\mathbb{Z}$ in degrees 0, 2, 4 and 0 otherwise. The same is therefore true in cohomology. What about the ring structure? We know that

$$H^\bullet(\mathbb{C}P^2; \mathbb{Z}) \cong \mathbb{Z}[\alpha]/(\alpha^3), \quad \deg \alpha = 2.$$

For the space $S^2 \vee S^4$, we have

$$\tilde{H}^\bullet(S^2 \vee S^4; \mathbb{Z}) \cong \tilde{H}^\bullet(S^2) \times \tilde{H}^\bullet(S^4) \cong \mathbb{Z}[x,y]/(x^2, y^2, xy).$$

These rings are not isomorphic, so the spaces $\mathbb{C}P^2$ and $S^2 \vee S^4$ are not homotopy equivalent.

The Künneth formula

We will prove a more general version of this in Section 3.4 below, but for now we consider the following situation. Suppose X and Y are spaces. Let $p_1 : X \times Y \to X$ and $p_2 : X \times Y \to Y$ be the projection maps. We then get an induced map, called the *cross product*,

$$H^\bullet(X; R) \times H^\bullet(Y; R) \longrightarrow H^\bullet(X \times Y; R),$$

$$\alpha \times \beta \longmapsto p_1^*(\alpha) \smile p_2^*(\beta).$$

Note that this map is bilinear, so we may replace $\times$ by $\otimes_R$. This map is a ring homomorphism if we set

$$(a \otimes b)(c \otimes d) = (-1)^{|b||c|}(ac \otimes bd).$$

Theorem 3.2.13 (Künneth Formula). *Suppose X and Y are cell complexes with $H^k(Y; R)$ a finitely generated free R-module for all k (e. g. R a field and Y finite). Then the cross product*

$$H^\bullet(X; R) \otimes_R H^\bullet(Y; R) \to H^\bullet(X \times Y; R)$$

is an isomorphism of rings.

So, for example, if $n_1, \ldots, n_\ell$ are odd integers, we have

$$H^\bullet(S^{n_1} \times \cdots S^{n_\ell}; \mathbb{Z}) \cong \bigwedge\nolimits_{\mathbb{Z}}[\alpha_1, \ldots, \alpha_n], \deg \alpha_i = n_i.$$

But since degrees of generators matter, we have, for example,

$$H^\bullet(S^2 \times S^3; \mathbb{Z}) \cong \mathbb{Z}[\alpha]/(\alpha^2) \otimes \bigwedge\nolimits_{\mathbb{Z}}[\beta], \deg \alpha = 2, \deg \beta = 3.$$

Note that in this ring

$$\alpha \smile \beta = (-1)^{2 \cdot 3}\beta \smile \alpha = \beta \smile \alpha \in H^5(S^2 \times S^3; \mathbb{Z}).$$

We also have a ring isomorphism

$$H^\bullet(T^n; \mathbb{Z}) = H^\bullet((S^1)^n; \mathbb{Z}) \cong \bigwedge\nolimits_{\mathbb{Z}}[\alpha_1, \ldots, \alpha_n], \deg \alpha_i = 1.$$

Finally, note that we also have

$$H^\bullet(\mathbb{R}P^n \times \mathbb{R}P^m; \mathbb{Z}_2) \cong \mathbb{Z}_2[\alpha, \beta]/(\alpha^{n+1}, \beta^{m+1}).$$

An application in algebra

A *division algebra structure* on $\mathbb{R}^n$ is a product $\mathbb{R}^n \times \mathbb{R}^n \to \mathbb{R}^n$ such that $a(b+c) = ab+ac$, and if $a \neq 0$, there is an $x \in \mathbb{R}^n$ with $ax = 1$, where 1 is a distinguished element satisfying $1 \cdot x = x \cdot 1 = x$ for all $x \in \mathbb{R}^n$. Note that such a product need not be commutative or even associative.

The most familiar examples are the cases $n = 1$, which is just $\mathbb{R}$ with the usual multiplication, and $n = 2$, which is $\mathbb{C}$ with complex multiplication. A nonexample is given by the cross product in $\mathbb{R}^3$. There is no unit element for the cross product, so inverses are not possible.

The obvious question is: for which n is there a division algebra structure on $\mathbb{R}^n$? This is a purely algebraic question, but it turns out that topology puts some limits on things.

Theorem 3.2.14. *If $\mathbb{R}^n$ has the structure of a division algebra over $\mathbb{R}$, then n must be a power of 2.*

Proof. The maps $x \mapsto ax$ and $x \mapsto xa$ are linear isomorphisms for any $a \neq 0$. It follows that a multiplication map $\mathbb{R}^n \times \mathbb{R}^n \to \mathbb{R}^n$ induces a map

$$h : \mathbb{R}P^{n-1} \times \mathbb{R}P^{n-1} \to \mathbb{R}P^{n-1},$$

which is a homeomorphism when restricted to each subspace $\mathbb{R}P^{n-1} \times \{y\}$ or $\{x\} \times \mathbb{R}P^{n-1}$. The map h is continuous since it is the quotient of a bilinear map.

Consider the induced map

$$H^\bullet(\mathbb{R}P^{n-1}; \mathbb{Z}_2) \xrightarrow{h^*} H^\bullet(\mathbb{R}P^{n-1} \times \mathbb{R}P^{n-1}; \mathbb{Z}_2)$$

$$\mathbb{Z}_2[\alpha]/(\alpha^n) \xrightarrow{h^*} \mathbb{Z}_2[\alpha_1, \alpha_2]/(\alpha_1^n, \alpha_2^n)$$

Note that $h^*(\alpha) = k_1\alpha_1 + k_2\alpha_2$ for some $k_j \in \mathbb{Z}_2$. Observe that the inclusion $\mathbb{R}P^{n-1} \hookrightarrow \mathbb{R}P^{n-1} \times \mathbb{R}P^{n-1}$ into the first factor sends α_1 to α and α_2 to 0 in cohomology, and since h restricts to a homeomorphism on the first factor, $k_1 \neq 0$. Similarly, $k_2 \neq 0$. Thus, $h^*(\alpha) = \alpha_1 + \alpha_2$. Since $\alpha^n = 0$, we must then have $(\alpha_1 + \alpha_2)^n = 0$. But then

$$0 = (\alpha_1 + \alpha_2)^n = \sum_{k=0}^{n} \binom{n}{k} \alpha_1^k \alpha_2^{n-k}$$

so that

$$\binom{n}{k} \equiv 0 \mod 2, \quad 0 < k < n.$$

This can happen only if $n = 2^d$ for some d. $\qquad\qquad\qquad\qquad\qquad\qquad$ □

Note that the same proof works for $\mathbb{C}$ instead of $\mathbb{R}$, so that if $\mathbb{C}^n$ is a division algebra over $\mathbb{C}$, then $\binom{n}{k} = 0$ for $0 < k < n$, which implies that $n = 1$. In particular, there are no finite field extensions of $\mathbb{C}$; that is, $\mathbb{C}$ is algebraically closed.

So, what are these division algebras, if they exist? We know the cases $d = 0$ ($\mathbb{R}$) and $d = 1$ ($\mathbb{C}$). When $d = 2$, we have the quaternions, $\mathbb{H}$, which is $\mathbb{R}^4$ with basis $1, i, j, k$ with the quaternionic multiplication: $ij = k, jk = i, ki = j$. In fact, we have quaternionic projective space $\mathbb{H}P^n$, consisting of "quaternionic lines" in $\mathbb{H}^{n+1} = \mathbb{R}^{4n+4}$. The space $\mathbb{H}P^n$ is a quotient of $S^{4n+3} \subset \mathbb{R}^{4n+4}$ with fiber S^3 and we can inductively build $\mathbb{H}P^n$ from $\mathbb{H}P^{n-1}$ by attaching a $4n$-cell via the quotient map. Note that $\mathbb{H}P^1 = S^4$ and we have the quotient $S^7 \to S^4$ called the *Hopf map* (compare with the quotient $S^3 \to S^2 = \mathbb{C}P^1$).

When $d = 3$, we have the *octonions*, or Cayley numbers, denoted by $\mathbb{O}$. This is $\mathbb{R}^8$ with a nonassociative product. Denote the basis by $\{e_0, e_1, \ldots, e_7\}$. Then the product structure is defined by

$$e_i e_j = -\delta_{ij} e_0 + \varepsilon_{ijk} e_k, \varepsilon_{ijk} = 1 \quad \text{for } ijk = 123, 145, 176, 246, 257, 347, 365$$
$$e_0 e_i = e_i e_0 = e_i$$

along with the requirement that the product is antisymmetric. This is a strange object, but it does work. We then get the corresponding projective spaces $\mathbb{O}P^n$, a quotient of $S^{8n+7} \subset \mathbb{R}^{8n+8} = \mathbb{O}^{n+1}$. Note that $\mathbb{O}P^1 = S^8$ and there is a Hopf map $S^{15} \to S^8$.

For $d > 3$, there are no such structures. We cannot provide a full proof of this fact here, as it involves some concepts that are outside our scope, but we can say a few things. Suppose $f : S^{2n-1} \to S^n$ is a continuous map. Choose generators $\zeta \in H^{2n-1}(S^{2n-1})$ and $\eta \in H^n(S^n)$ and let y be a cocycle representing η. Since $\eta \smile \eta = 0$, the chain $y \smile y$ is a coboundary; say $y \smile y = \delta u$ in singular cohomology. Also, since $f^*(\eta) = 0 \in H^n(S^{2n-1})$ (assume $n \geq 2$ here), there is an $(n-1)$-cochain x on S^{2n-1} such that $f^\sharp(y) = \delta x$. Note that $x \smile f^\sharp(y)$ and $f^\sharp(u)$ are both $(2n-1)$-cochains on S^{2n-1}. Now

$$\delta(x \smile f^\sharp(y) - f^\sharp(u)) = \delta(x \smile \delta x) - f^\sharp(y \smile y)$$
$$= \delta x \smile \delta x - f^\sharp(y) \smile f^\sharp(y)$$
$$= 0.$$

So, the cohomology class of this cocycle is a multiple $m\zeta$ of the generator of $H^{2n-1}(S^{2n-1})$. This integer m is independent of the choices made and is called the *Hopf invariant* of f.

We have not yet defined the higher homotopy groups (we will do so in Chapter 4), but for now, let us simply remark that for $k \geq 2$ there are abelian groups $\pi_k(X)$, defined for any topological space X, built from homotopy classes of maps $S^k \to X$. These are homotopy invariants of X. The Hopf invariant is in fact a homotopy invariant (i. e., if $f \simeq g$, then f and g have the same Hopf invariants), so we have a well-defined homomorphism

$$\gamma : \pi_{2n-1}(S^n) \to \mathbb{Z}, \quad [f] \mapsto m(f).$$

Here are some facts about γ:
(1) If n is odd, then $\gamma \equiv 0$.
(2) If n is even, then $2 \in \text{im } \gamma$.
(3) If $n = 2, 4, 8$ and f is the Hopf map $S^{2n-1} \to S^n$, then $m(f) = 1$.

The following result is due to J.F. Adams (Ann. Math. **72** (1960), 20–104).

Theorem 3.2.15. *Up to homotopy, the only maps of Hopf invariant 1 are the three maps above. Hence, the only real division algebras occur in dimensions 1, 2, 4, 8.*

Exercises

Exercise 3.2.5. Here is another calculation of the cohomology ring $H^\bullet(T)$. Consider the cellular decomposition of T shown in Figure 3.4. We use the homology classes of the loops a and b as the generators of $H_1(T)$, as before, and let α and β denote the dual basis of $H^1(T)$. In Example 3.2.7, we did not construct explicit cocycles representing these cohomology classes. Define $\varphi : C_1(T) \to \mathbb{Z}$ by setting $\varphi(e) = 1$ if and only if the vertical arc labeled with φ crosses the edge e, and $\varphi(e) = 0$ otherwise. Similarly, define $\psi : C_1(T) \to \mathbb{Z}$ by $\psi(e) = 1$ if and only if the horizontal arc labeled with ψ crosses the edge e, and $\psi(e) = 0$ otherwise. Prove that φ and ψ are cocycles representing α and β, re-

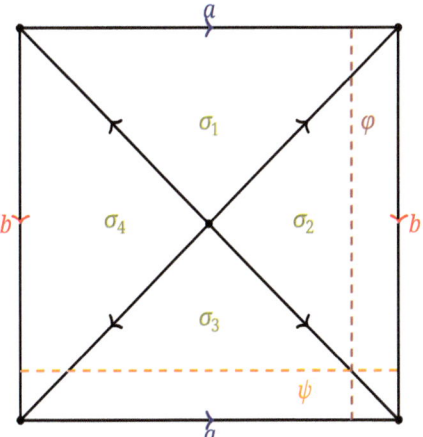

Figure 3.4: A cellular decomposition of the torus.

spectively. Moreover, show that $c = \sigma_4 + \sigma_3 - \sigma_2 - \sigma_1$ (the σ_i oriented as shown) generates $H_2(T)$, and that $\varphi \smile \psi$ represents the dual generator c^* of $H^2(T)$.

Exercise 3.2.6. Compute the cohomology ring $H^\bullet(K; \mathbb{Z})$, where K is Klein bottle.

Exercise 3.2.7. Compute the cohomology ring $H^\bullet(M_g; \mathbb{Z})$, where M_g is the closed orientable surface of genus g.

Exercise 3.2.8. Repeat the previous exercise as follows: Given M_g, pinch homologically trivial embedded circles to points to obtain a wedge of tori. Use naturality to deduce the ring structure.

Exercise 3.2.9. Suppose $f : S^{2n-1} \to S^n$ is continuous and forms the mapping cone Cf. Denote the inclusion $S^n \to Cf$ by i. Show that $H^k(Cf; \mathbb{Z}) \cong \mathbb{Z}$ for $k = 0, n, 2n$ and 0 otherwise.

Exercise 3.2.10. Compute the cohomology ring $H^\bullet(\mathbb{H}^n; \mathbb{Z})$, where $\mathbb{H}^n$ is the quaternionic projective space.

Exercise 3.2.11. Let $f : \mathbb{C}^{n+1} \to \mathbb{C}^{n+1}$ be the map raising each coordinate to the dth power, $d > 0$, and let $\bar{f} : \mathbb{C}P^n \to \mathbb{C}P^n$ be the induced map on projective spaces. Compute the induced map $\bar{f}^* : H^\bullet(\mathbb{C}P^n; \mathbb{Z}) \to H^\bullet(\mathbb{C}P^n; \mathbb{Z})$.

Exercise 3.2.12. Prove that $\mathbb{R}P^3$ is not homotopy equivalent to $\mathbb{R}P^2 \vee S^3$.

Exercise 3.2.13. Prove that the rings mentioned in Example 3.2.12 are not isomorphic.

3.3 Cohomology of manifolds and Poincaré duality

Manifolds have a particularly nice structure that influences their cohomology. The goal of this section is to prove various duality results relating the homology and cohomology of compact manifolds. These results have some powerful consequences; as one example, if M is an odd dimensional manifold, then $\chi(M) = 0$. The approach we will take here is somewhat formal, but in the project at the end of the chapter, we will explore the geometry behind duality in some detail.

3.3.1 Cap products and the duality theorem

Let X be any topological space and let R be a commutative ring with 1.

Definition 3.3.1. The *cap product* is a bilinear operation

$$\frown: S_k(X; R) \otimes_R S^\ell(X; R) \to S_{k-\ell}(X; R),$$

defined for $k \geq \ell$ by setting

$$\sigma \frown \varphi = \varphi(\sigma|_{\langle v_0, \ldots, v_\ell \rangle}) \sigma|_{\langle v_\ell, \ldots, v_k \rangle}$$

for $\sigma : \Delta^k \to X$, $\varphi \in S^\ell(X; R)$.

That is, the cap product of a k-simplex and an ℓ-cochain is the back $k - \ell$ face of σ, scaled by the value of φ on the front ℓ face of σ.

Lemma 3.3.2.

$$\partial(\sigma \frown \varphi) = (-1)^\ell (\partial\sigma \frown \varphi - \sigma \frown \delta\varphi).$$

Proof. This is just a direct computation:

$$\partial\sigma \frown \varphi = \sum_{i=0}^{\ell} (-1)^i \varphi(\sigma|_{\langle v_0, \ldots, \hat{v}_i, \ldots, v_{\ell+1} \rangle}) \sigma|_{\langle v_{\ell+1}, \ldots, v_k \rangle}$$

$$+ \sum_{i=\ell+1}^{k} (-1)^i \varphi(\sigma|_{\langle v_0, \ldots, v_\ell \rangle}) \sigma|_{\langle v_{\ell+1}, \ldots, v_k \rangle}$$

$$\sigma \frown \delta\varphi = \sum_{i=0}^{\ell+1} (-1)^i \varphi(\sigma|_{\langle v_0, \ldots, \hat{v}_i, \ldots, v_{\ell+1} \rangle}) \sigma|_{\langle v_{\ell+1}, \ldots, v_k \rangle}$$

while

$$\partial(\sigma \frown \varphi) = \sum_{i=\ell}^{k} (-1)^{i-\ell} \varphi(\sigma|_{\langle v_0, \ldots, v_\ell \rangle}) \sigma|_{\langle v_\ell, \ldots, \hat{v}_i, \ldots, v_k \rangle}$$

By inspection, we see that the result holds. □

As a consequence of Lemma 3.3.2, we see that the cap product of a cycle and a co-cycle is a cycle, and that the cap product of a cycle and a coboundary is a boundary. We therefore get an induced cap product in homology

$$\frown : H_k(X; R) \otimes_R H^\ell(X; R) \to H_{k-\ell}(X; R).$$

There are also relative versions, which we leave to the reader to formulate. Moreover, if $f : X \to Y$ is continuous, then we have

$$f_*(\alpha) \frown \varphi = f_*(\alpha \frown f^*(\varphi)).$$

Theorem 3.3.3 (Poincaré Duality). *If M is a closed R-orientable n-manifold with fundamental class $[M] \in H_n(M; R)$, then the map*

$$D : H^k(M; R) \longrightarrow H_{n-k}(M; R)$$

$$\alpha \longmapsto [M] \frown \alpha$$

is an isomorphism for all k.

Before proving this result, let us take a look at a pair of examples.

Example 3.3.4. Consider the cell decomposition of the torus shown in Figure 3.4. In one of the exercises in the previous section, you used this to compute the cup product structure on the integral cohomology. The classes of the edges a and b generate $H_1(T)$ and their Hom-duals are the classes of φ and ψ; these generate $H^1(T)$. A fundamental class for T is $[T] = [\sigma_4 + \sigma_3 - \sigma_2 - \sigma_1]$. Denote the intersection $\sigma_i \cap \sigma_j$ by e_{ij}. We then have the following

$$\partial\sigma_1 = a - e_{12} + e_{14}$$
$$\partial\sigma_2 = b - e_{23} + e_{12}$$
$$\partial\sigma_3 = a - e_{23} + e_{34}$$
$$\partial\sigma_4 = b - e_{34} + e_{14}$$

With this in hand, let us compute some cap products.

$$[T] \frown [\varphi] = \varphi(b)e_{14} + \varphi(a)e_{34} - \varphi(b)e_{12} - \varphi(a)e_{14}$$
$$= 0 + e_{34} - 0 - e_{14}$$
$$= b - \partial\sigma_4$$

Thus, $[T] \frown [\varphi] = [b]$. Note that $[\varphi]$ is the Hom-dual of $[a]$. Similarly, one checks that $[T] \frown [\psi] = -[a]$. So in this case, the map $D : H^1(T; \mathbb{Z}) \to H_1(T; \mathbb{Z})$ is the map $\mathbb{Z}^2 \to \mathbb{Z}^2$ with matrix

$$\begin{bmatrix} 0 & -1 \\ 1 & 0 \end{bmatrix}$$

This map is an isomorphism (but not the identity).

Example 3.3.5. Consider the cell decomposition of $X = \mathbb{R}P^2$ in Figure 3.3. We saw that the group $H_1(\mathbb{R}P^2; \mathbb{Z}_2)$ is generated by $[a + b]$, whose dual cocycle is φ shown in the diagram. A fundamental class is $[X] = [U + L]$. The cap product is then

$$[X] \frown [\varphi] = \varphi(b)c + \varphi(a)c$$
$$= c$$

It follows that $[X] \frown [\varphi] = [c] = [a + b]$, so that $D : H^1(\mathbb{R}P^2; \mathbb{Z}_2) \to H_1(\mathbb{R}P^2; \mathbb{Z}_2)$ is an isomorphism.

An algebraic detour

By a *directed set* we mean a partially ordered set I such that for any $i, i' \in I$, there is an $i'' \in I$ with $i \le i''$ and $i' \le i''$.

Example 3.3.6. Here are some examples of directed sets.

(1) $I = \mathbb{N}$ with the usual order.

(2) $I = \mathbb{N}^k$ with $(a_1, \dots, a_k) \le (b_1, \dots, b_k)$ if and only if $a_i \le b_i, 1 \le i \le k$.

(3) Let X be a set and let K be a subset. Let $I = \{A \subseteq X \mid K \subset A\}$. Define $A \le A'$ if $A \supseteq A'$. Then this is a directed set: given A, A', take $A'' = A \cap A'$.

(4) Let (X, x_0) be a pointed space and let I be the set of all covering spaces $p : (E, e_0) \to (X, x_0)$. Define $(E, e_0; p) \le (E', e_0'; p')$ if there exists an $f : (E', e_0') \to (E, e_0)$ with $pf = p'$. This set is directed: given E, E', define

$$E'' = \{(e, e') \in E \times E' \mid p(e) = p'(e')\}, \quad p''(e, e') = p(e) = p(e').$$

Definition 3.3.7. Suppose I is a directed set and $\{M_i\}_{i \in I}$ is a family of R-modules such that for any $i \le i'$, there is an R-module homomorphism $\varphi_{i',i} : M_i \to M_{i'}$ such that for $i \le i' \le i''$, $\varphi_{i'',i'} \circ \varphi_{i',i} = \varphi_{i'',i}$ and $\varphi_{i,i} = \mathrm{id}_{M_i}$. We call this a *direct system of R-modules*. A *direct limit* of this system is a module M with a family of homomorphisms $\varphi_i : M_i \to M$ such that $\varphi_{i'} \circ \varphi_{i',i} = \varphi_i$ for $i \le i'$, which satisfies the following universal mapping property. For any module N and family of maps $\psi_i : M_i \to N$ satisfying $\psi_{i'} \circ \varphi_{i',i} = \psi_i$ for $i \le i'$, there is a unique homomorphism $\psi : M \to N$ such that $\psi_i = \psi \circ \varphi_i$ for all i. We write $M = \varinjlim M_i$.

Another way to state this definition is that, viewing I as a category with a single morphism for each $i \le i'$, a direct system of modules is simply a functor $I \to R - \mathrm{mod}$. Because of the universal mapping property, the direct limit is unique if it exists.

Proposition 3.3.8. *The direct limit of a system of R-modules exists.*

Proof. We need to construct an R-module that admits maps from each M_i. There is an obvious candidate, namely the module $M^+ = \bigoplus_{i \in I} M_i$, with $\varphi_i^+ : M_i \to M$ the canonical inclusion. Let N be the submodule of M^+, generated by all

$$\varphi_{i'}^+(\varphi_{i',i}(x_i)) - \varphi_i^+(x_i), \quad i \le i', \quad x_i \in M_i.$$

In other words, we identify elements of the direct sum that map to a common element further down the line. Set $M = M^+/N$, let $\pi : M^+ \to M$ be the quotient map, and set $\varphi_i = \pi \circ \varphi_i^+$. Then it is straightforward to check that $(M, \{\varphi_i\})$ is the direct limit. ☐

Proposition 3.3.9. *Suppose I has a largest element m. Then the map $\varphi_m : M_m \to \varinjlim M_i$ is an isomorphism.*

Proof. For all i, we have $\varphi_{m,i} : M_i \to M_m$ compatible with $\psi_i : M_i \to M$. By the universal mapping property, we have a unique map $\psi : M = \varinjlim M_i \to M_m$ and $\psi_i = \psi \circ \varphi_i$. Thus, id $= \varphi_{m,m} = \psi \circ \varphi_m$, which implies that φ_m is injective and ψ is surjective. But also, φ_m is surjective and ψ is injective. ☐

Lemma 3.3.10. *Suppose that for each $i \in I$, $M_i = N_i \oplus P_i$ and that for $i \leq i'$, the map $\varphi_{i',i}$ decomposes accordingly as $\varphi_{i',i} = \psi_{i',i} \oplus \rho_{i',i}$. Let $N = \varinjlim N_i$ and $P = \varinjlim P_i$, so that we have induced maps $\psi : N \to M$ and $\rho : P \to M$ with $\psi \circ \psi_i = \varphi_i|_{N_i}$ and $\rho \circ \rho_i = \varphi_i|_{P_i}$. Then $\psi \oplus \rho : N \oplus P \to M$ is an isomorphism.*

Proof. Construct the inverse as follows. Given $x \in M$, choose $x_i \in M_i$ with $x = \varphi_i(x_i)$. Write $x_i = y_i + z_i$ uniquely with $y_i \in N_i$, $z_i \in P_i$. Define $\theta(x) = (\psi_i(y_i), \rho_i(z_i)) \in N \oplus P$. It is straightforward to check that $\theta(x)$ is independent of the choice of x_i and that θ is inverse to $\psi \oplus \rho$. ☐

Definition 3.3.11. A set $J \subset I$ is *final* if J is a directed set under the induced order and for any $i \in I$ there is a $j \in J$ with $i \leq j$.

If we form the limit over a final subset $J \subset I$ of a directed system $\{M_i\}_{i \in J}$, we have a canonical map

$$\lambda : \varinjlim_J M_j \to \varinjlim_I M_i.$$

Lemma 3.3.12. *The map λ is an isomorphism.*

We leave the proof of this as an exercise. The surjectivity of λ is clear, since if $x \in M = \varinjlim M_i$, we have $x = \varphi_i(x_i)$ for some $x_i \in M_i$. Then since J is final, there is a $j \geq i$ and $x = \varphi_j(\varphi_{j,i}(x_i))$. The injectivity is more difficult, but it is still relatively straightforward.

Suppose we have three direct systems on I with homomorphisms between them such that the following diagram commutes for each $i \leq i'$:

$$
\begin{array}{ccccc}
M_i' & \xrightarrow{\lambda_i} & M_i & \xrightarrow{\rho_i} & M_i'' \\
\downarrow{\varphi_{i',i}'} & & \downarrow{\varphi_{i',i}} & & \downarrow{\varphi_{i',i}''} \\
M_{i'}' & \xrightarrow{\lambda_{i'}} & M_{i'} & \xrightarrow{\rho_{i'}} & M_{i'}''
\end{array}
$$

Passing to the limit we obtain a sequence

$$M' \xrightarrow{\lambda} M \xrightarrow{\rho} M''$$

with $\lambda \varphi_i' = \varphi_i \lambda_i$, and $\rho \varphi_i = \varphi_i'' \rho_i$ for all $i \in I$.

Lemma 3.3.13. *If the rows are exact for all i, then the limit sequence is exact.*

Again, we leave the proof of this as an exercise. Note that, in particular, if each ρ_i is injective, then so is ρ, and if each λ_i is surjective, then so is λ.

Here is an important application.

Proposition 3.3.14. *Let x be any point in a Hausdorff space X. Then*

$$\varphi : \lim_{\substack{\longrightarrow \\ x \in U}} H_n(X, X - U) \to H_n(X, X - \{x\})$$

is an isomorphism, where the limit is taken over all open U containing x.

Proof. Since we have a map $H_n(X, X - U) \to H_n(X, X - \{x\})$ for any such U, we have the map φ. Represent an n-cycle in $(X, X - \{x\})$ by a finite sum of singular simplices. The image of this is compact and so lies in some $(X, X - U)$; this implies that φ is surjective. For injectivity, if a cycle in some $(X, X-U)$ is a boundary in $(X, X-\{x\})$, then compactness forces it to be a boundary in some $(X, X - V)$ with $U \subseteq V$. This implies that it is zero in the direct limit. □

The proof of Poincaré duality

We will actually prove a stronger statement. Suppose X is a space, not necessarily compact. The compact subspaces of X form a directed system under inclusion. The modules $H^k(X, X - K)$ then form a directed system since $K \subset K'$ implies that $X - K' \subset X - K$ and therefore we have a map $H^k(X, X - K) \to H^k(X, X - K')$.

Definition 3.3.15. The *kth cohomology with compact supports* is the direct limit

$$H_c^k(X) = \lim_{\substack{\longrightarrow \\ K}} H^k(X, X - K).$$

Note that if X is compact, then $H_c^k(X) = H^k(X)$.

The geometric interpretation here is that a cohomology class in $H_c^k(X)$ is represented by a cochain that vanishes off of some compact subset K; that is, it annihilates all chains with support in $X - K$.

Cohomology with compact supports has two versions of functoriality. Suppose $f : X \to Y$ is continuous. Then for any compact $K \subset X$, $f(K)$ is compact in Y. However, f might not map $X - K$ into $Y - f(K)$ and so we might not get an induced map on H_c^k. If we assume that f is *proper* ($L \subset Y$ compact implies that $f^{-1}(L)$ is compact in X), then f maps $X - f^{-1}(L)$ into $Y - L$. We therefore have induced maps

$$H^k(Y, Y - L) \to H^k(X, X - f^{-1}(L)) \to H_c^k(X),$$

and we then obtain a map on the direct limits

$$f_c^* : H_c^k(Y) \to H_c^k(X).$$

So cohomology with compact supports is contravariant with respect to proper maps.

There is another way to induce a map, however. Suppose $U \subset X$ is open. If $K \subset U$ is compact, we have the inverse of the excision isomorphism $H^k(U, U-K) \to H^k(X, X-K)$

and this is compatible with the inclusion homomorphisms. Passing to the limit gives a unique map making the following diagram commute for all K:

$$
\begin{array}{ccc}
H_c^k(U) & \longrightarrow & H_c^k(X) \\
\uparrow & & \uparrow \\
H^k(U, U - K) & \longrightarrow & H^k(X, X - K)
\end{array}
$$

So cohomology with compact supports is covariant with respect to inclusions of open sets.

Example 3.3.16. Let $U = \mathbb{R}^n$ considered as the open subset $S^n - \{x\} \subset S^n$. The sets $S^n - K$ form a system of neighborhoods of x which contains a final system of contractible neighborhoods. For these, the map $H^k(S^n, S^n - K) \to \tilde{H}^k(S^n)$ is an isomorphism. It follows that

$$
H_c^k(\mathbb{R}^n) \cong \tilde{H}^k(S^n), \quad k \geq 0.
$$

Now, if X is an n-dimensional manifold, fix an R-orientation of X. For each compact set $K \subset X$, restricting the R-orientation (regarded as a section of the R-orientation sheaf) to K gives a fundamental class $[X]_K \in H_n(X, X - K)$ via the isomorphism of this group with the group of sections ΓK. Note that if X happens to be compact with fundamental class $[X] \in H_n(X; R)$, then $[X]_K$ is the image of $[X]$ under the map $H_n(X) \to H_n(X, X - K)$. We have the relative cap product

$$
[X]_K \frown - : H^k(X, X - K) \to H_{n-k}(X).
$$

If $K \subset K'$, then the diagram

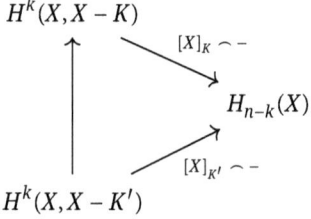

commutes. Passing to the limit, we obtain a homomorphism

$$
D : H_c^k(X; R) \to H_{n-k}(X; R).
$$

Theorem 3.3.17. *If X is an R-oriented n-manifold, then the map*

$$
D : H_c^k(X; R) \to H_{n-k}(X; R)
$$

is an isomorphism for all $k \geq 0$.

If X happens to be compact, then $H^k(X)$ coincides with $H^k_c(X)$ and we get the statement of Theorem 3.3.3.

The picture to keep in mind is that as the compact sets K enlarge to become all of X, the modules $H^k(X, X - K)$ "see" more of $H^k(X)$. Figure 3.5 shows a torus. Let K be the compact portion of the space above the dashed line. A representative of the dual of a 1-dimensional homology class is shown in red. Letting W be an open neighborhood of K, we have maps

$$H^1(T, T - K) \xrightarrow[\text{excision}]{\cong} H^1(W, W - K) \xrightarrow{[T]_K \frown} H_1(W)$$

So, corresponding to the cohomology of T seen by $H^1(T, T - K)$, there is homology of T whose support lies in W. Enlarging K by unions produces enlargements of W.

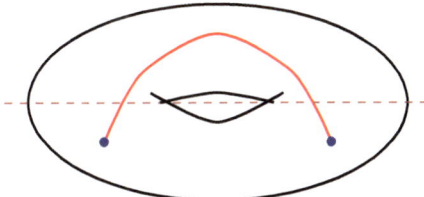

Figure 3.5: Duality on the torus.

Proof of Theorem 3.3.17. We proceed in steps.

Step 1. Suppose the theorem holds for open sets U, V, and $B = U \cap V$. Then it holds for $Y = U \cup V$. To see this, let $K \subset U$ and $L \subset V$ be compact. Use Mayer–Vietoris for the triple $(Y, Y - K, Y - L)$ to obtain a diagram:

$$\cdots \to H^k(B, B - K \cap L) \to H^k(U, U - K) \oplus H^k(V, V - L) \to H^k(Y, Y - K \cup L) \to H^{k+1}(B, B - K \cap L) \to \cdots$$
$$\downarrow \qquad\qquad \downarrow \qquad\qquad\qquad \downarrow \qquad\qquad \downarrow$$
$$\cdots \longrightarrow H_{n-k}(B) \longrightarrow H_{n-k}(U) \oplus H_{n-k}(V) \longrightarrow H_{n-k}(Y) \longrightarrow H_{n-k-1}(B) \longrightarrow \cdots$$

where the vertical maps are cap product with the relevant class $[X]_-$ and the horizontal maps have used excision isomorphisms of the form $(W, W - S) \subset (Y, Y - S)$. The bottom row is the Mayer–Vietoris sequence for the triple (Y, U, V). Commutativity of the two left squares follows from the naturality of the cap product. The right square is a consequence of the following.

Lemma 3.3.18. *The following diagram commutes up to* $(-1)^{k+1}$:

$$
\begin{array}{ccc}
H^k(Y, Y - K \cup L) & \xrightarrow{\;\delta\;} & H^{k+1}(Y, Y - K \cup L) \\
\Big\downarrow {\scriptstyle [X]_{K \cup L} \frown -} & & \Big\downarrow {\scriptstyle \cong} \\
& & H^{k+1}(B, B - K \cap L) \\
& & \Big\downarrow \\
H_{n-k}(Y) & \xrightarrow{\;\partial\;} & H_{n-k-1}(B)
\end{array}
$$

We leave the proof of this as an exercise. Assuming this, note that every compact set in Y has the form $K \cup L$. Passing to the limit gives a sign-commutative diagram

$$
\begin{array}{ccccccccc}
\cdots \to & H_c^k(B) & \to & H_c^k(U) \oplus H_c^k(V) & \to & H_c^k(Y) & \xrightarrow{\;\delta\;} & H^{k+1}(B) & \to \cdots \\
& \Big\downarrow{\scriptstyle D} & & \Big\downarrow{\scriptstyle D \oplus D} & & \Big\downarrow{\scriptstyle D} & & \Big\downarrow{\scriptstyle D} & \\
\cdots \to & H_{n-k}(B) & \to & H_{n-k}(U) \oplus H_{n-k}(V) & \to & H_{n-k}(Y) & \xrightarrow{\;\partial\;} & H_{n-k-1}(B) & \to \cdots
\end{array}
$$

By Five Lemma, if the theorem holds for U, V, and B, then it holds for Y.

Step 2. Let $\{U_i\}$ be a system of open sets totally ordered by inclusion and let $U = \bigcup U_i$. If the theorem holds for each U_i, then it holds for U. This amounts to verifying that we have isomorphisms

$$\psi_1 : \varinjlim H_{n-k}(U_i) \to H_{n-k}(U)$$

and

$$\psi_2 : \varinjlim H_c^k(U_i) \to H_c^k(U).$$

Note that for any compact $K \subset U$, we have $K \subset U_i$ for some i (because of the total order). It follows that ψ_1 is an isomorphism by considering the compact support of any chain. A similar argument applies to ψ_2.

Step 3. The theorem is true if U is contained in a coordinate neighborhood. Regard U as a subset of $\mathbb{R}^n$. If U is convex, then U is homeomorphic to the interior of the closed n-ball D^n. In computing the limit

$$\varinjlim_K H^k(\operatorname{int}(D^n), \operatorname{int}(D^n) - K)$$

it suffices to consider the final system of closed balls K_r of radius $r < 1$ centered at 0. For such K_r the modules are all 0 except for $k = n$ and

$$[X]_K \frown - : H^n(\operatorname{int}(D^n), \operatorname{int}(D^n) - K) \to H_0(\operatorname{int}(D^n)) \cong R$$

is certainly an isomorphism, since the generator of H^n takes value 1 on $\text{int}(D^n)$ and by the definition of cap product we get the vertex generating $H_0(\text{int}(D^n))$. It follows that the limiting map is an isomorphism.

If U is not convex, enumerate a dense set of points in U having rational coordinates and choose a convex set V_j contained in U about the jth point. Let $U_1 = V_1$ and $U_i = U_{i-1} \cup V_i$ for $i > 1$. The theorem holds for U_1. Assume inductively that it holds for a union of $k < i$ convex open sets. Note that $U_{i-1} \cap V_i$ is the union of at most $i-1$ convex open sets. By Step 1 and induction the theorem holds for U_i and then by Step 2 it holds for U.

Step 4. The theorem holds for X. By Zorn's Lemma, there is a maximal open $U \subseteq X$ for which the theorem is true. For any V contained in a coordinate neighborhood, the theorem is true for V. The theorem is then true for $U \cup V$ and by maximality we conclude that $U = X$. □

Corollary 3.3.19. *If X is connected and R-orientable, $H_c^n(X; R) \cong R$.*

Corollary 3.3.20. *If X is compact and orientable, then*

$$\beta_k = \beta_{n-k} \quad \text{for all } k$$
$$T_k \cong T_{n-k-1}, \ T_r = \text{torsion subgroup of } H_r$$

Corollary 3.3.21. *If X is odd dimensional, say $\dim X = n = 2k + 1$, then $\chi(X) = 0$.*

Proof. If X is orientable, then

$$\chi(X) = \sum_{i=0}^{n}(-1)^i \beta_i$$

$$= \sum_{i=0}^{k}(-1)^i \beta_i + \sum_{i=k+1}^{2k+1}(-1)^i \beta_i$$

$$= \sum_{i=0}^{k}(-1)^i \beta_i + \sum_{i=k+1}^{2k+1}(-1)^i \beta_{(2k+1)-i}$$

$$= \sum_{i=0}^{k}(-1)^i \beta_i + \sum_{i=0}^{k}(-1)^{i+1} \beta_i$$

$$= 0.$$

If X is nonorientable, the same argument works with $\mathbb{Z}_2$-coefficients to show that $\sum_i(-1)^i \dim H_i(X; \mathbb{Z}_2) = 0$. But then the universal coefficient theorem implies the result for integral homology. □

In the project at the end of the chapter we examine the geometry of Poincaré Duality.

Exercises

Exercise 3.3.1. Show that if (X,A) is a pair of spaces, then there are well-defined cap products

$$\frown: H_k(X,A;R) \otimes_R H^\ell(X;R) \to H_{k-\ell}(X,A;R)$$

and

$$\frown: H_k(X,A;R) \otimes_R H^\ell(X,A;R) \to H_{k-\ell}(X,A;R).$$

Exercise 3.3.2. Prove that if $f : X \to Y$ is continuous, then

$$f_*(\alpha) \frown \varphi = f_*(\alpha \frown f^*(\varphi)).$$

Exercise 3.3.3. Verify that the map θ defined in the proof of Lemma 3.3.10 is well-defined and inverse to $\psi \oplus \rho$.

Exercise 3.3.4. Complete the proof of Lemma 3.3.12 by showing that the map λ is injective.

Exercise 3.3.5. Prove Lemma 3.3.13.

Exercise 3.3.6. Prove Lemma 3.3.18.

3.3.2 Interaction with cup products

Since the cap product of a $(k + \ell)$-chain and a k-cochain is an ℓ-chain, we can apply an ℓ-cochain to the result. The formula for this is given by the following.

Proposition 3.3.22. *If $\psi \in S^\ell(X;R)$, $\varphi \in S^k(X;R)$, and $a \in S_{k+\ell}(X;R)$, then $\psi(a \frown \varphi) = (\varphi \smile \psi)(a)$.*

Proof. Suppose $\sigma : \Delta^{k+\ell} \to X$ is a singular simplex. Then

$$\psi(\sigma \frown \varphi) = \psi(\varphi(\sigma|_{\langle v_0,\dots,v_k\rangle})\sigma|_{\langle v_k,\dots,v_{k+\ell}\rangle})$$
$$= \varphi(\sigma|_{\langle v_0,\dots,v_k\rangle})\psi(\sigma|_{\langle v_k,\dots,v_{k+\ell}\rangle})$$
$$= (\varphi \smile \psi)(\sigma). \qquad \square$$

In other words, the map $\varphi \smile - : S^\ell \to S^{k+\ell}$ is Hom-dual to the map $- \frown \varphi$. Passing to homology, we get a commutative diagram

$$
\begin{array}{ccc}
H^\ell(X;R) & \xrightarrow{\;h\;} & \mathrm{Hom}_R(H_\ell(X;R),R) \\
\downarrow{\scriptstyle \varphi\smile -} & & \downarrow{\scriptstyle (-\frown\varphi)^*} \\
H^{k+\ell}(X;R) & \xrightarrow{\;h\;} & \mathrm{Hom}_R(H_{k+\ell}(X;R),R)
\end{array}
$$

Exercises

Exercise 3.3.7. Prove that the cap product makes $S_*(X)$ a right unitary $S^*(X)$-module.

Exercise 3.3.8. If X is path connected, so that $\partial_\sharp : H_0(X;R) \to R$ is an isomorphism, prove that for any $\alpha \in H_p(X;R)$ and $\beta \in H^p(X;R)$, $\partial_\sharp(\alpha \frown \beta) = \varphi(z)$, where z represents α and φ represents β.

3.3.3 Consequences of duality

Definition 3.3.23. Suppose A and B are R-modules. A bilinear pairing $p : A \times B \to R$ is *nondegenerate* if the following maps are both isomorphisms.

$$A \longrightarrow \mathrm{Hom}_R(B,R) \qquad\qquad B \longrightarrow \mathrm{Hom}_R(A,R)$$
$$\text{and}$$
$$a \mapsto \{b \overset{f}{\mapsto} p(a,b)\} \qquad\qquad b \mapsto \{a \overset{g}{\mapsto} p(a,b)\}$$

The cup product is a bilinear operation. If M is a closed R-oriented n-manifold with fundamental class $[M] \in H_n(M;R)$, consider the pairing

$$H^k(M;R) \times H^{n-k}(M;R) \longrightarrow R$$

$$(\varphi,\psi) \longmapsto (\varphi \smile \psi)([M])$$

Proposition 3.3.24. *If R is a field, then the cup product pairing is nondegenerate. If $R = \mathbb{Z}$, then it is nondegenerate modulo torsion.*

Proof. Consider the composition

$$H^{n-k}(M;R) \overset{h}{\to} \mathrm{Hom}_R(H_{n-k}(M;R),R) \overset{D^*}{\to} \mathrm{Hom}_R(H^k(M;R),R)$$

We have

$$(D^*h)(\psi) = \{\varphi \mapsto \psi([M] \frown \varphi)\} = (\varphi \smile \psi)([M]).$$

Under either hypothesis on R, the map h is an isomorphism. Nondegeneracy in one of the variables is equivalent to D^* being an isomorphism, which follows from Poincaré Duality. Nondegeneracy in the other follows from commutativity of the cup product. $\square$

We can use this result to compute the cohomology rings of projective spaces. We first have the following consequence.

Corollary 3.3.25. *Suppose M is closed, connected, and orientable. An element $\alpha \in H^k(M;\mathbb{Z})$ generates an infinite cyclic summand if and only if there is a $\beta \in H^{n-k}(M;\mathbb{Z})$ such that*

$\alpha \smile \beta$ generates $H^n(M; \mathbb{Z}) \cong \mathbb{Z}$. With field coefficients, this result holds for any $\alpha \neq 0$, where "infinite cyclic summand" means a one-dimensional subspace.

Proof. Such an α exists if and only if there is a $\varphi : H^k(M; \mathbb{Z}) \to \mathbb{Z}$ with $\varphi(\alpha) = \pm 1$. By Proposition 3.3.24, φ is realized by taking the cup product with some $\beta \in H^{n-k}(M; \mathbb{Z})$ and evaluating on $[M]$. So the existence of β with $\alpha \smile \beta$, generating $H^n(M; \mathbb{Z})$ is equivalent to the existence of φ with $\varphi(\alpha) = \pm 1$. The field case is entirely similar. □

Corollary 3.3.26. $H^{\bullet}(\mathbb{C}P^n; \mathbb{Z}) \cong \mathbb{Z}[\alpha]/(\alpha^{n+1})$, where $\deg \alpha = 2$.

Proof. The inclusion $\mathbb{C}P^{n-1} \hookrightarrow \mathbb{C}P^n$ induces an isomorphism on cohomology H^{2i} for $i < n$ and so H^{2i} is generated by α^i for $i < n$ (the induction begins at $n = 1$, where $\mathbb{C}P^1 = S^2$). By Corollary 3.3.25, there exists $m \in \mathbb{Z}$ with $\alpha \smile m\alpha^{n-1} = m\alpha^n$ generating $H^{2n}(\mathbb{C}P^n; \mathbb{Z})$. But then $m = \pm 1$ and the result follows. □

Corollary 3.3.27. $H^{\bullet}(\mathbb{R}P^n; \mathbb{Z}_2) \cong \mathbb{Z}_2[\alpha]/(\alpha^{n+1})$, where $\deg \alpha = 1$.

Proof. Exercise. □

Theorem 3.3.28 (Borsuk–Ulam). *If $n > m \geq 1$, there is no map $g : S^n \to S^m$ commuting with the antipodal maps.*

Proof. Such a map g would induce, by passage to the quotient, a map $f : \mathbb{R}P^n \to \mathbb{R}P^m$ making the following diagram commute.

$$\begin{array}{ccc} S^n & \xrightarrow{\;g\;} & S^m \\ \downarrow{\scriptstyle p'} & \overset{f'}{\nearrow} & \downarrow{\scriptstyle p} \\ \mathbb{R}P^n & \xrightarrow{\;f\;} & \mathbb{R}P^m \end{array}$$

Lemma 3.3.29. *There exists a lift $f' : \mathbb{R}P^n \to S^m$ with $pf' = f$.*

Proof. We use the lifting criterion for covering spaces. If $m = 1$, then since the only map $\pi_1(\mathbb{R}P^n) \to \pi_1(\mathbb{R}P^1) \cong \mathbb{Z}$ is trivial, we have a lift f'. Suppose $m > 1$. The map $f^* : H^{\bullet}(\mathbb{R}P^m; \mathbb{Z}_2) \to H^{\bullet}(\mathbb{R}P^n; \mathbb{Z}_2)$ is a ring homomorphism. But

$$0 = f^*(\alpha^{m+1}) = (f^*(\alpha))^{m+1}$$

implies that $f^*(\alpha) = 0$, since $n > m$. Let $i : \mathbb{R}P^1 \to \mathbb{R}P^n$ and $j : \mathbb{R}P^1 \to \mathbb{R}P^m$ be the inclusions obtained by setting all but the first two homogeneous coordinates equal to 0. We know that $j^* : H^1(\mathbb{R}P^m; \mathbb{Z}_2) \to H^1(\mathbb{R}P^1; \mathbb{Z}_2)$ is an isomorphism and so $j^*(\alpha) \neq 0$. This implies that $j^* \neq (fi)^*$ and so $fi \neq j$. But i and j are generators of the fundamental groups and so it follows that $f_* : \pi_1(\mathbb{R}P^n) \to \pi_1(\mathbb{R}P^m)$ is the zero map and the lifting criterion applies. □

Now we have $pf'p' = pg$. If $x \in S^n$, either $g(x) = f'p'(x)$ or $g(-x) = f'p'(x) = f'p'(-x)$. Thus, $f'p'$ and g are two lifts of fp', which agree at a point and by the uniqueness of lifts we must have $f'p' = g$. But $g(x) \neq g(-x) = -g(x)$ while $p'(x) = p'(-x)$, a contradiction. $\qquad\square$

This result implies the following more familiar statement of the Borsuk–Ulam Theorem, which appeared earlier as Theorem 2.5.4. That proof involved the transfer homomorphism and results about the degree of a map on S^n. The proof above made use of the cohomology ring of projective space.

Corollary 3.3.30. *If* $f : S^n \to \mathbb{R}^n$ *is continuous, then there is a point* $x \in S^n$ *with* $f(x) = f(-x)$.

Proof. Suppose not. Then we can define a map $g : S^n \to S^{n-1}$ by setting $g(x)$ to be the point at which the vector from 0 through $f(x) - f(-x)$ intersects S^{n-1}. The map g commutes with the antipodal map, contradicting the previous result. $\qquad\square$

This formulation of the Borsuk–Ulam Theorem has many consequences. For example, when $n = 2$ it implies various physical phenomena such as the existence of two antipodal points on the Earth's surface at which the temperature and barometric pressure agree. It also implies a result due to Lyusternik and Shnirlman that asserts that any cover of S^n by $n + 1$ open sets has at least one set containing a pair of antipodal points. In the exercises, we give one other application.

Exercises

Exercise 3.3.9. Prove Corollary 3.3.27.

Exercise 3.3.10. Compute the cohomology ring $H^\bullet(\mathbb{H}P^n; \mathbb{Z})$.

Exercise 3.3.11. Suppose M is an n-manifold. Prove that if M is orientable, then $H_{n-1}(M; \mathbb{Z})$ is torsion-free. Prove that if M is nonorientable, then $H_n(M; \mathbb{Z}_k) = 0$ for k is odd, the torsion subgroup of $H_{n-1}(M; \mathbb{Z}_2)$ is cyclic of order 2, and $H_1(M; \mathbb{Z}_2) \neq 0$.

Exercise 3.3.12. Prove that if M is an orientable 3-manifold with $H_1(M; \mathbb{Z}) = 0$, then M has the homology of S^3.

Exercise 3.3.13. Prove that if M is a nonorientable 3-manifold, then $H_1(M; \mathbb{Z})$ is infinite.

Exercise 3.3.14. Prove that the cup product pairing $H^p(M; F) \otimes H^p(M; F) \to H^n(M; F)$, $p + q = n$, has the property that for a fixed $x \in H^p(M; F)$, $x \smile y = 0$ for all $y \in H^q(M; F)$ only if $x = 0$.

Exercise 3.3.15. Prove that $\mathbb{C}P^{2n}$ does not admit an orientation reversing homotopy equivalence.

Exercise 3.3.16. Prove the Ham Sandwich Theorem: Suppose that $U_1, U_2, \ldots, U_n$ are measurable subsets of $\mathbb{R}^n$ with positive measure. Then there is a hyperplane H that divides each U_i in half.

3.4 Universal coefficient and Künneth theorems

In the final section of this chapter we collect some miscellaneous theorems that allow us to compute (co)homology with various coefficients and also to compute the (co)homology of a product of two spaces.

3.4.1 Universal coefficient theorems

We have seen that there is an exact sequence of the form

$$0 \to \mathrm{Ext}(H_{n-1}(X), G) \to H^n(X; G) \to \mathrm{Hom}(H_n(X), G) \to 0,$$

so it is an obvious question to ask if there is a corresponding result for homology with coefficients $H_n(X; G)$. The answer is yes, of course, and it involves a new construction.

Consider the short exact sequence of chain complexes

$$
\begin{array}{ccccccccc}
0 & \longrightarrow & Z_n & \longrightarrow & C_n & \xrightarrow{\partial_n} & B_{n-1} & \longrightarrow & 0 \\
& & \downarrow{\scriptstyle \partial_n \equiv 0} & & \downarrow{\scriptstyle \partial_n} & & \downarrow{\scriptstyle \partial_{n-1} \equiv 0} & & \\
0 & \longrightarrow & Z_{n-1} & \longrightarrow & C_{n-1} & \xrightarrow{\partial_{n-1}} & B_{n-2} & \longrightarrow & 0
\end{array}
$$

Note that for each n, $C_n \cong Z_n \oplus B_{n-1}$, but as chain complexes $C_\bullet \neq Z_\bullet \oplus B_\bullet$. If G is an abelian group, apply the functor $- \otimes G$ to this to obtain a diagram

$$
\begin{array}{ccccccccc}
0 & \longrightarrow & Z_n \otimes G & \longrightarrow & C_n \otimes G & \xrightarrow{\partial_n \otimes 1} & B_{n-1} \otimes G & \longrightarrow & 0 \\
& & \downarrow{\scriptstyle \partial_n \otimes 1} & & \downarrow{\scriptstyle \partial_n \otimes 1} & & \downarrow{\scriptstyle \partial_{n-1} \otimes 1} & & \\
0 & \longrightarrow & Z_{n-1} \otimes G & \longrightarrow & C_{n-1} \otimes G & \xrightarrow{\partial_{n-1} \otimes 1} & B_{n-2} \otimes G & \longrightarrow & 0
\end{array}
$$

The rows are still exact since $\otimes$ distributes over direct sums. The long exact sequence in homology then gives us

$$\cdots \to B_n \otimes G \xrightarrow{i_n \otimes 1} Z_n \otimes G \to H_n(C_\bullet; G) \to B_{n-1} \otimes G \xrightarrow{i_{n-1} \otimes 1} Z_{n-1} \otimes G \to \cdots,$$

where i_n is induced by the inclusion $B_n \hookrightarrow Z_n$. This breaks into short exact sequences

$$0 \to \mathrm{coker}(i_n \otimes 1) \to H_n(C_\bullet; G) \to \ker(i_{n-1} \otimes 1) \to 0.$$

Lemma 3.4.1. *If* $A \xrightarrow{i} B \xrightarrow{j} C \to 0$ *is an exact sequence, then the sequence* $A \otimes G \xrightarrow{i \otimes 1} B \otimes G \xrightarrow{j \otimes 1} C \to 0$ *is also exact.*

Proof. Exercise. □

Corollary 3.4.2. *For each n, we have* $\operatorname{coker}(i_n \otimes 1) \cong H_n(C_\bullet) \otimes G$.

Proof. We have an exact sequence $B_n \xrightarrow{i_n} Z_n \to H_n(C) \to 0$ and so the sequence $B_n \otimes G \to Z_n \otimes G \to \operatorname{coker}(i_n \otimes 1) \to 0$ is also exact. It follows that $\operatorname{coker}(i_n \otimes 1) \cong H_n(C_\bullet) \otimes G$. □

It remains to compute $\ker(i_n \otimes 1)$. Applying $- \otimes G$ to the exact sequence in the proof of Lemma 3.4.2, we obtain an exact sequence

$$0 \to \ker(i_n \otimes 1) \to B_n \otimes G \to Z_n \otimes G \to H_n(C_\bullet) \otimes G \to 0.$$

To compute this kernel, we need to generalize.

Let H be an abelian group and construct a *free resolution* of H:

$$\cdots \to F_2 \to F_1 \to F_0 \to H \to 0.$$

This is an exact sequence with each F_i a free abelian group. Apply $- \otimes G$ to this resolution to get a chain complex

$$\cdots \to F_2 \otimes G \to F_1 \otimes G \to F_0 \otimes G \to H \otimes G \to 0.$$

Definition 3.4.3. The ith homology of this group is called the ith Tor-*group*, denote by $\operatorname{Tor}_i(H, G)$. Note that $\operatorname{Tor}_0(H, G) = H \otimes G$.

Here are two important facts about this definition:
(1) If we choose another free resolution $F'_\bullet \to H$, the resulting Tor-groups do not change.
(2) We can always find a resolution of the form $0 \to F_1 \to F_0 \to H \to 0$ so that $\operatorname{Tor}_i(H, G) = 0$ for $i > 1$. We therefore drop the subscript and simply refer to $\operatorname{Tor}(H, G)$.

Remark 3.4.4. Of course, this definition makes sense in the category of R-modules, for R a commutative ring. In this more general setting, free resolutions may have length longer than 1 (and indeed, may even be infinite) and so the higher Tor-modules might not vanish.

Returning to homology with coefficients, we see that $\ker(i_n \otimes 1) = \operatorname{Tor}(H_n(C_\bullet), G)$. We therefore have the following result.

Theorem 3.4.5 (Universal Coefficient Theorem for homology). *If $C_\bullet$ is a chain complex of free abelian groups, then there are natural short exact sequences*

$$0 \to H_n(C_\bullet) \otimes G \to H_n(C_\bullet; G) \to \operatorname{Tor}(H_{n-1}(C_\bullet), G) \to 0.$$

These split, but not naturally.

It remains to determine how to compute Tor.

Proposition 3.4.6. *The following hold.*

(1) $\mathrm{Tor}(B, A) \cong \mathrm{Tor}(A, B)$.

(2) $\mathrm{Tor}(\bigoplus_i A_i, B) \cong \bigoplus_i \mathrm{Tor}(A_i, B)$.

(3) $\mathrm{Tor}(A, B) = 0$ *if A or B is free (or, more generally, torsion-free).*

(4) $\mathrm{Tor}(A, B) = \mathrm{Tor}(T(A), B)$, *where T(A) is the torsion subgroup of A.*

(5) $\mathrm{Tor}(\mathbb{Z}_n, A) = \ker(A \overset{\times n}{\to} A)$.

Proof. For (1), note that if $0 \to F_1 \to F_0 \to B \to 0$ is a free resolution of B, then $\mathrm{Tor}(B, A) = \ker(F_1 \otimes A \to F_0 \otimes A)$. But the tensor product is commutative and so this kernel is the same as $\ker(A \otimes F_1 \to A \otimes F_0)$, which is simply $\mathrm{Tor}(A, B)$. Statement (2) follows from the commutativity of $\otimes$ with direct sums. For (3), consider first $A = \mathbb{Z}$. A resolution of $\mathbb{Z}$ is $0 \to 0 \to \mathbb{Z} \to \mathbb{Z} \to 0$. Applying $- \otimes B$ we obtain

$$0 \to 0 \otimes B \to \mathbb{Z} \otimes B \to \mathbb{Z} \otimes B \to 0,$$

from which it follows that $\mathrm{Tor}(\mathbb{Z}, B) = 0$. For a finitely generated free abelian group, the result follows from (2). Since $\otimes$ commutes with direct limits, we obtain the result for an arbitrary torsion-free A. Statement (4) follows from (2) and (3) in the finitely generated case and by passage to the limit in general. Finally, for (5), consider the resolution of $\mathbb{Z}_n$:

$$0 \to \mathbb{Z} \overset{\times n}{\to} \mathbb{Z} \to \mathbb{Z}_n \to 0.$$

Applying $- \otimes A$, we obtain the complex

$$
\begin{array}{ccc}
0 \longrightarrow \mathbb{Z} \otimes A & \overset{\times n \otimes 1}{\longrightarrow} & \mathbb{Z} \otimes A \\
\downarrow{\cong} & & \downarrow{\cong} \\
A & \overset{\times n}{\longrightarrow} & A
\end{array}
$$

We therefore conclude the result. □

So, for example, we have $\mathrm{Tor}(\mathbb{Z}_m, \mathbb{Z}_n) \cong \mathbb{Z}_{\gcd(m,n)}$.

Corollary 3.4.7. *For a space X, we have $H_n(X; \mathbb{Q}) \cong H_n(X; \mathbb{Z}) \otimes \mathbb{Q}$, and hence $\beta_n = \dim_\mathbb{Q} H_n(X; \mathbb{Q})$.*

Proof. Since $\mathbb{Q}$ is torsion-free, (3) implies that $\mathrm{Tor}(H_{n-1}(X; \mathbb{Z}), \mathbb{Q}) = 0$ for all n. □

Exercises

Exercise 3.4.1. Prove Lemma 3.4.1.

Exercise 3.4.2. Suppose that $H_*(X; \mathbb{Z})$ is finitely generated. Prove that for any coefficient field F $\chi(X) = \sum_i (-1)^i \dim H_i(X; F)$.

Exercise 3.4.3. Prove that if $\tilde{H}_n(X;\mathbb{Q})$ and $\tilde{H}_n(X;\mathbb{Z}_p)$ are zero for all n and all primes p, then $\tilde{H}_n(X;\mathbb{Z}) = 0$.

3.4.2 The Künneth formula

There is one more question we need to answer: Given spaces X and Y, how is $H_*(X \times Y)$ related to the homology of X and Y? The dream answer would be that the homology of the product is just the product of the homology groups, but alas things are not quite that neat.

To motivate the main result, let us work with a pair of cell complexes and their cellular chain complexes. We begin by defining a cross product of homology

$$H_i(X;R) \times H_j(Y;R) \rightarrow H_{i+j}(X \times Y;R).$$

Consider the cellular chain complex $C_*(X \times Y)$. If e^i is a cell in X and e^j is a cell in Y, then $e^i \times e^j$ is an $(i+j)$-cell in $X \times Y$. All cells of $X \times Y$ occur in this way, and we must compute the boundary map in $C_*(X \times Y)$ in terms of these cells. Recall that the cellular chain group $C_i(X)$ is defined to be $H_i(X^{(i)}, X^{(i-1)})$, with basis $\{e^i\}$ consisting of the i-cells in X. There is a sign ambiguity about the basis cell e^i, depending on the choice of corresponding generator of $H_i(X^{(i)}, X^{(i-1)})$. Such a choice is called *choosing an orientation* for e^i.

The question is then how the orientations of e^i and e^j determine one for $e^i \times e^j$. The proper way to think about this is to think of an orientation of e^i corresponding to a choice of ordered basis for $\mathbb{R}^i$ by identifying e^i with the standard cube I^i. An orientation of $I^i \times I^j$ is obtained by choosing an ordered basis for I^i and an ordered basis for I^j and then concatenating them. Reversing the orientation of either factor reverses the product orientation, so we need only worry about the order of factors.

Proposition 3.4.8. *The boundary map in $C_*(X \times Y)$ is determined by those in $C_*(X)$ and $C_*(Y)$ via the formula*

$$d(e^i \times e^j) = de^i \times e^j + (-1)^i e^i \times de^j.$$

Proof. This is a surprisingly technical result. Let's first prove it for a cube I^n. Give I the cell structure with two 0-cells and one edge. In the ith copy of I, denote these by v_i^0, v_i^1, and e_i. The boundary map is $de_i = v_i^1 - v_i^0$. The n-cell in I^n is $e_1 \times \cdots \times e_n$ and its boundary is

$$d(e_1 \times \cdots \times e_n) = \sum_i (-1)^{i+1} e_1 \times \cdots \times de_i \times \cdots \times e_n.$$

Now if we write $I^n = I^i \times I^j$, $i+j = n$, then letting $e^i = e_1 \times \cdots \times e_i$ and $e^j = e_{i+1} \times \cdots \times e_n$, we deduce

$$d(e^i \times e^j) = de^i \times e^j + (-1)^i e^i \times de^j.$$

The general result follows from naturality. Indeed, if $f : X \to Z$ and $g : Y \to W$ are cellular maps, then at the level of cellular chains we have $(f \times g)_* : C_*(X \times Y) \to C_*(Z \times W)$ and this map agrees with $f_* \times g_*$. Assuming this, we can finish the proof as follows. Let $\varphi_\alpha : I^i \to X^{(i)}$ and $\psi_\beta : I^j \to Y^{(j)}$ be the characteristic maps for the cells e_α^i and e_β^j, respectively. We may assume these maps are cellular, so that the product map $\varphi_\alpha \times \psi_\beta$ is also cellular. It therefore induces a map on cellular chain complexes:

$$\varphi_{\alpha*} \times \psi_{\beta*} : C_*(I^i \times I^j) \to C_*(X \times Y).$$

Let e^i be the i-cell in I^i and e^j the j-cell in I^j. Then $\varphi_{\alpha*}(e^i) = e_\alpha^i$ and $\psi_{\beta*}(e^j) = e_\beta^j$. We also have $(\varphi_\alpha \times \psi_\beta)_*(e^i \times e^j) = e_\alpha^i \times e_\beta^j$. It follows that

$$d(e_\alpha^i \times e_\beta^j) = d((\varphi_\alpha \times \psi_\beta)_*(e^i \times e^j)).$$

But $(\varphi_\alpha \times \psi_\beta)_*$ is a chain map and therefore commutes with d. Since we have computed d on the cube, we have

$$(\varphi_\alpha \times \psi_\beta)_* d(e^i \times e^j) = (\varphi_\alpha \times \psi_\beta)_*(de^i \times e^j + (-1)^i e^i \times de^j).$$

Now, since $(\varphi_\alpha \times \psi_\beta)_* = \varphi_{\alpha*} \times \psi_{\beta*}$, we can distribute over the product:

$$(\varphi_\alpha \times \psi_\beta)_*(de^i \times e^j + (-1)^i e^i \times de^j) = \varphi_{\alpha*}(de^i) \times \psi_{\beta*}(e^j) + (-1)^i \varphi_{\alpha*}(e^i) \times \psi_{\beta*}(de^j).$$

Finally, since both $\varphi_{\alpha*}$ and $\psi_{\beta*}$ are chain maps, they commute with d and we obtain the result. □

We now have a well-defined bilinear map

$$H_i(X; R) \times H_j(Y; R) \to H_{i+j}(X \times Y; R),$$

and therefore a map, for each $n \geq 0$

$$\bigoplus_i (H_i(X; R) \otimes H_{n-i}(Y; R)) \to H_n(X \times Y; R).$$

One might hope that this map is an isomorphism, but it is not. However, we have shown that

$$C_n(X \times Y) = \bigoplus_i (C_i(X) \otimes C_{n-i}(Y)), \quad d(e^i \times e^{n-i}) = de^i \otimes e^{n-i} + (-1)^i e^i \otimes de^{n-i}.$$

So, let us generalize a bit. Suppose C_* and C'_* are chain complexes of R-modules. The *tensor product* is the chain complex $C_* \otimes_R C'_*$ defined by

$$(C_* \otimes_R C'_*)_n = \bigoplus_i (C_i \otimes_R C'_{n-i}), \quad d(c \otimes c') = dc \otimes c' + (-1)^{|c|} c \otimes dc',$$

where $|c|$ is the degree of c. This is indeed a chain complex:

$$d^2(c \otimes c') = d(dc \otimes c' + (-1)^{|c|}c \otimes dc')$$
$$= d^2c \otimes c' + (-1)^{|c|-1}dc \otimes dc' + (-1)^{|c|}dc \otimes dc' + (-1)^{2|c|}c \otimes d^2c'$$
$$= 0 + (-1)^{|c|-1}dc \otimes dc' + (-1)^{|c|}dc \otimes dc' + 0$$
$$= 0.$$

As in the cellular homology case, we get a map on homology

$$\bigoplus_i (H_i(C_\bullet) \otimes_R H_{n-i}(C'_\bullet)) \to H_n(C_\bullet \otimes_R C'_\bullet).$$

Theorem 3.4.9. *If R is a principal ideal domain and the R-modules C_i are free, then there are natural short exact sequences*

$$0 \to \bigoplus_i (H_i(C_\bullet) \otimes_R H_{n-i}(C'_\bullet)) \to H_n(C_\bullet \otimes_R C'_\bullet) \to \bigoplus_i \mathrm{Tor}_1^R(H_i(C_\bullet), H_{n-i-1}(C'_\bullet)) \to 0.$$

Proof. The proof is similar to the proof of Theorem 3.4.5. Indeed, the case $C'_\bullet = G$ in degree 0 is exactly that theorem. This more general result follows similarly by using the fact that $C_\bullet$ consists of free modules and that, since R is a PID, any submodule of C_i is free. It follows that $0 \to B_i \to Z_i \to H_i(C_\bullet) \to 0$ is a free resolution for all i and this allows one to compute various kernels and cokernels. The rest is purely formal and we leave it to the reader. □

Applying this to cell complexes, we obtain the following.

Theorem 3.4.10 (Künneth Formula). *If X and Y are cell complexes and R is a principal ideal domain, then there are natural short exact sequences*

$$0 \to \bigoplus_i (H_i(X;R) \otimes_R H_{n-i}(Y;R)) \to H_n(X \times Y;R) \to \bigoplus_i \mathrm{Tor}_1^R(H_i(X;R), H_{n-i-1}(Y;R)) \to 0.$$

These split, but not naturally.

Corollary 3.4.11. *If F is a field, then the map*

$$\bigoplus_i (H_i(X;F) \otimes_F H_{n-i}(Y;F)) \to H_n(X \times Y;F)$$

is an isomorphism for all $n \geq 0$.

Proof. All Tor-modules vanish over a field since any F-module is free. □

Corollary 3.4.12. *Assume R is a principal ideal domain. Then*

$$\chi(X \times Y;R) = \chi(X;R)\chi(Y;R).$$

Example 3.4.13. Consider the 5-manifolds $\mathbb{R}P^2 \times S^3$ and $\mathbb{R}P^3 \times S^2$. These spaces have the same fundamental group, and even all the same higher homotopy groups (see Chapter 4). Yet, they are not homotopy equivalent. The Künneth formula gives us the following:

$$H_n(\mathbb{R}P^2 \times S^3; \mathbb{Z}) = \begin{cases} \mathbb{Z} & n = 0 \\ \mathbb{Z}_2 & n = 1 \\ 0 & n = 2 \\ \mathbb{Z} & n = 3 \\ \mathbb{Z}_2 & n = 4 \\ 0 & n \geq 5 \end{cases} \qquad H_n(\mathbb{R}P^3 \times S^2; \mathbb{Z}) = \begin{cases} \mathbb{Z} & n = 0 \\ \mathbb{Z}_2 & n = 1 \\ \mathbb{Z} & n = 2 \\ \mathbb{Z} \oplus \mathbb{Z}_2 & n = 3 \\ 0 & n = 4 \\ \mathbb{Z} & n = 5 \\ 0 & n \geq 6. \end{cases}$$

Note that the Euler characteristic of both of these spaces is 0, as one can compute directly or via Corollary 3.4.12.

We make one final note about products. The Künneth formula works perfectly well for cohomology since one can view a chain complex as a cochain complex by negating indices and vice versa. It is relatively straightforward, but tedious, to check that all of this plays nicely with respect to cup products. We simply state the following result without proof.

Theorem 3.4.14. *Suppose one of $H^\bullet(X)$ or $H^\bullet(Y)$ is torsion-free (or coefficients are in a field). Then there is a canonical isomorphism of rings*

$$H^\bullet(X) \otimes H^\bullet(Y) \to H^\bullet(X \times Y).$$

Example 3.4.15. $H^\bullet(\mathbb{C}P^n \times \mathbb{C}P^m) \cong \mathbb{Z}[\alpha, \beta]/(\alpha^{n+1}, \beta^{m+1})$, $\deg \alpha = 2 = \deg \beta$.

Exercises

Exercise 3.4.4. Prove the assertion in the proof of Proposition 3.4.8 that the induced chain map on products is the product of the induced chain maps.

Exercise 3.4.5. Complete the proof of Theorem 3.4.9.

Exercise 3.4.6. Consider the spaces $S^1 \times S^2$ and $S^1 \vee S^2 \vee S^3$. Show that these spaces have the same integral homology groups, and therefore the same integral cohomology groups. Show, however, that the cohomology *rings* are not isomorphic.

Exercise 3.4.7. Show that the cohomology rings of $\mathbb{C}P^{n(n+1)/2}$ and $S^2 \times S^4 \times \cdots \times S^{2n}$ are not isomorphic.

Exercise 3.4.8. Prove that $S^n \vee S^m$ is not a retract of $S^n \times S^m$.

Project: the geometry of Poincaré Duality

Our proof of Poincaré Duality was very high-powered and abstract. In the end, we appealed to Zorn's Lemma, etc. In this project we explore the geometry behind Poincaré Duality, which is how it came to be discovered in the first place.

Suppose X is a locally finite simplicial complex and let $\mathrm{sd}(X)$ be the first barycentric subdivision. The simplices of $\mathrm{sd}(X)$ are of the form $\langle \hat{\sigma}_{i_1}, \hat{\sigma}_{i_2} \cdots, \hat{\sigma}_{i_k} \rangle$, where $\sigma_{i_1} > \sigma_{i_2} > \cdots > \sigma_{i_k}$ are simplices in X and $\hat{\sigma}_{i_j}$ is the barycenter of σ_{i_j}. Impose a partial order on the vertices of $\mathrm{sd}(X)$ by decreasing dimension of the simplices of X of which they are the barycenters. This induces a linear ordering on the vertices of each simplex of $\mathrm{sd}(X)$.

Definition 3.4.16. Given a simplex σ in X, set $D(\sigma)$ to be the union of all open simplices of $\mathrm{sd}(X)$ of which $\hat{\sigma}$ is the final vertex. The set $D(\sigma)$ is called the *block dual* to σ.

Figure 3.6 shows the duals of some simplices in a complex.

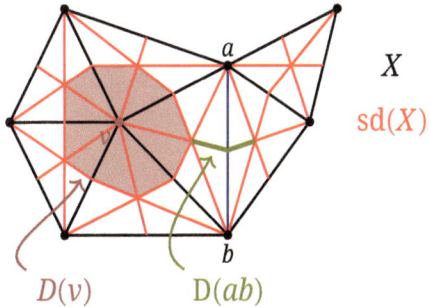

Figure 3.6: The duals of various simplices.

Suppose X is a locally finite simplicial complex consisting entirely of n-simplices and their faces. Prove the following facts.
(1) The dual blocks are disjoint and their union is $|X|$.
(2) $\overline{D(\sigma)}$, for a k-simplex σ, is a polytope of dimension $n - k$.
(3) $\mathrm{int}(D(\sigma))$ is the union of all blocks $D(\tau)$ for $\tau < \sigma$.
(4) If $H_i(X, X - \hat{\sigma}) \cong \mathbb{Z}$ for $i = n$ and 0 otherwise, then $(\overline{D(\sigma)}, \mathrm{int}(D(\sigma)))$ has the homology of $(D^{n-k}, \partial D^{n-k})$.

The collection of dual blocks is called the *dual block decomposition* of X. The union of blocks of dimension at most p is denoted by X_p and is called the *dual p-skeleton* of X. The *dual chain complex* $D_\bullet(X)$ is defined by $D_p(X) = H_p(X_p, X_{p-1})$, with ∂ the connecting homomorphism of the triple (X_p, X_{p-1}, X_{p-2}).

Prove that if X is a homology n-manifold, (i. e., $\tilde{H}_k(X, X - \{x\}) \cong \mathbb{Z}$ if and only if $k = n$ for all $x \in X$), then

$$H_\bullet(D_\bullet(X)) \cong H_\bullet(X).$$

Recall that if X is a compact triangulated n-manifold, then it is possible to orient the n-simplices of X so that $\gamma = \sum \sigma_i$ is an n-cycle. Moreover, γ represents the fundamental class of X. Now prove Poincaré Duality in this context–$H^p(X;G) \cong H_{n-p}(X;G)$ (if X is nonorientable, this holds for $G = \mathbb{Z}_2$)–as follows.

(1) There is an obvious one-to-one correspondence $C_p(X) \to D_{n-p}(X)$, taking σ to $D(\sigma)$. Since $C^p(X) = \mathrm{Hom}(C_p(X), \mathbb{Z})$, we obtain an isomorphism $C^p(X) \cong D_{n-p}(X)$ via $\sigma^* \overset{\varphi}{\mapsto} D(\sigma)$, a generator of $H_{n-p}(\overline{D(\sigma)}, \mathrm{int}(D(\sigma)))$. We need this to be a chain map. Prove by induction in items (2) and (3) below that it is possible to choose the sign of the generator so that $\varphi : C^p \to D_{n-p}$ makes the following diagram commute.

$$
\begin{array}{ccc}
C^{p-1}(X) & \xrightarrow{\ \varphi\ } & D_{n-p+1}(X) \\
\downarrow{\scriptstyle \delta} & & \downarrow{\scriptstyle \partial} \\
C^{p}(X) & \xrightarrow{\ \varphi\ } & D_{n-p}(X)
\end{array}
$$

(2) Begin with the n-simplices, the orientation of X gives you an orientation of each n-simplex that makes γ a cycle. Since $D(\sigma) = \hat{\sigma}$ for an n-simplex σ, there is an obvious choice of φ here.

(3) For an $(n-1)$-simplex s, you must define $\varphi(s^*)$ so that $\partial\varphi(s^*) = \varphi\delta(s^*)$. Note that s is a face of exactly two n-simplices σ_0 and σ_1, which have a prescribed orientation. Use the fact that γ is a cycle to compute what $\delta(s^*)$ must be, and since $\overline{D(s)}$ is the union of line segments from $\hat{s}$ to $\hat{\sigma}_0$ and $\hat{\sigma}_1$, you can define $\varphi(s^*)$. Proceed inductively to define φ in all dimensions.

Bibliographic notes
The discussion of orientability follows that of Greenberg & Harper. We admit that it is very technical, but it is quite powerful and streamlines some of the later proofs involving the homology of manifolds. The direct calculation of the cohomology rings of projective spaces is borrowed from Hatcher [2], who did the difficult work of avoiding Poincaré duality to prove this result. The geometry project is inspired by Munkres's discussion of duality in his book [5], which focuses on simplicial complexes. The proof, involving cohomology with compact supports and direct limits, is beautiful, but the intuition gets lost quickly.

4 Homotopy and spectral sequences

4.1 Homotopy groups

4.1.1 Definitions and basic calculations

If X and Y are spaces, we denote the set of homotopy classes of maps $X \to Y$ by $[X, Y]$. There are relative versions of this idea as well.

Definition 4.1.1. Let $x_0 \in X$. The *nth homotopy group* is the set

$$\pi_n(X, x_0) = [(S^n, p), (X, x_0)] = [(I^n, \partial I^n), (X, x_0)],$$

where $p \in S^n$ can be any point. The group operation is defined as follows for $n \geq 1$ (π_0 is only a set). If $[f]$ and $[g]$ denote homotopy classes of maps $(I^n, \partial I^n) \to (X, x_0)$, then $[f] \cdot [g] = [f \cdot g]$, where

$$(f \cdot g)(t_1, \ldots, t_n) = \begin{cases} f(2t_1, t_2, \ldots, t_n) & 0 \leq t_1 \leq 1/2 \\ g(2t_1 - 1, t_2, \ldots, t_n) & 1/2 \leq t_1 \leq 1. \end{cases}$$

We introduced $\pi_1(X, x_0)$ in Chapter 1. The fact that this binary operation on the set of homotopy classes induces a group structure was proved there and the same proof works in this case as well. Moreover, if $f : (X, x_0) \to (Y, y_0)$ is continuous, we have an induced map $f_* : \pi_n(X, x_0) \to \pi_n(Y, y_0)$ defined in the obvious way.

Example 4.1.2. Note that $\pi_0(X, x_0) = [(S^0, 1), (X, x_0)]$ is the set of path components of X.

Proposition 4.1.3. *The group $\pi_n(X, x_0)$ is abelian for $n \geq 2$.*

Proof. This is always demonstrated using the following diagram

The point is that having a second coordinate in I^n allows us to push the maps f and g past each other, up to homotopy, by using the constant map in the first two coordinates as needed. □

We also have *relative homotopy groups* for $x_0 \in A \subset X$, defined by

$$\pi_n(X, A, x_0) = [(I^n, \partial I^n, \overline{\partial I^n - I^{n-1} \times \{1\}}), (X, A, x_0)],$$

with the same group operation. These are only sets for $n = 0, 1$, and groups for $n \geq 2$. These groups are abelian for $n \geq 3$ (we need a third coordinate to move maps past each other).

https://doi.org/10.1515/9783111014852-004

The long exact sequence

Consider the inclusions $i : (A, x_0) \hookrightarrow (X, x_0)$ and $j : (X, x_0, x_0) \hookrightarrow (X, A, x_0)$. Note that $\pi_n(X, x_0, x_0) = \pi_n(X, x_0)$. We have a sequence (basepoints omitted)

$$\cdots \xrightarrow{\partial} \pi_n(A) \xrightarrow{i_*} \pi_n(X) \xrightarrow{j_*} \pi_n(X, A) \xrightarrow{\partial} \pi_{n-1}(A) \to \cdots \xrightarrow{\partial} \pi_1(A) \xrightarrow{i_*} \pi_1(X),$$

where $\partial : \pi_n(X, A, x_0) \to \pi_{n-1}(A, x_0)$ is defined as follows. If $[f] \in \pi_n(X, A, x_0)$, then $f : (I^n, \partial I^n, \overline{\partial I^n - I^{n-1} \times \{1\}}) \to (X, A, x_0)$. Set

$$\partial[f] = [f|_{\partial I^n}] : (\partial I^n, \overline{\partial I^n - I^{n-1} \times \{1\}}) \to (A, x_0).$$

This is well defined because $\overline{\partial I^n - I^{n-1} \times \{1\}}$ is contractible.

Proposition 4.1.4. *This sequence is exact.*

Proof. We show that $\operatorname{im}(j_*) = \ker(\partial)$ and leave the others as an exercise. Suppose $\alpha = [f] \in \pi_n(X, x_0)$. Then f represents $j_*(\alpha)$. By definition, f maps I^{n-1} into x_0 and so $\partial(j_*(\alpha)) = [f|_{I^{n-1}}] = 0 \in \pi_{n-1}(A, x_0)$. Thus, $\operatorname{im}(j_*) \subseteq \ker(\partial)$. Conversely, if $\partial([f]) = 0$, then $\partial(f)$ is nullhomotopic and so $[f] = 0 \in \pi_n(X, A, x_0)$. $\square$

Note that homotopy groups are much easier to define than homology groups. They are much more difficult to compute, however. In fact, we still do not know all the homotopy groups of S^n, $n \geq 2$.

Proposition 4.1.5. $\pi_k(S^n) = 0$ *for* $k < n$ *and* $\pi_3(S^2) \neq 0$.

Proof. Let $f : S^k \to S^n$ be a continuous map with $k < n$. We may assume that f is cellular and since the k-skeleton of S^n may be taken to be a point, this deformation carries f to a constant map. (In general, this shows that $\pi_k(X^{(m-1)}) \xrightarrow{\cong} \pi_k(X)$ for $k < m - 1$ and $\pi_k(X^{(k)}) \to \pi_k(X)$ is surjective.) To see that $\pi_3(S^2) \neq 0$, consider $\mathbb{C}P^2 = \mathbb{C}P^1 \cup_f e^4$, where $f : S^3 \to \mathbb{C}P^1 = S^2$ is the Hopf map. The homotopy type depends only on the homotopy class of f. So if $\pi_3(S^2) = 0$, we would have $\mathbb{C}P^2 \simeq S^2 \vee S^4$. But then the cup product $\alpha \smile \alpha$, where α generates $H^2(\mathbb{C}P^2)$ would be 0, a contradiction. $\square$

Similarly, we can use $\mathbb{H}P^n$ to show that $\pi_{4n-1}(S^{2n}) \neq 0$.

Proposition 4.1.6. *If* $\tilde{X} \to X$ *is the universal cover, then* $\pi_i(X) \cong \pi_i(\tilde{X})$ *for* $i \geq 2$.

Proof. Consider the following diagram

$$\begin{array}{ccc} & & \tilde{X} \\ & \nearrow{\scriptstyle \tilde{f}} & \downarrow{\scriptstyle p} \\ S^n & \xrightarrow{f} & X \end{array}$$

Since S^n is simply connected for $n \geq 2$, the lift $\tilde{f}$ exists and hence $p_* : \pi_n(\tilde{X}) \rightarrow \pi_n(X)$ is surjective for $n \geq 2$. Suppose $\tilde{f} : S^n \rightarrow \tilde{X}$ satisfies $p \circ \tilde{f} = f \simeq \eta_{x_0}$. Then using homotopy lifting we get a homotopy $\tilde{f}_t$. By uniqueness of lifts, $\tilde{f} \simeq \eta_{\tilde{x}_0}$ and so p_* is injective. □

Now, if $f : S^n \rightarrow X$, we have an induced map $f_* : H_n(S^n) \rightarrow H_n(X)$. Since $H_n(S^n) \cong \mathbb{Z}$, we have the element $f_*(1) \in H_n(X)$. This determines a map $H : \pi_n(X) \rightarrow H_n(X)$ defined by

$$H : [f] \mapsto f_*(1).$$

This map is called the *Hurewicz homomorphism*. Note that if $X = S^n$, then $H(f) = \deg f$.

Theorem 4.1.7. *The map $H : \pi_n(S^n) \rightarrow H_n(S^n)$ is an isomorphism.*

Proof. It is not at all obvious that H is a homomorphism. We leave this as an exercise (with generous hints).

Since $H(\mathrm{id}_{S^n}) = 1$, H is surjective. Injectivity is much trickier. Suppose $H([f]) = 0$; that is, $\deg f = 0$. We must show that f is homotopic to a constant map. We may deform f until it is smooth and let $p \in S^n$ be a regular value. Thanks to an exercise in Section 2.4.1, we know that

$$\deg f = \sum_{x \in f^{-1}(p)} \deg f|_{x_i},$$

where $\deg f|_{x_i}$ is the local degree at x_i. By assumption, this sum is 0. We proceed by induction on the number of points in $f^{-1}(p)$. Assume $n \geq 2$. Suppose first that $f^{-1}(p) = \{x\} \cup \{y\}$, with $\deg f|_x = 1$ and $\deg f|_y = -1$. In $S^n \times I$, choose an embedded arc A connecting x and y, meeting $S^n \times \{0\}$ transversely at x and y (see Figure 4.1). We can choose small discs around x and y and construct a tubular neighborhood N of A. Note that $N \approx e^n \times I$. The map f is defined on $e^n \times \{0\}$ and $e^n \times \{1\}$ and since the local degrees of f at x and y have opposite signs, the degrees of f on $S^{n-1} \times \{0\}$ and $S^{n-1} \times \{1\}$ are the same. Thus,

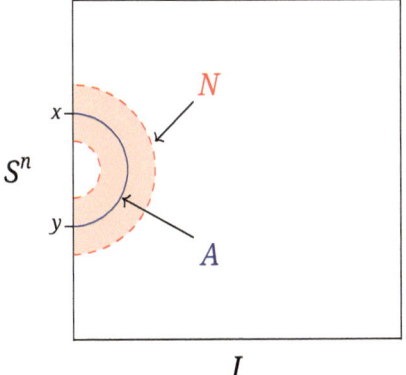

Figure 4.1: The embedded arc in $S^n \times I$.

f extends to $F : S^{n-1} \times I \to \partial D^n$. We may extend F to a map $e^n \times I \to D^n$ and so f extends to $\tilde{f} : S^n \times \{0\} \cup N \to S^n$ so that $\tilde{f} : \partial N \to \partial D^n \subset S^n$. Since $S^n - \text{int} D^n$ is contractible, we can extend $\tilde{f}$ to a map on the rest of $S^n \times I$. This map has p as a regular value with preimage A. The resulting map on $S^n \times \{1\}$ will miss p and so is nullhomotopic.

Now if $f^{-1}(p)$ has more than 2 points and $n \geq 2$, this argument allows us to deform f to "cancel" two of the points with opposite local degree. Repeat this process until $f^{-1}(p)$ is empty and then $\deg f = 0$ implies that f is nullhomotopic. If $n = 1$, more care is required, but we know that $\pi_1(S^1) \cong \mathbb{Z}$. □

A similar argument proves the following.

Theorem 4.1.8. *The map* $H : \pi_n(\bigvee_i S^n) \to H_n(\bigvee_i S^n) \cong \bigoplus_i \mathbb{Z}$ *is an isomorphism for* $n > 1$.

Corollary 4.1.9. *Excision fails for homotopy groups.*

Proof. Consider the pair (D^2, S^1) and the corresponding long exact sequence in homotopy

$$\pi_3(D^2) \to \pi_3(D^2, S^1) \xrightarrow{\cong} \pi_2(S^1) \to \pi_2(D^2)$$
$$\| \qquad\qquad\qquad\qquad\quad \| \qquad\qquad$$
$$0 \qquad\qquad\qquad\qquad 0 \qquad\quad 0$$

We conclude that $\pi_3(D^2, S^1) = 0$. But a similar calculation shows that $\pi_3(S^2, D^2) \cong \pi_3(S^2) \neq 0$. Consider the excisive triple (X, A, Z) shown in Figure 4.2. We have $\bar{Z} \subset \text{int}(A) \subset X$ and

$$(X - Z, A - Z) \longleftrightarrow (X, A)$$
$$\downarrow{\cong} \qquad\qquad\qquad \|$$
$$(D^2, S^1) \longleftrightarrow (S^2, D^2)$$

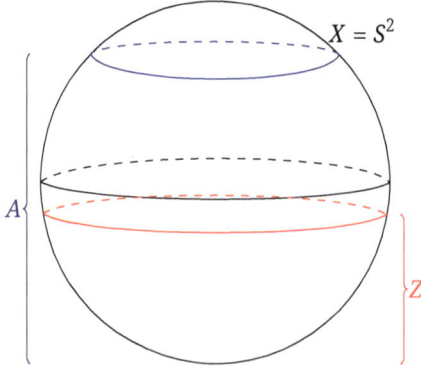

Figure 4.2: The excisive triple in S^2.

So if excision held for homotopy, we would have $\pi_3(D^2, S^1) \cong \pi_3(S^2, D^2)$, which is not the case. ☐

Exercises

Exercise 4.1.1. Prove that $\pi_n(X \times Y) \cong \pi_n(X) \times \pi_n(Y)$.

Exercise 4.1.2. Write down an explicit homotopy that proves that $\pi_n(X)$ is abelian for $n \geq 2$.

Exercise 4.1.3. Prove that $\pi_n(X, A, x_0)$ is abelian for $n \geq 3$.

Exercise 4.1.4. Complete the proof of Proposition 4.1.4.

Exercise 4.1.5. Prove that $\pi_{4n-1}(S^{2n}) \neq 0$.

Exercise 4.1.6. This exercise concerns the Hurewicz map $H : \pi_n(X) \to H_n(X)$. Denote the fundamental class of S^n by $[S^n] \in H_n(S^n; \mathbb{Z})$. Recall that the map H is defined as $H([f]) = f_*([S^n])$. Show that the operation in $\pi_n(X)$ can be realized as follows. Given two maps $f, g : S^n \to X$ consider the composite

$$S^n \xrightarrow{\text{pinch}} S^n \vee S^n \xrightarrow{f \vee g} X \vee X \xrightarrow{\text{fold}} X,$$

where "pinch" is the map that collapses the equator $S^{n-1} \subset S^n$ to a point, and "fold" is the map arising from the following pushout diagram

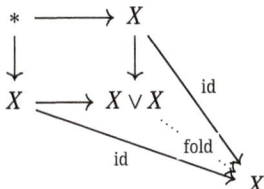

Show that this composite represents the group operation in $\pi_n(X)$. Deduce that H is a homomorphism.

Exercise 4.1.7. Consider the Hurewicz homomorphism $H : \pi_1(X) \to H_1(X)$. This map is an isomorphism for $X = S^1$, but in general it will not be (e. g., if $\pi_1(X)$ is nonabelian). Prove that H is always surjective and factors through the abelianization of $\pi_1(X)$. With more work, one can show that $H_1(X)$ is the abelianization of $\pi_1(X)$ (try it!).

Exercise 4.1.8. Let $X = S^1 \vee S^2$. Show that $\pi_2(X) \cong \mathbb{Z} \oplus \mathbb{Z} \oplus \cdots$ (an infinite direct sum of copies of $\mathbb{Z}$). (Hint: what is the universal cover of X?)

4.1.2 Whitehead's theorem

We have seen that it is possible for spaces that are not homotopy equivalent to have the same homology groups. It is even possible for such spaces to have the same homotopy groups (e. g. $\mathbb{R}P^2 \times S^3$ and $S^2 \times \mathbb{R}P^3$). While the algebraic invariants may be the same, the fact that the spaces are not homotopy equivalent means that there is no map between them that induces the isomorphisms on the algebra side.

Theorem 4.1.10 (Whitehead's Theorem). *Let X and Y be cell complexes with basepoints x_0 and y_0 0-cells. Let $f : (X, x_0) \to (Y, y_0)$ be a map inducing isomorphisms*

$$f_* : \pi_n(X, x_0) \to \pi_n(Y, y_0)$$

for all n. Suppose that Y is connected. Then f is a homotopy equivalence.

Proof. We first prove the following. Suppose $\dim X < \infty$ and $\pi_n(X, x_0) = 0$ for all $n \geq 0$. We claim that X is contractible (this is the special case $x_0 \to X$). Since $\pi_0(X, x_0)$ is a one-point set, the 0-skeleton can be deformed to $x_0 \in X$ (i. e., for each 0-cell x, choose a path from x to x_0; this gives a homotopy from $X^{(0)}$ to x_0). Use the homotopy extension property to obtain a continuous family $f_t : X \to X, 0 \leq t \leq 1$ with $f_0 = \mathrm{id}_X$ and $f_1(X^{(0)}) = \{x_0\}$. Now consider $X^{(1)}$. The image under f_1 of any edge e in $X^{(1)}$ is a loop at x_0. Since $\pi_1(X, x_0) = 0$, we can deform $f_1(e)$ through loops based at x_0 to the constant loop. Doing this for each 1-cell gives a homotopy from $f_1|_{X^{(1)}}$ to the constant map $f_2 : X^{(1)} \to \{x_0\}$. Use the HEP to extend this to a homotopy $f_t : X \to X, 1 \leq t \leq 2$ with $f_2(X^{(2)}) = \{x_0\}$. Continue in this fashion, using the fact that $\pi_n(X, x_0) = 0$ for all n to get a homotopy from id_X to the constant map at x_0. We note the following.

(1) The same argument shows that if $\pi_n(X, x_0) = 0$ for $n < N$, then there is a homotopy $f_t : X \to X, 0 \leq t \leq 1$, with $f_0 = \mathrm{id}_X$ and $f_1(X^{(N-1)}) = \{x_0\}$. It follows that $\tilde{H}_k(X) = 0$ for $k < N$.

(2) If $\dim X = \infty$, make the first homotopy last for $0 \leq t \leq 1/2$, the second for $1/2 \leq t \leq 3/4$, etc., and then define $f_1(X) = \{x_0\}$. Since X has the weak topology, this homotopy is continuous.

(3) There is a relative version: if $\pi_n(X, A, x_0) = 0$ for all n, then there is a deformation retraction $f_1 : X \to A$.

To prove the general case, given $f : X \to Y$, consider the mapping cylinder M_f. Since M_f retracts to Y, we have $\pi_k(M_f) \cong \pi_k(Y)$. If $f_* : \pi_k(X) \to \pi_k(Y)$ is an isomorphism for all k, then so is $i_* : \pi_k(X) \to \pi_k(M_f)$. It follows that $\pi_k(M_f, X) = 0$ for all k and so there is a deformation retraction $r : M_f \to X$. The composition $Y \hookrightarrow M \xrightarrow{r} X$ is the required homotopy inverse to f. □

Example 4.1.11. Consider S^k as $I^k/\partial I^k$. There is a product map

$$I^2/\partial I^2 \times I^2/\partial I^2 \longrightarrow I^4/\partial I^4$$

$$\overline{((x_1, x_2), (x_3, x_4))} \longmapsto \overline{(x_1, x_2, x_3, x_4)}$$

This is a map $f : S^2 \times S^2 \to S^4$. Note that $f|_{S^2 \times \{p\}}$ and $f|_{\{q\} \times S^2}$ are nullhomotopic for any p or q, since $\pi_2(S^4) = 0$. Since $\pi_k(S^2 \times S^2) \cong \pi_k(S^2) \times \pi_k(S^2)$ for any k, we see that $f_* : \pi_k(S^2 \times S^2) \to \pi_k(S^4)$ is trivial for all k. But f is not homotopic to a constant map, since one can show that $f_* : H_4(S^2 \times S^2) \to H_4(S^4)$ is an isomorphism.

Exercises

Exercise 4.1.9. Prove that the map f in Example 4.1.11 induces an isomorphism on H_4.

Exercise 4.1.10. Suppose the cell complex X is the union of an increasing sequence of subcomplexes $X_1 \subset X_2 \subset \cdots$ such that each inclusion $X_i \to X_{i+1}$ is nullhomotopic. Prove that X is contractible. Thus, S^∞ is contractible.

Exercise 4.1.11. Let X be the union of the y-axis in the plane and the set of points $\{(x, \sin(1/x)) \mid x > 0\}$. Show that $\pi_0(X)$ has two elements, but X is connected as a topological space. Let $Y = \{a, b\}$ be a space with two distinct points and let $f : Y \to X$ be the map $f(a) = (0, 0)$ and $f(b) = (1, \sin(1))$. Show that f is an isomorphism on homotopy groups, but f is not a homotopy equivalence. Thus, Whitehead's theorem is false for non-CW complexes.

4.1.3 Hurewicz theorem

The main goal of this section is to prove the following result and derive some consequences.

Theorem 4.1.12 (Hurewicz). *Let X be a cell complex. If $\pi_k(X) = 0$ for $k < n$, then*
(1) *$\tilde{H}_k(X) = 0$ for $k < n$, and*
(2) *$H : \pi_n(X) \to H_n(X)$ is an isomorphism, provided $n > 1$.*

Proof. The first statement follows from the Whitehead theorem. We need only prove the second.

 Step 1. We first show that H is surjective. Since $\pi_k(X) = 0$ for $k < n$, there is a map $f : X \to X$ such that (a) $f \simeq \mathrm{id}_X$ and (b) $f(X^{(n-1)}) = \{x_0\}$. Let $\alpha \in H_n(X)$ and let $\sum m_i e_i^n$ be a cycle representing α. We know that $f|_{e_i^n}$ takes the pair $(e_i^n, \partial e_i^n)$ to (X, x_0); it follows that $f|_{e_i^n}$ represents an element of $\pi_n(X)$. Since $f \simeq \mathrm{id}_X$, we see that

$$a = f_*(\alpha) = \left[\sum m_i f(e_i^n)\right].$$

Since each $f(e_i^n)$ is represented by a sphere, this latter class is in the image of H.

Step 2. We now show that H is injective. Let $f : S^n \to X$ be a map such that $f_*([S^n]) = 0$ in $H_n(X)$. We can deform f until $f(S^n) \subseteq X^{(n)}$ and f is transverse to a point p_i in the interior of each cell e_i^n (this means that the image of f intersects a point in the interior of each n-cell rather than hitting the boundary or intersecting tangentially). Let

$$\lambda_i = \sum_{x \in f^{-1}(p_i)} \deg f|_{p_i}.$$

The chain $\sum \lambda_i e_i^n$ is a cycle in $C_n(X)$ representing $f_*([S^n])$ in $H_n(X)$, and so this is a boundary. That is, there exist μ_j so that $\sum \lambda_i e_i^n = \partial(\sum \mu_j e_j^{n+1})$. Adding to $f : S^n \to X^{(n)}$ the chain $\sum \mu_j \partial e_j^{n+1}$ makes each $\lambda_i = 0$ and does not change the homotopy class of f. We know there is a map $\psi : X \to X$ with $\psi \simeq \mathrm{id}_X$ and $\psi(X^{(n-1)}) = \{x_0\}$. This means that ψ factors through a wedge of spheres:

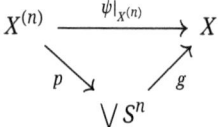

Since $f : S^n \to X^{(n)}$ has each $\lambda_i = 0$, the composite $p \circ f : S^n \to \bigvee S^n$ is homologous to 0, but then $p \circ f$ is nullhomotopic, since $\pi_n(\bigvee S^n) \cong H_n(\bigvee S^n)$. Thus, $g \circ p \circ f = \psi \circ f$ is nullhomotopic, and since $\psi \simeq \mathrm{id}_X$, f is nullhomotopic. □

Remark 4.1.13. There is a relative version: if (X, A) is a pair with $\pi_1(A) = 0$ and $\pi_k(X, A) = 0$ for $k < n$, $n \geq 2$, then $H_k(X, A) = 0$ for $k < n$ and $H : \pi_n(X, A) \to H_n(X, A)$ is an isomorphism.

Corollary 4.1.14. *If X and Y are simply connected cell complexes and $f : X \to Y$ induces an isomorphism on homology, then f is a homotopy equivalence.*

Proof. Let M_f be the mapping cylinder. Then $Y \hookrightarrow M_f$ is a deformation retract and we have $\pi_i(Y) \cong \pi_i(M_f)$ and $H_i(Y) \cong H_i(M_f)$ for all $i \geq 0$. The inclusion $X \hookrightarrow M_f$ gives exact sequences

$$\cdots \to \pi_i(X) \xrightarrow{f_*} \pi_i(Y) \to \pi_i(M_f, X) \to \pi_{i-1}(X) \xrightarrow{f_*} \cdots$$

$$\cdots \to H_i(X) \xrightarrow{f_*} H_i(Y) \to H_i(M_f, X) \to H_{i-1}(X) \xrightarrow{f_*} \cdots$$

Since f_* is an isomorphism on homology, we see that $H_i(M_f, X) = 0$ for all $i \geq 0$. By the Whitehead theorem, M_f deformation retracts to X. □

Corollary 4.1.15. *If X has the homotopy type of an n-dimensional cell complex and if $\pi_i(X) = 0$ for $i \leq n$, then X is contractible.*

Proof. Since $\pi_i(X) = 0$ for $i \leq n$, we have $\tilde{H}_i(X) = 0$ for $i \leq n$. But also, $\tilde{H}_i(X) = 0$ for $i > n$ and so $\pi_i(X) = 0$ for all $i \geq 0$ by Theorem 4.1.12. The result then follows from the Whitehead Theorem. □

Corollary 4.1.16. *If X has the homotopy type of an n-dimensional cell complex and $\pi_i(X) = 0$ for $i \leq n - 1$, then $X \simeq \bigvee S^n$. In particular, if $H_n(X) \cong \mathbb{Z}$, then $X \simeq S^n$ (thus, a simply-connected homology sphere is a homotopy sphere).*

Proof. We may assume that $X = X^{(n)}$ and that we have

$$X \xrightarrow{g} \bigvee S^n \xrightarrow{f} X.$$

It follows that $f_* : H_n(\bigvee S^n) \to H_n(X)$ is surjective and we may choose a basis for $H_n(\bigvee S^n) = \bigoplus H_n(S^n)$ and for $H_n(X)$ so that the matrix of f_* is given by

$$\left[\begin{array}{cccc|c}
a_1 & & & & 0 \;\; 0 \\
& a_2 & & & \vdots \\
& & \ddots & & \vdots \\
& & & a_b & 0
\end{array}\right]$$
$$\underbrace{}_{c=\operatorname{rank} H_n(\bigvee S^n)}$$

where $b = \operatorname{rank} H_n(X)$ and $a_i = \pm 1$ for all i. Since $\pi_n(\bigvee S^n) \cong H_n(\bigvee S^n)$, we may find a map fitting into the diagram

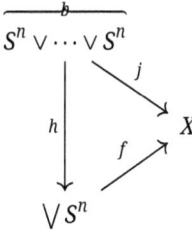

such that h_* has matrix

$$\left[\begin{array}{ccc}
a_1 & & \\
& \ddots & \\
& & a_b \\
\hline
& 0 &
\end{array}\right]$$
$$\underbrace{}_{b}$$

where there are $c - b$ rows of 0's. It follows that j_* is an isomorphism on homology and so is a homotopy equivalence. □

Exercises

Exercise 4.1.12. Prove that if X is a simply connected cell complex and $H_i(X) = 0$ for $i > n$, then X has the homotopy type of an $(n + 1)$-dimensional complex. If, in addition, $H_n(X)$ is free, then X has the homotopy type of an n-dimensional complex.

Exercise 4.1.13. In this exercise, we will make use of relative homotopy groups to give another proof of the cellular approximation theorem (Theorem 1.2.25). Suppose that X and Y are (connected) CW complexes and that $f : X \to Y$ is continuous. Show that f is homotopic to a cellular map $g : X \to Y$, as follows. First, for simplicity, assume that $\dim X < \infty$. Show that $f \simeq g_0$, where $g_0(X^{(0)}) \subset Y^{(0)}$ (this is easy). Assume inductively that we have defined a cellular map $g_n : X \to Y$ with $g_n(X^{(n)}) \subset Y^{(n)}$ and $f \simeq g_n$. Show that we have an exact sequence

$$\pi_{n+1}(Y^{(n+1)}, Y^{(n)}) \to \pi_{n+1}(Y, Y^{(n)}) \to 0,$$

and that $\pi_{n+1}(Y^{(n+1)}, Y^{(n)})$ is a free group on the $(n+1)$-cells in Y. Now let e^{n+1} be an $(n+1)$-cell in X. Then $g_n : (e^{n+1}, \partial e^{(n+1)}) \to (Y, Y^{(n)})$ gives an element in $\pi_{n+1}(Y, Y^{(n+1)})$. By the previous result, g_n may be deformed into $\tilde{g}_n : (e^{n+1}, \partial e^{(n+1)}) \to (Y^{(n+1)}, Y^{(n)})$. Show that it now suffices to show that, during the homotopy $g_n \simeq \tilde{g}_n$, the map may be kept constant on ∂e^{n+1}. Prove the more general statement: If $A \subset B \subset Y$ and if $\alpha : (e^{n+1}, S^n) \to (Y, A)$ is homotopic to $\beta : (e^{n+1}, S^n) \to (B, A)$, then α is homotopic to $\gamma : (e^{n+1}, S^n) \to (B, A)$, where all maps in the homotopy are constant on S^n. (Hint: given a homotopy α_t from α to β, put a collar on S^n and define γ_t by α_{st}, $0 \leq s \leq 1$ on the collar and then use α_t on the top of the cell.). Proceeding inductively, we then obtain $g_{n+1} : X^{(n+1)} \to Y^{(n+1)}$, $g_{n+1}|_{X^{(n)}}$ and $g_{n+1} \simeq f$ on $X^{(n+1)}$. Apply HEP to extend g_{n+1} to all of X. What modification is necessary if $\dim X = \infty$? Note the similarity of this proof with that of Theorem 1.2.25.

4.1.4 The long exact sequence of a fibration

Suppose that $\pi : E \to B$ is a fibration and let $\gamma : I \to B$ be a path from b_0 to b_1. Consider the following diagram

Apply homotopy lifting to get a map $\pi^{-1}(b_0) \times I \to E$ covering $\gamma \circ p_2$. This implies that $\pi^{-1}(b_0) \times \{1\} \simeq \pi^{-1}(b_1)$ and so all fibers are homotopic if B is path connected.

Now, if γ is a loop at b_0, the resulting homotopy equivalence $\pi^{-1}(b_0) \times \{1\} \to \pi^{-1}(b_0)$ is a homotopy automorphism of $\pi^{-1}(b_0)$. Its homotopy class depends only on the homotopy class of γ in $\pi_1(B, b_0)$ and so we have a homomorphism

$$\pi_1(B, b_0) \to \mathrm{Aut}(\pi^{-1}(b_0)),$$

where the target is the group of homotopy classes of homotopy equivalences of $\pi^{-1}(b_0)$. This is called the *action of $\pi_1(B, b_0)$ on the fiber*.

Theorem 4.1.17. *Let $\pi : E \to B$ be a fibration and let $F = \pi^{-1}(b_0)$ be the fiber. Then there is an exact sequence*

$$\cdots \to \pi_{n+1}(B, b_0) \xrightarrow{\partial} \pi_n(F, e_0) \xrightarrow{i_*} \pi_n(E, e_0) \xrightarrow{\pi_*} \pi_n(B, b_0) \xrightarrow{\partial} \cdots$$

where $i : F \hookrightarrow E$ is the inclusion.

Proof. Consider the long exact sequence of the pair $i : F \hookrightarrow E$:

$$\cdots \to \pi_{n+1}(E, F) \xrightarrow{\partial} \pi_n(F) \to \pi_n(E) \to \pi_n(E, F) \xrightarrow{\partial} \pi_{n-1}(F) \to \cdots$$

Consider the map $\pi_* : \pi_n(E, F) \to \pi_n(B, b_0)$. We claim that this map is an isomorphism. To see this, let $f : (I^n, \partial I^n) \to (B, b_0)$ be a map. There is a lift $g : I^n \to E$ and since I^n is contractible, $f \simeq \eta_{b_0} : I^n \to \{b_0\}$ (the homotopy will deform ∂I^n off b_0 in general). By homotopy lifting, any map which is homotopic to a map that lifts must itself lift. The map $g : (I^n, \partial I^n) \to (E, F)$ determines an element of $\pi_n(E, F)$ and $\pi_*([g]) = [f] \in \pi_n(B, b_0)$, so that π_* is surjective. To prove that π_* is injective, we show that if $g_0, g_1 : (I^n, \partial I^n) \to (E, F)$ are such that $\pi \circ g_0 \simeq \pi \circ g_1$, then $g_0 \simeq g_1$. This follows from the following lemma by taking $X = I^{n-1}$.

Lemma 4.1.18. *Given $H : X \times I \to B$ and two lifts $\tilde{H}_1, \tilde{H}_2 : X \times I \to E$ which agree on $X \times \{0\}$, there is a homotopy of lifts of H connecting $\tilde{H}_1$ and $\tilde{H}_2$, all of which agree on $X \times \{0\}$.*

Proof. Let $\tilde{h} = H|_{X \times \{0\}}$. We have a commutative diagram

$$
\begin{array}{ccc}
Z & \xrightarrow{\tilde{H}_1 \cup (\tilde{h} \times I) \cup \tilde{H}_2} & E \\
\downarrow & & \downarrow{\scriptstyle \pi} \\
X \times I \times I & \xrightarrow{J} & B
\end{array}
$$

where $Z = (X \times I \times \{0\}) \cup (X \times \{0\} \times I) \cup (X \times \{1\} \times I)$. Since $X \times I \times I$ deforms to Z, we can extend $\tilde{H}_1 \cup (\tilde{h} \times I) \cup \tilde{H}_2$ to a lifting of J. $\quad\square$

$\square$

Example 4.1.19. Consider the fibration

$$S^1 \longrightarrow S^{2n+1}$$
$$\downarrow$$
$$\mathbb{C}P^n$$

Since $\pi_i(S^1) = 0$ for $i > 1$, we have $\pi_i(\mathbb{C}P^n) \cong \pi_i(S^{2n+1})$ for $i \neq 2$ and $\pi_2(\mathbb{C}P^n) \cong \mathbb{Z}$:

$$\cdots \to \pi_2(S^{2n+1}) \to \pi_2(\mathbb{C}P^n) \xrightarrow{\cong} \pi_1(S^1) \to \pi_1(S^{2n+1})$$
$$\parallel \qquad\qquad\qquad\qquad \downarrow\cong \qquad\qquad \parallel$$
$$0 \qquad\qquad\qquad\qquad \mathbb{Z} \qquad\qquad\quad 0$$

Example 4.1.20. Let ΩB be the loop space on B. We have a fibration

$$\Omega B \longrightarrow \mathcal{P}B$$
$$\downarrow$$
$$B$$

and since $\mathcal{P}B$ is contractible, we see $\pi_{i-1}(\Omega B) \cong \pi_i(B)$. Iterating this constructing, we have the higher loop spaces $\Omega^i B$ and $\pi_i(B) \cong \pi_0(\Omega^i B)$.

Example 4.1.21. Recall that an r-frame in $\mathbb{C}^N$ is given by r mutually orthogonal unit vectors $(e_1, \ldots, e_r)$. The set of r-frames in $\mathbb{C}^N$ is denoted by $S(r, N)$ and is called the Stiefel manifold. (It is a manifold since it is a closed subset of the product $\underbrace{S^{2N-1} \times \cdots \times S^{2N-1}}_{r}$ with the induced topology.) We claim that $\pi_i(S(r, N)) = 0$ for $i < 2N - 2r + 1$ and $\pi_{2N-2r+1}(S(r, N)) \cong \mathbb{Z}$. To see this, note that $S(1, N) = S^{2N-1}$ and the result holds for $r = 1$. Assume inductively that the result is true for $r - 1$ and consider the fibration

$$S^{(2N-1)-(2r-2)} \longrightarrow S(r, N)$$
$$\downarrow \pi$$
$$S(r-1, N)$$

where $\pi(e_1, \ldots, e_r) = (e_1, \ldots, e_{r-1})$. Note that $(2N - 1) - (2r - 2) = 2N - 2r + 1$. Why is this sphere the fiber? Choose a unit vector in $\mathbb{C}^N$ orthogonal to the $(r-1)$-plane determined by $(e_1, \ldots, e_{r-1})$. The number of real dimensions available is $(2N - 1) - 2(r - 1)$ and so we obtain this sphere as the fiber. Consider the long exact homotopy sequence

$$\cdots \to \pi_{i+1}(S(r-1, N)) \to \pi_i(S^{2N-2r+1}) \to \pi_i(S(r, N)) \to \pi_i(S(r-1, N)) \to \cdots$$

By the inductive hypothesis,

$$\pi_i(S(r-1, N)) = 0, \quad i < 2N - 2(r-1) + 1 = 2N - 2r + 3$$

and $\pi_{2N-2r+3}(S(r-1,N)) \cong \mathbb{Z}$. We then have

$$\cdots \to \pi_{2N-2r+2}(S(r-1,N)) \to \pi_{2N-2r+1}(S^{2N-2r+1}) \xrightarrow{\cong} \pi_{2N-2r+1}(S(r,N)) \to \pi_{2N-2r+1}(S(r-1,N))$$

$$\| \qquad\qquad\qquad\qquad\quad \downarrow^{\cong} \qquad\qquad\qquad\qquad\qquad\qquad\qquad \|$$

$$0 \qquad\qquad\qquad\qquad\quad \mathbb{Z} \qquad\qquad\qquad\qquad\qquad\qquad\qquad 0$$

Thus, $\pi_{2N-2r+1}(S(r,N)) \cong \mathbb{Z}$ and $\pi_i(S(r,N)) = 0$ for $i < 2N - 2r + 1$.

Example 4.1.22. Let $U(r)$ be the group of unitary $r \times r$ complex matrices ($A\overline{A}^T = I$). Note that columns of a unitary matrix give an orthonormal basis of $\mathbb{C}^r$ and so $U(r) = S(r,r)$. Recall the Grassmann manifold $G(r,N)$ of r-planes in $\mathbb{C}^N$ and define a map $\pi : S(r,N) \to G(r,N)$ by $\pi(e_1,\ldots,e_r) = e_1 \wedge \cdots \wedge e_r$, the r-plane spanned by $e_1,\ldots,e_r$. To compute the fiber of π, fix an r-plane and note that the set of orthonormal bases of the plane is $U(r)$. We therefore have a fibration

$$U(r) \longrightarrow S(r,N)$$
$$\downarrow^{\pi}$$
$$G(r,N)$$

Via the previous example, we conclude that $\pi(G(r,N)) \cong \pi_{i-1}(U(r))$ for $i < 2N - 2r + 1$.

Example 4.1.23. Consider the inclusion $U(n) \hookrightarrow U(n+1)$ given by

$$A \mapsto \left[\begin{array}{c|c} A & 0 \\ \hline 0 & 1 \end{array}\right]$$

Define $\pi : U(n+1) \to S^{2n+1}$ by $\pi(A) = $ last column of $A \in S^{2n+1}$. Note that

$$\pi^{-1}\begin{pmatrix} 0 \\ \vdots \\ 0 \\ 1 \end{pmatrix} = \left[\begin{array}{c|c} A & 0 \\ \hline 0 & 1 \end{array}\right], \quad A \in U(n).$$

We therefore have a fibration

$$U(n) \longrightarrow U(n+1)$$
$$\downarrow^{\pi}$$
$$S^{2n+1}$$

The long exact sequence then implies that $\pi_i(U(n)) \cong \pi_i(U(n+1))$ for $i < 2n$. It follows that the homotopy groups of $U(n)$ are stable:

$$\pi_i(U(n)) \cong \pi_i(U(n+m)), \quad i < 2n, m \geq 0.$$

In fact, we have the following result, whose proof is beyond the scope of this text.

Theorem 4.1.24 (Bott Periodicity). *Let* $U = \varinjlim U(n)$. *Then* $\pi_{2i+1}(U) \cong \mathbb{Z}$ *and* $\pi_{2i}(U) = 0$.

Eilenberg–MacLane spaces

Let π be a group and let n be a positive integer. If $n > 1$, assume that π is abelian. We claim that there is a space $K(\pi, n)$ satisfying

(1) $\pi_n(K(\pi, n)) \cong \pi$;

(2) $\pi_k(K(\pi, n)) = 0$ for $k \neq n$.

The construction is straightforward. Choose a presentation of π. Take a point, and for each generator of π attach an n-cell e^n to obtain a wedge of spheres, one for each generator. This gives the n-skeleton $X^{(n)}$. A relation among the generators is a word and so this provides an attaching map $e^{n+1} \to X^{(n)}$. Attach a cell for each relation to obtain the $(n + 1)$-skeleton $X^{(n+1)}$. We now have $\pi_n(X^{(n+1)}) \cong \pi$ and $\pi_k(X^{(n+1)}) = 0$ for $k < n$. Now attach $(n + 2)$-cells to $X^{(n+1)}$ to kill π_{n+1}, etc. The end result is a space $K(\pi, n)$ with the requisite properties

There may be other spaces satisfying this, but they are all homotopy equivalent.

Homology of groups

Since Eilenberg–MacLane spaces are unique up to homotopy, we can define the homology of a group G as follows:

$$H_\bullet(G) = H_\bullet(K(G, 1)).$$

Moreover, we may use any abelian group of coefficients for homology (or, more generally, a module over the group ring $\mathbb{Z}G$, but we will not pursue that here). The exercises at the end of the section give some examples of $K(G, 1)$ for various groups, and we will use these now to compute some homology.

Example 4.1.25. We have $K(\mathbb{Z}, 1) = S^1$. From this we deduce that

$$H_k(\mathbb{Z}) = \begin{cases} \mathbb{Z} & k = 0, 1 \\ 0 & \text{otherwise.} \end{cases}$$

Example 4.1.26. We have $K(\mathbb{Z}_2, 1) = \mathbb{R}P^\infty$. This space has a cell decomposition

$$\mathbb{R}P^\infty = e^0 \cup e^1 \cup e^2 \cup \cdots$$

with one cell in each positive dimension. The corresponding boundary maps in the cellular chain complex are alternately 0 and multiplication by 2. As a result we compute

$$H_k(\mathbb{Z}_2; \mathbb{Z}) = \begin{cases} \mathbb{Z} & k = 0 \\ \mathbb{Z}_2 & k \text{ odd} \\ 0 & \text{otherwise.} \end{cases}$$

Finding explicit models of $K(G, 1)$ can be tricky, so we need to find an alternate approach. The universal cover E of a $K(G, 1)$ is contractible and admits a free action of the group $G = \pi_1(G, 1)$ as the group of deck transformations. This implies that the cellular chain complex $C_{\bullet}(E)$ is a complex of free $\mathbb{Z}G$-modules. Recall that the *group ring* $\mathbb{Z}G$ is the free abelian group on the elements of G with product

$$\left(\sum_{g \in G} n_g g\right)\left(\sum_{h \in G} m_h h\right) = \sum_{g, h \in G} n_g m_h(gh),$$

(here, we write the group operation multiplicatively). For example, if $G = \langle t \mid t^n = 1 \rangle$ is the cyclic group of order n, then

$$\mathbb{Z}G = \mathbb{Z}[t]/\langle t^n - 1 \rangle,$$

the truncated polynomial ring. The group ring is a noncommutative ring in general (unless G is abelian), so we need to take care when talking about $\mathbb{Z}G$-modules, specifying whether they are left or right modules.

Now, consider the augmented cellular chain complex $C_{\bullet}(E)$:

$$\cdots \to C_k(E) \xrightarrow{d} C_{k-1}(E) \xrightarrow{d} \cdots \xrightarrow{d} C_0(E) \xrightarrow{\varepsilon} \mathbb{Z} \to 0.$$

Since E is contractible, this chain complex is exact, and since G acts freely (on the right, say) on these modules, this is a *free resolution* of the trivial module $\mathbb{Z}$ over $\mathbb{Z}G$. Moreover, the chain complex $C_{\bullet}(K(G, 1))$ is obtained by factoring out this action; that is, we can compute $H_{\bullet}(G, \mathbb{Z})$ by applying the functor $-\otimes_{\mathbb{Z}G} \mathbb{Z}$ to $C_{\bullet}(E)$. This suggests that to compute $H_{\bullet}(G)$ it suffices to find a free (or, more generally, projective) resolution $F_{\bullet}$ of $\mathbb{Z}$ over the group ring $\mathbb{Z}G$ and then apply $-\otimes_{\mathbb{Z}G} \mathbb{Z}$ to it to yield a complex whose homology is that of G (since any two free resolutions are homotopy equivalent).

Example 4.1.27. Consider the following resolution of $\mathbb{Z}$ over the group ring $\mathbb{Z}G$ for $G = \mathbb{Z}_n = \langle t \mid t^n = 1 \rangle$:

$$\cdots \to \mathbb{Z}G \xrightarrow{\Delta} \mathbb{Z}G \xrightarrow{t-1} \cdots \xrightarrow{\Delta} \mathbb{Z}G \xrightarrow{t-1} \mathbb{Z}G \xrightarrow{\varepsilon} \mathbb{Z} \to 0,$$

where the map Δ is multiplication by the element $1 + t + \cdots + t^{n-1}$. Since $\Delta(t - 1) = 0$, this is at least a chain complex and it is easy to see that it is exact. Now, applying $-\otimes_{\mathbb{Z}G} \mathbb{Z}$, we obtain a chain complex

$$\cdots \to \mathbb{Z} \xrightarrow{\times n} \mathbb{Z} \xrightarrow{0} \cdots \xrightarrow{0} \mathbb{Z} \xrightarrow{\times n} \mathbb{Z} \xrightarrow{0} \mathbb{Z}.$$

The maps are as indicated since t acts trivially on $\mathbb{Z}$, and therefore $t-1$ is the zero map, while Δ becomes multiplication by n. Note that when $n = 2$ this is precisely the cellular chain complex for $\mathbb{R}P^\infty$. We therefore conclude

$$H_k(\mathbb{Z}_n) = \begin{cases} \mathbb{Z} & k = 0 \\ \mathbb{Z}_n & k \text{ odd} \\ 0 & k > 0 \text{ even.} \end{cases}$$

Note that we could use any left $\mathbb{Z}G$-module M in place of $\mathbb{Z}$ and thereby obtain $H_\bullet(G; M)$. We will restrict our attention to the trivial module in our discussion, but the interested reader is encouraged to investigate this idea further.

Now, suppose G is an amalgamated free product: $G = G_1 *_A G_2$. Then we have a diagram of associated Eilenberg–MacLane spaces:

$$\begin{array}{ccc} K(A,1) & \longrightarrow & K(G_1,1) \\ \downarrow & & \downarrow \\ K(G_2) & \longrightarrow & X \end{array}$$

We claim that $X = K(G_1,1) \cup_{K(A,1)} K(G_2,1)$ is a $K(G,1)$-space. The Seifert–Van Kampen theorem implies that the fundamental group of X is indeed G. We need only show that the universal cover of X is contractible, and for that it suffices to show that it has no nontrivial homology in positive degrees. Consider the universal covers of each piece:

$$\begin{array}{ccc} E_A & \longrightarrow & E_{G_1} \\ \downarrow & & \downarrow \\ E_{G_2} & \longrightarrow & \tilde{X} \end{array}$$

The Mayer–Vietoris sequence associated to this diagram shows that the homology of $\tilde{X}$ vanishes in positive degrees. We therefore obtain a Mayer–Vietoris sequence in group homology:

$$\cdots \to H_n(A) \to H_n(G_1) \oplus H_n(G_2) \to H_n(G) \xrightarrow{\partial} H_{n-1}(A) \to \cdots,$$

where the maps are induced by the various inclusions and ∂ is the usual connecting homomorphism.

Example 4.1.28. Suppose $G = G_1 * G_2$ is the free product of G_1 and G_2. Then the Mayer–Vietoris sequence shows that

$$\tilde{H}_n(G) \cong \tilde{H}_n(G_1) \oplus \tilde{H}_n(G_2),$$

since the homology of the trivial group is clearly trivial in all positive degrees.

Exercises

Exercise 4.1.14. Show that $K(\mathbb{Z}, 1) = S^1$.

Exercise 4.1.15. Show that $K(\mathbb{Z}, 2) = \mathbb{C}P^\infty$.

Exercise 4.1.16. Show that $K(\mathbb{Z}_2, 1) = \mathbb{R}P^\infty$.

Exercise 4.1.17. Show that the infinite lens space L_m is a $K(\mathbb{Z}_m, 1)$ space.

Exercise 4.1.18. Show that the orientable surface M_g is a $K(\pi, 1)$ for $\pi = \pi_1(M_g)$.

Exercise 4.1.19. Use the Mayer–Vietoris sequence to compute $H_\bullet(\mathrm{SL}_2(\mathbb{Z}))$.

4.2 The Leray–Serre spectral sequence

Recall the long exact homology sequence of the pair (X, A):

$$\cdots \longrightarrow H_n(A) \longrightarrow H_n(X) \longrightarrow H_n(X, A) \longrightarrow H_{n-1}(A) \longrightarrow \cdots$$

This arises from a more formal set-up on the chain level:

$$0 \longrightarrow C_\bullet(A) \longrightarrow C_\bullet(X) \longrightarrow C_\bullet(X, A) \longrightarrow 0.$$

Moreover, this formalism works with cohomology. But this really has to do with a *filtration* $\emptyset \subseteq A \subseteq X$ of the space X and the corresponding filtration on the (co)chains of X. The long exact sequence allows us to compute $H_\bullet(X)$ (or $H^\bullet(X)$) from the (co)homology of A and X/A.

More generally, a filtration

$$\emptyset \subset X_0 \subset X_1 \subset \cdots \subset X_r = X$$

gives rise to a *spectral sequence* for computing the (co)homology of X. The reader may note that we have seen filtrations before (Section 2.7), and there is a relationship between spectral sequences and persistent homology. Our goal here, however, is to compute the (co)homology of the space X, not its persistence diagram.

4.2.1 Cohomology of fibrations

Suppose we have a fibration

$$
\begin{array}{ccc}
F & \longrightarrow & E \\
& & \downarrow{\scriptstyle \pi} \\
& & B
\end{array}
$$

where B is connected. We are interested in the cohomology of the space E and we hope that it can be computed in terms of the cohomology of F and B. Filter B by its skeleta; denote the p-skeleton by $B^{(p)}$. Let $E^{(p)} = \pi^{-1}(B^{(p)})$. Choose $b_0 \in B$ and let $F = \pi^{-1}(b_0)$.

Proposition 4.2.1. *If B is path connected and if $\pi_1(B, b_0)$ acts trivially on $H^{\bullet}(F)$, then there are isomorphisms*

$$H^n(E^{(p)}, E^{(p-1)}) \cong \prod_{e^p \in B} H^n(\pi^{-1}(e^p), \pi^{-1}(\partial e^p)) \cong C^p(B; H^{n-p}(F)).$$

Proof. Let $\varphi : \coprod e^p \to B$ be the map of all p-cells into B. Consider the induced fibration $\varphi^* E \to \coprod e^p$ (Definition 1.3.9). Since the map

$$\left(\coprod e^p, \coprod \partial e^p \right) \to (B^{(p)}, B^{(p-1)})$$

is a relative homeomorphism, so is the map

$$(\varphi^* E, \varphi^* E|_{\coprod \partial e^p}) \to (E^{(p)}, E^{(p-1)}).$$

Thus, by excision,

$$H^{\bullet}(E^{(p)}, E^{(p-1)}) \cong H^{\bullet}(\varphi^* E, \varphi^* E|_{\coprod \partial e^p})$$
$$\cong \prod_{e^p \in B} H^{\bullet}(\varphi^* E|_{e^p}, \varphi^* E|_{\partial e^p}).$$

Now, suppose $\eta : E' \to e^p$ is any fibration. Let $F_0 = \eta^{-1}(0)$. We have a diagram

$$
\begin{array}{ccc}
F_0 \times \{0\} & \lhook\joinrel\longrightarrow & E' \\
\downarrow & & \downarrow{\scriptstyle \eta} \\
F_0 \times e^p & \xrightarrow{\ p_2\ } & e^p
\end{array}
$$

Since e^p is contractible, p_2 lifts to $\tilde{p} : F_0 \times e^p \to E'$ extending the inclusion $F_0 \times \{0\} \to E'$. This gives a fiberwise map

$$\tilde{p} : (F_0 \times e^p, F_0 \times \partial e^p) \to (E', E'|_{\partial e^p}).$$

Comparing the long exact sequences in homotopy, we see that $\tilde{p}$ induces an isomorphism in relative homotopy and so it is a homotopy equivalence of pairs (i. e., $\pi_{\bullet}(F_0) \cong \pi_{\bullet}(E')$, since e^p is contractible):

$$
\begin{array}{ccccccc}
\cdots \to & \pi_k(F_0 \times \partial e^p) & \to & \pi_k(F_0 \times e^p) & \to & \pi_k(F_0 \times e^p, F_0 \times \partial e^p) & \to \cdots \\
& {\scriptstyle \tilde{p}_*}\downarrow{\scriptstyle \cong} & & {\scriptstyle \tilde{p}_*}\downarrow{\scriptstyle \cong} & & {\scriptstyle \tilde{p}_*}\downarrow & \\
\cdots \longrightarrow & \pi_k(E'|_{\partial e^p}) & \longrightarrow & \pi_k(E') & \longrightarrow & \pi_k(E', E'|_{\partial e^p}) & \longrightarrow \cdots
\end{array}
$$

Thus,

$$H^{\bullet}(\varphi^* E|_{e^p}, \varphi^* E|_{\partial e^p}) \cong H^{\bullet}(F_0) \otimes H^{\bullet}(e^p, \partial e^p).$$

If $\pi_1(B, b_0)$ acts trivially on $H^{\bullet}(F)$, then choosing a path from b_0 to $0 \in e^p \subset B^{(p)}$ gives an identification of $H^{\bullet}(F)$ with $H^{\bullet}(F_0)$ *independent of path*. Thus,

$$H^{\bullet}(F_0) \otimes H^{\bullet}(e^p, \partial e^p) \cong H^{\bullet}(F) \otimes H^{\bullet}(e^p, \partial e^p)$$

and so

$$H^{\bullet}(E^{(p)}, E^{(p-1)}) \cong \prod_{e^p \in B} H^{\bullet}(e^p, \partial e^p) \otimes H^{\bullet}(F) \cong C^p(B; H^{\bullet}(F)). \qquad \square$$

Now, the idea is that
- $H^{\bullet}(E^{(p)}, E^{(p-1)})$ is a first approximation to $H^{\bullet}(E^{(p)})$;
- $H^{\bullet}(E^{(p)}, E^{(p-2)})$ is a second, and better, approximation to $H^{\bullet}(E^{(p)})$;
- etc.

These are related by exact sequences in which the missing terms are given by Proposition 4.2.1. So we get a whole chain of exact sequences, each giving a better approximation to $H^{\bullet}(E)$.

Definition 4.2.2. A (*first quadrant*) *spectral sequence* consists of abelian groups $E_r = \{E_r^{p,q}\}_{p,q \geq 0}$, $r \geq 0$ and maps $d_r : E_r^{p,q} \to E_r^{p+r,q-r+1}$, with $d_r^2 = 0$, such that $H(E_r, d_r) = E_{r+1}$. That is, the E_{r+1} page is the homology of the E_r page:

$$E_{r+1}^{p,q} = \frac{\ker d_r : E_r^{p,q} \to E_r^{p+r,q-r+1}}{\operatorname{im} d_r : E_r^{p-r,q+r-1} \to E_r^{p,q}}.$$

The schematic of the maps on the E_r page of a spectral sequence is shown at the right. The map d_r goes "over r and down $r - 1$." An element $\alpha \in E_r^{p,q}$ is said to *live to infinity* if $d_r\alpha = 0$, $d_{r+1}\alpha = 0$, etc. An element $\beta \in E_r^{p,q}$ is said to be *killed* if $d_r\beta = d_{r+1}\beta = \cdots = d_{s-1}\beta = 0$, but $\beta = d_s\gamma$ for some $\gamma \in E_s^{p+s,q-s+1}$. A spectral sequence is said to *degenerate* or *collapse* at E_r if $E_r \cong E_{r+1} \cong \cdots$.

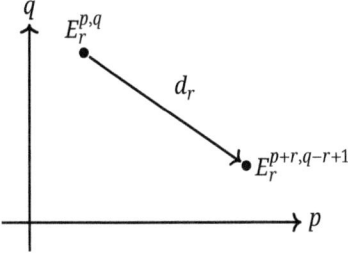

Lemma 4.2.3. *If $N > p, q + 1$, then $E_N^{p,q} \cong E_{N+1}^{p,q}$.*

Proof. Since the map d_N mapping into $E_N^{p,q}$ has domain in a spot outside the first quadrant, it is the zero map, and similarly the map out of $E_N^{p,q}$ lands outside the first quadrant and is therefore also the zero map. The result follows. ☐

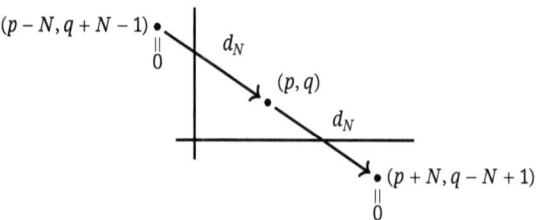

This leads to the following definition of the E_∞ page of a spectral sequence.

Definition 4.2.4. Set $E_\infty^{p,q} = E_N^{p,q}$ for $N > p, q + 1$, and

$$E_\infty^n = \bigoplus_{p+q=n} E_\infty^{p,q}.$$

Lemma 4.2.5. *Given a spectral sequence $\{E_r^{p,q}\}$, the sequence*

$$0 \to E_2^{1,0} \to E_\infty^1 \to E_2^{0,1} \xrightarrow{d_2} E_2^{2,0} \to E_\infty^2$$

is exact.

Proof. Note that $E_2^{1,0} = E_\infty^{1,0}$ and

$$E_\infty^1 = E_2^{1,0} \oplus E_3^{0,1} = E_2^{1,0} \oplus (\ker d_2 : E_2^{0,1} \to E_2^{2,0}).$$

The result follows. ☐

Theorem 4.2.6 (Leray–Serre). *Let $F \to E \to B$ be a fibration in which B is a connected cell complex with $\pi_1(B)$ acting trivially on $H^\bullet(F)$. Then there is a spectral sequence $\{E_r^{p,q}\}$ with $E_1^{p,q} = C^p(B; H^q(F))$ in which d_1 is the coboundary map in $C^\bullet(B; H^q(F))$. Moreover, $E_2^{p,q} = H^p(B; H^q(F))$ and for $N > p, q + 1$, we have*

$$E_N^{p,q} = \frac{\ker(H^{p+q}(E) \to H^{p+q}(E^{(p-1)}))}{\mathrm{im}(H^{p+q}(E) \to H^{p+q}(E^{(p)}))}.$$

Remark 4.2.7. Define

$$\mathcal{F}^p H^{p+q}(E) = \ker(H^{p+q}(E) \to H^{p+q}(E^{(p-1)})).$$

Then we have a filtration

$$H^n(E) = \mathcal{F}^0 H^n(E) \supseteq \mathcal{F}^1 H^n(E) \supseteq \cdots \supseteq \mathcal{F}^n H^n(E) \supseteq \mathcal{F}^{n+1} H^n(E) = 0.$$

The *associated graded module*, defined as

$$\bigoplus_{i \geq 0} \mathcal{F}^i / \mathcal{F}^{i+1},$$

is isomorphic to the direct sum

$$\bigoplus_{p+q=n} E_\infty^{p,q}.$$

We write

$$E_r^{p,q} \Longrightarrow H^{p+q}(E)$$

and say that the spectral sequence *converges* or *abuts* to $H^\bullet(E)$.

Warning
To say that the spectral sequence converges to $H^\bullet(E)$ does not mean that $H^n(E) \cong \bigoplus_{p+q=n} E_\infty^{p,q}$. It says only that there is a filtration of $H^n(E)$ whose associated graded module is the direct sum on the right side of the above equation. In general, there may be an extension problem. However, if we take cohomology with field coefficients, then everything works and we may conclude that $H^n(E) \cong \bigoplus E_\infty^{p,q}$.

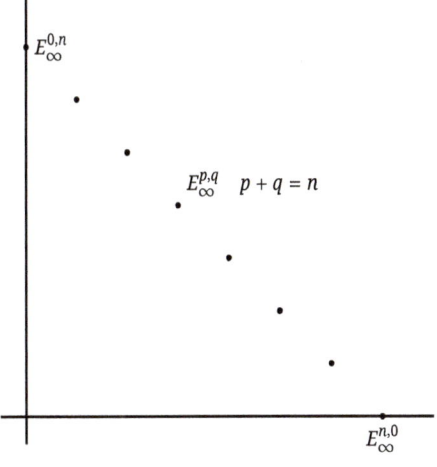

Example 4.2.8. Consider the spectral sequence whose E_∞-term is

We have a filtration

$$\mathcal{F}^0 H^2(E) \supseteq \mathcal{F}^1 H^2(E) \supseteq \mathcal{F}^2 H^2(E) \supseteq \mathcal{F}^3 H^2(E) = 0,$$

with graded quotients

$$\mathcal{F}^0/\mathcal{F}^1 \cong \mathbb{Z}_2, \quad \mathcal{F}^1/\mathcal{F}^2 = 0, \quad \mathcal{F}^2/\mathcal{F}^3 \cong \mathbb{Z}_2.$$

Thus, we have an exact sequence

$$0 \to \mathbb{Z}_2 \to H^2(E) \to \mathbb{Z}_2 \to 0,$$

but we cannot distinguish whether $H^2(E)$ is $\mathbb{Z}_4$ or $\mathbb{Z}_2 \oplus \mathbb{Z}_2$ without additional information.

Proof of Theorem 4.2.6. Let $S^\bullet(E)$ be the singular cochains of E. Define

$$F^p(S^\bullet(E)) = \ker(S^\bullet(E) \to S^\bullet(E^{(p-1)})).$$

Note that $\delta F^p \subseteq F^p$ and so $S^\bullet(E)$ is a cochain complex with a decreasing filtration preserved by δ. In this case, it is a purely algebraic fact that the spectral sequence exists, converging to $H^\bullet(S^\bullet(E))$ (see, for example, Chapter 5 of [7]). The terms are $E_r^{p,q} = Z_r^{p,q}/B_r^{p,q}$, where

$$Z_r^{p,q} = \{\mu \in F^p(S^{p+q}(E)) \mid \delta\mu \in F^{p+r}(S^{p+q+1}(E))\}$$
$$B_r^{p,q} = Z_r^{p,q} \cap F^{p+1} + \delta F^{p-r+1}(S^{p+q-1}(E)) \cap F^p.$$

The map d_r is induced by the coboundary operator in $S^\bullet(E)$ and one checks easily that $H^\bullet(E_r)$, with respect to d_r is E_{r+1}. By definition

$$E_0^{p,q} = F^p(S^{p+q}(E))/F^{p+1}(S^{p+q}(E)) \cong S^{p+q}(E^{(p)}, E^{(p-1)}),$$

with d_0 the usual singular coboundary map. So

$$E_1^{p,q} = H^{p+q}(E^{(p)}, E^{(p-1)}) \cong C^p(B; H^q(F)).$$

Under this identification, d_1 becomes $\delta_B \otimes 1$. Thus, $E_2^{p,q} \cong H^p(B; H^q(F))$. □

Remark 4.2.9. (1) There is a homology spectral sequence $\{E_{p,q}^r\}$ with $E_{p,q}^2 = H_p(B; H_q(F))$ converging to $H_\bullet(E)$. Over a field, the two spectral sequences are dual. The differentials are maps $d_{p,q}^r : E_{p,q}^r \to E_{p-r,q+r-1}^r$; these go "left" by r and "up" by $r - 1$.

(2) If $\pi_1(B)$ acts nontrivially on $H^\bullet(F)$, there is still a spectral sequence. In this case, $E_2^{p,q} = H^p(B; H^q(F))$, but with twisted coefficients. This is more complicated to compute, but it is possible.

(3) There is a ring structure on $E_r^{\bullet,\bullet}$ such that d_r is a *derivation*. Over a field (or over $\mathbb{Z}$ if all cohomology groups are free)

$$E_2^{p,q} \cong H^p(B) \otimes H^q(F) \cong E_2^{p,0} \otimes E_2^{0,q}$$

and

$$d_2 = d_2 \otimes 1 + (-1)^p 1 \otimes d_2 = (-1)^p 1 \otimes d_2,$$

the latter equality following from the fact that $d_2 = 0$ on $E_2^{p,0}$.

This last fact is part of what makes the cohomology spectral sequence so useful. In the next section, we compute several examples.

4.2.2 Calculations

4.2.2.1 $\mathbb{C}P^n$

Consider the fibration $S^1 \to S^{2n+1} \to \mathbb{C}P^n$. This gives rise to a spectral sequence with

$$E_2 = H^\bullet(S^1) \otimes H^\bullet(\mathbb{C}P^n) \Longrightarrow H^\bullet(S^{2n+1}).$$

Since $H^q(S^1) = 0$ for $q > 1$, we see that $E_2^{p,q} = 0$ for $q > 1$ and $d_3 = d_4 = \cdots = 0$. Thus $E_3 \cong E_4 \cong \cdots \cong E_\infty$. Note that

$$E_\infty^{p,q} = E_3^{p,q} = 0, \quad p + q \neq 0, 2n + 1.$$

Let $\alpha \in E_2^{0,1} \cong H^1(S^1)$ be a generator. Consider $d_2 : E_2^{0,1} \to E_2^{2,0}$. Since $E_3^{p,q} = 0$ for $1 \le p + q \le 2$, this map must be an isomorphism. Set $\beta = d_2\alpha$ so that $E_2^{2,1} \cong H^2(\mathbb{C}P^n) \otimes H^1(S^1) = \mathbb{Z}(\beta \otimes \alpha)$. Then

$$d_2(\beta \otimes \alpha) = (-1)^2 \beta \otimes d_2\alpha$$
$$= \beta \otimes \beta \in E_2^{4,0} \cong H^4(\mathbb{C}P^n).$$

Continuing in this way, we see that $H^{2q}(\mathbb{CP}^n) \cong \mathbb{Z}$ generated by β^q, $q \leq n$, and $H^{2q+1}(\mathbb{CP}^n) = 0$. Moreover, since $E_2^{2n,1} = H^{2n}(\mathbb{CP}^n) \otimes H^1(S^1)$ is the only possible nonzero term on the diagonal $p + q = 2n + 1$, it must be equal to $H^{2n+1}(S^{2n+1}) \cong \mathbb{Z}$ and the map $d_2 : E_2^{2n,1} \to E_2^{2n+2,0}$ must be the zero map. Thus

$$H^{\bullet}(\mathbb{CP}^n) \cong \mathbb{Z}[\beta]/(\beta^{n+1}).$$

Note that since d_2 is a derivation, we actually recover the ring structure in cohomology.

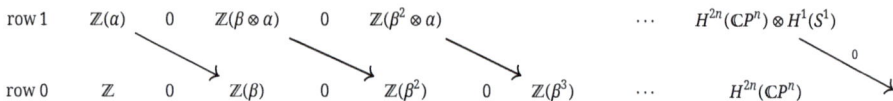

| row 1 | $\mathbb{Z}(\alpha)$ | 0 | $\mathbb{Z}(\beta \otimes \alpha)$ | 0 | $\mathbb{Z}(\beta^2 \otimes \alpha)$ | | $\cdots$ | $H^{2n}(\mathbb{CP}^n) \otimes H^1(S^1)$ | |
| row 0 | $\mathbb{Z}$ | 0 | $\mathbb{Z}(\beta)$ | 0 | $\mathbb{Z}(\beta^2)$ | 0 | $\mathbb{Z}(\beta^3)$ | $\cdots$ | $H^{2n}(\mathbb{CP}^n)$ |

4.2.2.2 $H^{\bullet}(K(\mathbb{Z}, n); \mathbb{Q})$

We note the following

$$K(\mathbb{Z}, 1) \simeq S^1$$
$$K(\mathbb{Z}, 2) \simeq \mathbb{CP}^\infty.$$

Theorem 4.2.10. *We have ring isomorphisms*

$$H^{\bullet}(K(\mathbb{Z}, 2k); \mathbb{Q}) \cong \mathbb{Q}[\alpha], \deg \alpha = 2k$$
$$H^{\bullet}(K(\mathbb{Z}, 2k + 1); \mathbb{Q}) \cong \bigwedge\nolimits_{\mathbb{Q}}[\beta], \deg \beta = 2k + 1.$$

Proof. By induction on n, the cases $n = 1, 2$ being clear. Consider the inductive step from $(2k - 1)$ to $2k$. We have the path fibration

$$K(\mathbb{Z}, 2k - 1) \longrightarrow P$$
$$\downarrow$$
$$K(\mathbb{Z}, 2k)$$

Since P is contractible, the long exact homotopy sequence implies that the fiber has the homotopy type of $K(\mathbb{Z}, 2k - 1)$. The corresponding spectral sequence has $E_\infty^{p,q} = 0$ for $(p, q) \neq (0, 0)$. Consider the E_2-term

0	
$H^{\bullet}(K(\mathbb{Z}, 2k); \mathbb{Q})$	row $2k - 1$
0	
$H^{\bullet}(K(\mathbb{Z}, 2k); \mathbb{Q})$	row 0

By induction $H^{\bullet}(K(\mathbb{Z}, 2k - 1); \mathbb{Q}) \cong \bigwedge_{\mathbb{Q}}[\beta]$, $\deg \beta = 2k - 1$. Note that $E_2 \cong E_3 \cong \cdots \cong E_{2k}$ and $E_{2k+1} = E_\infty$. Since the map

$$d_{2k} : H^0(K(\mathbb{Z}, 2k); \mathbb{Q}) \otimes H^{2k-1}(K(\mathbb{Z}, 2k - 1); \mathbb{Q}) \to H^{2k}(K(\mathbb{Z}, 2k); \mathbb{Q})$$

must be an isomorphism, it follows that

$$H^i(K(\mathbb{Z}, 2k); \mathbb{Q}) \cong \begin{cases} 0 & i < 2k \\ \mathbb{Q} & i = 2k. \end{cases}$$

Let $\alpha = d_{2k}\beta$. Then α is a generator of $H^{2k}(K(\mathbb{Z}, 2k); \mathbb{Q})$. Suppose inductively that we have shown that $\mathbb{Q}[\alpha] \to H^\bullet(K(\mathbb{Z}, 2k); \mathbb{Q})$ is an isomorphism in all degrees $\leq t(2k)$. Then we have the E_{2k}-term

	0		$\cdots$	0			
β	0	$\cdots$	0	$\beta \otimes \alpha$	$\cdots$	$\beta \otimes \alpha^2$	$\cdots$ $\beta \otimes \alpha^t$
	0			$\cdots$	0		
1	0	$\cdots$	0	α	$\cdots$	α^2	$\cdots$ α^t
0			$2k$		$2(2k)$		$t(2k)$

Since $E_{2k+1}^{p,q} = 0$ for all $(p, q) \neq (0, 0)$, we have $H^i(K(\mathbb{Z}, 2k); \mathbb{Q}) = 0$ for $2kt < i < 2k(t + 1)$ and $H^{2k(t+1)}(K(\mathbb{Z}, 2k); \mathbb{Q}) \cong \mathbb{Q}$, generated by $d_{2k}(\beta \otimes \alpha^t) = \alpha \otimes \alpha^t = \alpha^{t+1}$. This completes the inductive step $(2k - 1) \to 2k$.

For the inductive step $2k \to (2k + 1)$, consider the fibration $K(\mathbb{Z}, 2k) \to \mathcal{P} \to K(\mathbb{Z}, 2k + 1)$. The E_2-term of the associated spectral sequence looks as follows.

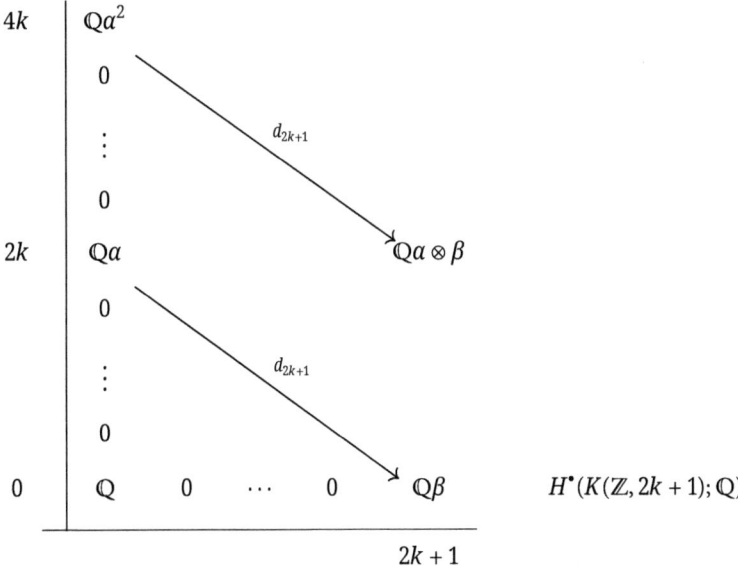

Again, we see that $E_2 = E_3 = \cdots E_{2k+1}$ and $E_{2k+2} = E_\infty$, with $E_\infty^{p,q} = 0$ for $(p, q) \neq (0, 0)$. It follows that $d_{2k+1} : E_{2k+1}^{0,2k} \to E_{2k+1}^{2k+1,0}$ is an isomorphism. Set $\beta = d_{2k+1}\alpha$. By the derivation property,

$$d_{2k+1}(a^2) = 2a \otimes d_{2k+1}a = 2a \otimes \beta.$$

It follows that $d_{2k+1} : E_{2k+1}^{0,4k} \to E_{2k+1}^{2k+1,2k}$ is an isomorphism (this is where we need $\mathbb{Q}$ coefficients). It follows that $H^{4k+2}(K(\mathbb{Z}, 2k + 1); \mathbb{Q}) = 0$ and continuing in this way, the result follows. □

4.2.2.3 Infinite Grassmannians

Let $G(k, N)$ be the Grassmann manifold of *complex k-planes* in $\mathbb{C}^N$. Let $G(k) = \varinjlim_N G(k, N)$ be the infinite Grassmannian.

Theorem 4.2.11. *There are classes $c_j \in H^{2j}(G(k); \mathbb{Z})$, $1 \leq j \leq k$ such that*

$$H^\bullet(G(k); \mathbb{Z}) \cong \mathbb{Z}[c_1, c_2, \ldots, c_k].$$

Proof. By induction on k, the case $k = 1$ being $G(1) = \mathbb{C}P^\infty$, whose cohomology algebra is $\mathbb{Z}[c]$, $c \in H^2(\mathbb{C}P^\infty)$ (Theorem 3.2.10). Assume the result for $k - 1$ and consider the fibrations

$$S^{2k-1} \longrightarrow F \qquad\qquad S^\infty \longrightarrow F$$
$$\downarrow \qquad\qquad\qquad\qquad \downarrow$$
$$G(k) \qquad\qquad\qquad\qquad G(k-1)$$

where

$$F = \{(v, \pi) \mid \pi \in G(k) \text{ and } v \in \pi \text{ is a unit vector}\}.$$

The map $F \to G(k)$ is the projection onto the second factor; its fiber is S^{2k-1} since π is a subspace of real dimension $2k$. The map $F \to G(k - 1)$ sends (v, π) to the $(k - 1)$-plane in π orthogonal to v. The fiber is S^∞, which is contractible. Thus $F \simeq G(k - 1)$, and the first fibration gives us a spectral sequence

$$E_2^{p,q} = H^p(G(k); H^q(S^{2k-1})) \Longrightarrow H^{p+q}(G(k-1)).$$

This sequence has only two nonzero rows:

0	
$H^\bullet(G(k)) \otimes H^{2k-1}(S^{2k-1})$	row $2k - 1$
0	
$H^\bullet(G(k)) \otimes H^0(S^{2k-1})$	row 0

It follows that $E_2 \cong E_3 \cong \cdots \cong E_{2k}$ and $E_{2k+1} = E_\infty$. Note that $E_{2k}^{0,2k-1} = H^0(G(k)) \otimes$ $H^{2k-1}(S^{2k-1}) \cong \mathbb{Z} \otimes \mathbb{Z}$, generated by $1 \otimes \alpha$. Let $c_k = d_{2k}\alpha$. For each i, we have exact sequences

$$H^{i+2k-1}(G(k-1)) \xrightarrow{d_{2k}^{i,2k-1}} H^i(G(k)) \xrightarrow{\smile c_k} H^{i+2k}(G(k)) \longrightarrow H^{i+2k}(G(k-1)).$$

By induction, $H^\bullet(G(k-1))$ is the polynomial algebra on $c_1, \ldots, c_{k-1}$ and these classes lift to $H^\bullet(G(k))$. It follows that $H^\bullet(G(k)) \to H^\bullet(G(k-1))$ is surjective and we obtain a short exact sequence

$$0 \longrightarrow H^i(G(k)) \xrightarrow{\smile c_k} H^{i+2k}(G(k)) \longrightarrow H^{i+2k}(G(k-1)) \longrightarrow 0.$$

It is now easy to prove that $H^\bullet(G(k)) \cong \mathbb{Z}[c_1, \ldots, c_k]$. $\qquad\square$

4.2.2.4 Unitary groups

Let $U(n)$ be the group of $n \times n$ complex unitary matrices $(A\overline{A}^T = I)$.

Theorem 4.2.12. *There are classes $x_k \in H^{2k-1}(U(n); \mathbb{Z})$ such that*

$$H^\bullet(U(n); \mathbb{Z}) \cong \bigwedge\nolimits_\mathbb{Z}[x_1, x_3, \ldots, x_{2n-1}].$$

Consequently, there is an isomorphism of rings $H^\bullet(U(n)) \cong H^\bullet(S^1 \times S^3 \times \cdots \times S^{2n-1})$.

Proof. If $A \in U(n)$, write $A = (e_1, \ldots, e_n)$, $e_i \in \mathbb{C}^n$, giving a unitary frame. Define $\pi : U(n) \to S^{2n-1}$ by $\pi(e_1, \ldots, e_n) = e_1$. This is a fibration with fiber $U(n-1)$. Assume inductively that $H^\bullet(U(n-1)) \cong \bigwedge\nolimits_\mathbb{Z}[x_1, \ldots, x_{2n-3}]$, the base case being $U(1) = S^1$. The spectral sequence of the fibration has two nonzero columns

$H^0(S^{2n-1}) \otimes H^\bullet(U(n-1))$	0	$H^{2n-1}(S^{2n-1}) \otimes H^\bullet(U(n-1))$	0
0		$2n-1$	

The only possible nontrivial differentials are then

$$d_{2n-1} : E_{2n-1}^{0,q} \to E_{2n-1}^{2n-1,q-2n+2}$$

and $E_2 \cong E_3 \cong \cdots \cong E_{2n-1}$ and $E_{2n} = E_\infty$. For all q, we then obtain a commutative diagram

$$0 \longrightarrow E_\infty^{0,q} \longrightarrow E_{2n-1}^{0,q} \xrightarrow{\ d_{2n-1}\ } E_{2n-1}^{2n-1,q-2n+2} \longrightarrow E_\infty^{2n-1,q-2n+2} \longrightarrow 0$$

$$H^q(U(n)) \xrightarrow{i^*} H^q(U(n-1)) \to H^{q-2n+1}(U(n-1)) \to H^{q+1}(U(n))$$

Note that $d_{2n-1}x_{2p-1} = 0$ for $1 \le p \le n-1$, since $x_{2p-1} \in H^{2p-1}(U(n-1))$ and

$$d_{2n-1}x_{2p-1} \in H^{2p-1-2n+2}(U(n-1)) = H^{2(p-n)+1}(U(n-1)) = 0, \quad p \le n-1.$$

Since d_{2n-1} is a derivation, we see that the spectral sequence collapses at E_2. But

$$E_2 \cong \bigwedge{}_{\mathbb{Z}}[x_1, \ldots, x_{2n-3}] \otimes \bigwedge{}_{\mathbb{Z}}[x_{2n-1}]$$
$$\cong \bigwedge{}_{\mathbb{Z}}[x_1, \ldots, x_{2n-1}]. \qquad \square$$

4.2.2.5 $\pi_4(S^3)$

A complete computation of the homotopy groups of any sphere (except S^1) still eludes us. To get a sense of just how complicated this can be, we consider a relatively low-dimensional example here. Consider the space $K(\mathbb{Z}, 3)$. We can build this by starting with S^3 and attaching cells to kill the higher homotopy groups. We therefore have an inclusion $S^3 \hookrightarrow K(\mathbb{Z}, 3)$. But we can replace this by a fibration

$$F \longrightarrow S^3$$
$$\downarrow$$
$$K(\mathbb{Z}, 3)$$

Using the long exact sequence in homotopy, we see that $\pi_i(F) \cong \pi_i(S^3)$ for $i \ge 4$, and by the Hurewicz theorem

$$\pi_4(S^3) \cong \pi_4(F) \cong H_4(F),$$

since $\pi_i(F) = 0$ for $i \le 3$.

Write $H_4(F) = \mathbb{Z}^r \oplus T$, where T is a finite torsion abelian group. We know that we have an exact sequence

$$0 \to \mathrm{Ext}(H_{i-1}(F), \mathbb{Z}) \to H^i(F; \mathbb{Z}) \to \mathrm{Hom}(H_i(F), \mathbb{Z}) \to 0.$$

For $i = 4$, we know that $H_3(F) = 0$ so that

$$H^4(F) \cong \mathrm{Hom}(H_4(F), \mathbb{Z}) \cong \mathbb{Z}^r.$$

Thus the ranks of $H_4(F)$ and $H^4(F)$ are equal. Taking $i = 5$, we see that

$$H^5(F) \cong \text{Ext}(H_4(F), \mathbb{Z}) \oplus \text{Hom}(H_5(F), \mathbb{Z})$$
$$\cong T \oplus \text{Hom}(H_5(F), \mathbb{Z})$$

So we can deduce the free part of $H_4(F)$ from $H^4(F)$ and the torsion part from $H^5(F)$. Consider the map $F \to S^3$. Replace this by a fibration; the result is

$$\Omega K(\mathbb{Z}, 3) \longrightarrow F$$
$$\downarrow$$
$$S^3$$

(we leave this as an exercise). Note that $\Omega K(\mathbb{Z}, 3) \simeq K(\mathbb{Z}, 2) \simeq \mathbb{C}P^\infty$. So the corresponding spectral sequence has the form

$$E_2^{p,q} = H^p(S^3; H^q(\mathbb{C}P^\infty)) \Longrightarrow H^{p+q}(F).$$

We show the groups and corresponding generators below. The maps shown are the first nontrivial differentials, d_3.

6	$\mathbb{Z}$	0	0	$\mathbb{Z}$	0	0
5	0	0	0	0	0	0
4	$\mathbb{Z}$	0	0	$\mathbb{Z}$	0	0
3	0	0	0	0	0	0
2	$\mathbb{Z}$	0	0	$\mathbb{Z}$	0	0
1	0	0	0	0	0	0
0	$\mathbb{Z}$	0	0	$\mathbb{Z}$	0	0
	0	1	2	3	4	5

	0	1	2	3	4	5
6	z^3	0	0	z^3x	0	0
5	0	0	0	0	0	0
4	z^2	0	0	z^2x	0	0
3	0	0	0	0	0	0
2	z	0	0	zx	0	0
1	0	0	0	0	0	0
0	1	0	0	x	0	0

We know that $H^2(F) = H^3(F) = 0$ and so we must have $d_3 : z \mapsto x$. Then $d_3(z^2) = 2zx$ so that $E_\infty^{0,4} = 0$ ($d_3^{0,4}$ is injective). Since $E_\infty^{p,q} = 0$ for $p + q = 4$, we see that $H^4(F) = 0$; in particular, the free part of $H^4(F)$ is trivial and hence the rank of $H_4(F)$ is 0. Observe further that $d_3(z^p) = pzx$ and hence for i even the $p+q = i$ diagonal consists of 0's. Thus, $H^i(F) = 0$ for i even.

Now we have a filtration

$$H^5(F) \supset F^0 H^5 \supset F^1 H^5 \supset \cdots \supset F^5 H^5 \supset F^6 H^5 = 0.$$

We know that $E_\infty^{3,2} \cong \mathbb{Z}_2$ and $E_\infty^{p,5-p} = 0$ for $p \neq 3$. Thus, $H^5(F) \cong \mathbb{Z}_2$. (Actually, we can extend this to show that $H^i(F) \cong \mathbb{Z}_{(i-1)/2}$ for i odd.) We therefore conclude that the torsion part of $H_4(F)$ is $\mathbb{Z}_2$ and hence

$$\pi_4(S^3) \cong \mathbb{Z}_2.$$

4.2.2.6 ΩS^2

We have the fibration

$$\Omega S^2 \longrightarrow P$$
$$\downarrow$$
$$S^2$$

The corresponding spectral sequence is then

$$E_2^{p,q} = H^p(S^2; H^q(\Omega S^2)) \Longrightarrow H^{p+q}(\mathcal{P}).$$

The E_2-term looks as follows:

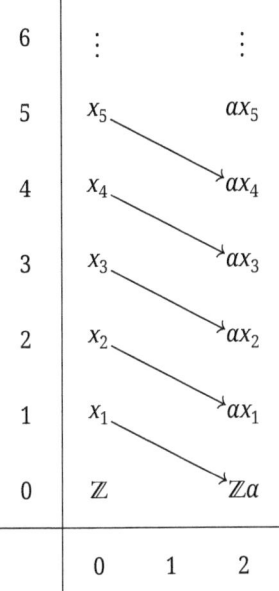

Since $H^q(\mathcal{P}) = 0$ for $q > 0$, we have $E_\infty^{p,q} = 0$ for $(p,q) \neq (0,0)$. It follows that d_2 : $E_2^{0,q} \to E_2^{2,q-1}$ is an isomorphism for $q \geq 1$. Since $E_2^{2,0} \cong \mathbb{Z}$, generated by a, we see that $H^1(\Omega S^2) = E_2^{0,1} \cong \mathbb{Z}$, generated by x_1 with $d_2 x_1 = a$. Then $ax_1 \in E_2^{2,1}$ and $H^2(\Omega S^2) = E_2^{0,2} \cong \mathbb{Z}$ generated by x_2 with $d_2 x_2 = ax_1$. Continuing in this manner, we see that $H^k(\Omega S^2) \cong \mathbb{Z}$ generated by x_k with $d_2 x_k = ax_{k-1}$.

What is the ring structure? By anti-commutativity, $x_1^2 = 0$. Thus, $x_1 x_k = N_k x_{k+1}$ for some integer N_k. But d_2 is a derivation, so

$$d_2(x_1 x_k) = (d_2 x_1) x_k - x_1 d_2 x_k$$
$$= ax_k - x_1 ax_{k-1}$$
$$= ax_k - N_{k-1} ax_k$$
$$= (1 - N_{k-1}) ax_k.$$

Now, $N_1 = 0$ (since $x_1^2 = 0$) and so $N_k \equiv (k+1) \bmod 2$. Thus, $x_1 x_k = x_{k+1}$ if k is even, and $x_1 x_k = 0$ if k is odd. Note that $x_2 \in H^2(\Omega S^2)$ commutes with everything and

$$d_2(x_2^k) = (d_2 x_2) x_2^{k-1} + x_2 (d_2 x_2^{k-1}) = \cdots = k x_2^{k-1} x_1 a.$$

Also, $x_2^k = M_k x_{2k}$ for some integer M_k. But then $d_2(x_2^k) = m_k d_2 x_{2k} = M_k x_{2k-1} a$. Thus

$$M_k x_{2k-1} a = k x_2^{k-1} x_1 a = k M_{k-1} x_{2(k-1)} x_1 a.$$

But $x_{2k-1} = x_1 x_{2(k-1)}$ and so

$$M_k x_1 x_{2(k-1)} a = k M_{k-1} x_{2(k-1)} x_1 a.$$

Inductively, we see that $M_k = k!$. To summarize:

$$x_1 x_k = \begin{cases} x_{k+1} & k \equiv 0 \mod 2 \\ 0 & k \equiv 1 \mod 2 \end{cases} \quad \text{and} \quad x_2^k = k! x_{2k}.$$

Thus,

$$H^\bullet(\Omega S^2) \cong \bigwedge\nolimits_{\mathbb{Z}}[x_1] \otimes \Gamma_{\mathbb{Z}}[x_2],$$

where $\Gamma_{\mathbb{Z}}$ denotes the *divided power algebra* on the element of indicated degree.

Remark 4.2.13. Using Morse theory, we can actually prove that ΩS^2 has the homotopy type of a cell complex with one cell in each dimension:

$$\Omega S^2 \simeq e^0 \cup e^1 \cup e^2 \cup \cdots.$$

Of course, this space is not homotopy equivalent to $\mathbb{R}P^\infty$, as the above calculation shows.

Exercises

Exercise 4.2.1. The calculation of the cohomology of the unitary group $U(n)$ is a special case of a more general result. Suppose $F \xrightarrow{i} E \to S^n$ is a fibration, with $n \geq 2$. Prove that there is an exact sequence

$$\cdots \to H^q(E) \xrightarrow{i^*} H^q(F) \xrightarrow{D} H^{q-n+1}(F) \to H^{q+1}(E) \xrightarrow{i^*} H^{q+1}(F),$$

where $D(uv) = D(u)v \pm u D(v)$. This is called the *Wang sequence*.

Exercise 4.2.2. Suppose $F \to E \to B$ is a fibration. Show that, if we replace the inclusion $F \to E$ by a fibration, then, up to homotopy, the result is $\Omega B \to F \to E$.

Exercise 4.2.3. Generalize the argument of Section 4.2.2.6 to compute the cohomology ring $H^\bullet(\Omega S^n; \mathbb{Z})$ for $n > 2$.

Exercise 4.2.4. Suppose $F \to E \to B$ is a fibration with $\pi_1(B)$ acting trivially on $H^\bullet(F)$. Suppose that $H^\bullet(B)$ and $H^\bullet(F)$ are finitely generated. Prove that $H^\bullet(E)$ is finitely generated and that $\chi(E) = \chi(B)\chi(F)$.

Exercise 4.2.5. Compute $H^\bullet(K(\mathbb{Z}_2, 2); \mathbb{Z}_2)$ and $H^\bullet(K(\mathbb{Z}_2, 2); \mathbb{Z})$ up to dimension 6.

Project: obstruction theory and the Hopf index formula

We met a special case of obstruction theory when we proved the Whitehead theorem. Recall that we showed (essentially) that if $f : X \to Y$ is map with $\pi_k(Y) = 0$ for all k, then f is homotopic to the basepoint map $X \to \{y_0\}$. We did this inductively over the skeleta of X, so we might ask the following two questions:

(1) If we have a map $f : X^{(n)} \to Y$, is it possible to extend it to a map $f : X^{(n+1)} \to Y$?
(2) If we have a homotopy $H : X^{(n)} \times I \to Y$ from $f_0 : X^{(n)} \to Y$ to $f_1 : X^{(n)} \to Y$, is it possible to extend it to a homotopy on $X^{(n+1)}$?

In this project, we will attack these two questions and work toward a proof of the Hopf index formula.

If Y is a space, denote by $[S^n, Y]$ the set of free homotopy classes of maps $S^n \to Y$ (*free* means no basepoint condition). Prove that if Y is simply connected, then there is a natural bijection

$$\pi_n(Y) \leftrightarrow [S^n, Y].$$

(Hint: there is always a function $\pi_n(Y) \to [S^n, Y]$. Show that if $\pi_0(Y) = 0$, then it is surjective by applying the homotopy extension property to $S^n \times \{0\} \cup * \times I \subset S^n \times I$, where $*$ is the basepoint of S^n. If $\pi_1(Y) = 0$, then the map is injective by applying HEP to $(S^n \times \{0,1\} \cup * \times I) \times I \subset (S^n \times I) \times I$.)

From now on, assume that Y is simply connected. Let (X,A) be a CW pair and denote by $X^{(n)} \cup A$ the union of A and all cells of dimension at most n in $X \setminus A$. Suppose we have a map $f_n : X^{(n)} \cup A \to Y$. Define the *obstruction cochain*

$$\theta(f_n) \in C^{n+1}(X, A; \pi_n(Y))$$

as follows. If e_α^{n+1} is an oriented $(n+1)$-cell of (X,A), then its attaching map $\varphi_\alpha : S^n \to X^{(n)} \cup A$ composed with f_n gives a map $f_n \circ \varphi_\alpha : S^n \to Y$; this determines an element of $\pi_n(Y)$. If we reverse the orientation on e_α^{n+1}, then the resulting element in $\pi_n(Y)$ changes sign. This gives a well-defined homomorphism $\theta(f_n) : C_{n+1}(X,A) \to \pi_n(Y)$, called the obstruction cochain.

Prove the following:

(1) $\theta(f_n)$ is an invariant of the homotopy class of f_n.
(2) $\theta(f_n)$ is zero if and only if f_n extends to a map $f_{n+1} : X^{(n+1)} \cup A \to Y$.
(3) $\theta(f_n)$ is a cocycle; that is $\delta\theta(f_n) : C_{n+2}(X,A) \to \pi_n(Y)$ is zero.
(4) If $g_n : X^{(n)} \cup A \to Y$ agrees with f_n on $X^{(n-1)} \cup A$, then $\theta(g_n) - \theta(f_n)$ is a coboundary.
(5) By varying the homotopy class of f_n, relative to $X^{(n-1)} \cup A$, we can change $\theta(f_n)$ by an arbitrary coboundary.

The upshot of these is the following result.

Theorem. *Suppose Y is simply connected and that $f_n : X^{(n)} \cup A \to Y$ is a map. Then there is a cohomology class $\Theta(f_n) \in H^{n+1}(X, A; \pi_n(Y))$, called the obstruction class, that vanishes if and only if $f_n|_{X^{(n-1)} \cup A}$ can be extended to a map $f_{n+1} : X^{(n+1)} \cup A \to Y$.*

If $f, g : X \to Y$ are given and $H : (X^{(n)} \cup A) \times I \to Y$ is a homotopy from $f|_{X^{(n)} \cup A}$ to $g|_{X^{(n)} \cup A}$, then the obstruction to extending H over $(X^{(n+1)} \cup A) \times I$ lies in

$$H^{n+2}(X \times I, ((X \times \{0,1\}) \cup A \times I); \pi_{n+1}(Y)),$$

which is isomorphic to the group

$$H^{n+1}(X, A; \pi_{n+1}(Y)).$$

Thus, the obstructions to constructing a homotopy between two maps $f, g : X \to Y$, given a fixed homotopy on A, lie in $H^n(X, A; \pi_n(Y))$.

So, as long as $H^i(X, \pi_i(Y))$ is 0, there is no obstruction to finding a homotopy between the restrictions of f and g to $X^{(i)}$. Suppose that $H^n(X; \pi_n(Y))$ is the first nonzero group. Prove that, given $f, g : X \to Y$, the first obstruction to finding a homotopy between them, $\Theta \in H^n(X; \pi_n(Y))$ is well-defined; that is, it does not depend on the step-by-step homotopy constructed from $f|_{X^{(n-1)}}$ to $g|_{X^{(n-1)}}$.

Prove the following:

(1) If $f : S^n \to Y$ is continuous, then the obstruction to f being nullhomotopic lies in $H^n(S^n; \pi_n(Y)) = \pi_n(Y)$. Show that the obstruction class is the homotopy class of f in $\pi_n(Y)$.

(2) Let $g : (\mathbb{CP}^2)^{(3)} \to S^2$ be the identity map. The obstruction to extending g over $\mathbb{CP}^2$ lies in $H^4(\mathbb{CP}^2; \pi_3(S^2)) = \pi_3(S^2)$. Prove that the obstruction is the Hopf map $S^3 \to S^2$, the attaching map for the 4-cell.

(3) More generally, if $X' = X \cup_\varphi e^n$, then the obstruction to extending id : $X \to X$ over X' is $[f] \in \pi_{n-1}(X)$.

Let us apply this idea to constructing sections of fibrations. Let $p : E \to B$ be a fibration with B a path connected cell complex, and let $F = p^{-1}(b_0)$. Recall that a *section* is a map $s : B \to E$ with $p \circ s = \mathrm{id}_B$. Assume that $\pi_1(B)$ acts trivially on F and that $\pi_1(F) = 0$. The obstructions to constructing a section lie in $H^i(B; \pi_{i-1}(F))$. Given two sections, the obstructions to constructing a homotopy between them lie in $H^i(B; \pi_i(F))$. Let us be explicit about this.

Suppose $s : B^{(n-1)} \to E$ is a section over the $(n-1)$-skeleton. For each n-cell of B, we have a trivialization of $p^{-1}(e^n) \to e^n$ as $e^n \times F$. This is constructed by connecting e^n to the basepoint b_0 be a path in B. Since $\pi_1(B)$ acts trivially on the fiber, this is well-defined up to homotopy, independent of the path. A section s on $B^{(n-1)}$ induces a section of $p^{-1}(e^n) \to e^n$ over the boundary ∂e^n. This gives a map $\hat{s} : \partial e^n \to e^n \times F$, and projecting this onto the second factor yields an element in $\pi_{n-1}(F)$. The obstruction cocycle $\theta_n(s)$:

$C_n(B) \to \pi_{n-1}(F)$ sends e^n to this element. The arguments above may be adapted to show that the class $\Theta(s) \in H^n(B; \pi_{n-1}(F))$ is the obstruction to extending $s|_{B^{(n-2)}}$ over $B^{(n)}$.

Now let $E^n \to B$ be an n-dimensional real vector bundle. A nowhere-vanishing section of this is the same as a section of the associated sphere bundle $S^{n-1}(E) \to B$. The first obstruction to finding a nowhere-vanishing section is in

$$H^n(B; \pi_{n-1}(S^{n-1})) \cong H^n(B; \mathbb{Z}).$$

It is called the *Euler class* of the bundle E^n and is denoted by $e(E)$. Note that if $\dim B < n$, then $E^n \to B$ always has a non-zero section.

Suppose $F \to E \xrightarrow{p} B$ is a bundle and that $\pi_i(F) = 0$ for $i < n - 1$. Assume B is a cell complex. Prove the following:

(1) There exists a section $s : B^{(n-1)} \to E$.
(2) Any two sections $s_1, s_2 : B^{(n-2)} \to E$ are homotopic.
(3) The obstruction to finding a section $s : B^{(n)} \to E$ is given by a cohomology class $c(E) \in H^n(B; H_{n-1}(F))$.

In the case of a sphere bundle, this class is the Euler class defined above, and the preceding exercises show that $S^{n-1} \to E \to B$ has a section over $B^{(n)}$ if and only if $e(E) = 0$. Now, if B is an oriented n-manifold, we may define the *Euler number* e by evaluating the class $e(E)$ on the fundamental class $[B]$. Then $S^{n-1} \to E \to B$ has a section if and only if $e = 0$. Prove that to compute e, one does the following: Find a nonzero section $s : B^{(n-1)} \to E^*$, where $E^* = E \setminus \{\text{zero section}\}$ (assume B has been triangulated). For each n-cell e_α^n, we have $E|_{e_\alpha^n} \approx e_\alpha^n \times \mathbb{R}^n$, so that s gives a map

$$s_\alpha : \partial e_\alpha^n \to S^{n-1}.$$

Show that

$$e = \sum_\alpha \deg(s_\alpha).$$

Exercise

Consider the *universal line bundle*

$$\mathbb{C} \longrightarrow E$$
$$\downarrow \pi$$
$$\mathbb{CP}^1$$

where $\pi^{-1}[z_0 : z_1]$ is the line spanned by $(z_0, z_1) \in \mathbb{C}^2$. Using the euclidean metric on $\mathbb{C}^2$, the associated sphere bundle is the Hopf fibration

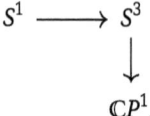

Over $\mathbb{C}P^1 - \{\infty\}$ (where $\infty = [0:1]$), take the section

$$s([z_0 : z_1]) = (1, z_1/z_0)$$

and use this to show that $e(E) = -1$.

A *vector field* in a neighborhood V of the origin in $\mathbb{R}^n$ is a smooth map

$$V = \sum_{i=1}^{n} v_i \frac{\partial}{\partial x_i} : U \to \mathbb{R}^n,$$

where each v_i is a smooth map on U. If V has an isolated zero at the origin, we define the *index* of V at 0 as follows. In a small sphere $\|x\| = \varepsilon$, the function $v(x) = (v_1, \ldots, v_n)$ is nonzero and so induces a map $f : S^{n-1} \to S^{n-1}$. We set

$$\mathrm{ind}_0(V) = \deg f.$$

Now suppose that B is a compact oriented manifold and that V is a vector field on B having only isolated zeros (that is, V is a section of the tangent bundle TB with only isolated zeros). Prove that

$$\sum_{V(x)=0} \mathrm{ind}_x(V) = e(TB) \in H^n(B; H_{n-1}(S^{n-1})) \cong H^n(B; \mathbb{Z}) \cong \mathbb{Z}.$$

(The previous exercise is useful here.)

Note that this proves that this sum $\sum_{V(x)=0} \mathrm{ind}_x(V)$ is *independent of V*. Following Steenrod [6], we now give a heuristic explanation of how one might actually compute this. Assume B is triangulated as a simplicial complex K, and let K' and K'' be the first and second barycentric subdivisions of K, respectively. A vertex v of K'' lies in the interior of exactly one simplex of K; denote the barycenter of that simplex by $\varphi(v)$. This assignment yields a unique simplicial map $\varphi : K'' \to K'$ whose fixed points are the barycenters of the simplices of K (prove this).

If x is any point in K', then x and $\varphi(x)$ lie in a single simplex of K' and are joined by a unique line segment in that simplex. Assuming the triangulation is smooth, this segment has a tangent direction at each point; define $f(x)$ to be the unit tangent vector at x. Then f is defined and continuous, except at the barycenters of the simplices of K.

Let L' be the cellular decomposition of B dual to K'; for each k-simplex σ of K' there is a dual $(n-k)$-cell of L' which is the union of the simplices of K'' having the barycenter of σ as the vertex of least order. Then the singularities of f occur at the centers of the cells of L' and so f is a nonwhere-vanishing section over the $(n-1)$-skeleton of L'. We therefore have the obstruction $\theta(f)$ in $C^n(L'; H_{n-1}(S^{n-1}))$.

Now, L' has one n-cell for each simplex of K, namely the dual σ^* of the barycenter of σ. Let z be a cellular cycle in L' representing the fundamental class $[B]$. This is a sum of all the n-cells with coefficients ± 1. We claim the following formula:

$$\theta(f)(\sigma^*) = (-1)^{\dim \sigma}.$$

To prove this, let x_σ be the barycenter of σ and let F_σ be the fiber of the sphere bundle $S(TB)$ over x_σ. For $v \in F_\sigma$, let $g(v)$ be the point in which the line segment from x_σ in the direction of v hits $\partial\sigma^*$. If the triangulation is fine enough and smooth, then g gives a continuous map $F_\sigma \to \partial\sigma^*$, and g^{-1} represents the generator σ^* of $\pi_{n-1}(F_\sigma)$. If $f' :$ $\partial\sigma^* \to F_\sigma$ is homotopic to $f|_{\partial\sigma^*}$, then $\theta(f)(\sigma^*)$ is simply the degree of the map gf' of $\partial\sigma^*$ to itself (i. e., the index of the singularity). Let $k = \dim \sigma$. The cell σ^* is the product of the k-cell $\sigma \cap \sigma^*$ and the $(n-k)$-cell $\tilde{\sigma} \cap \sigma^*$, where $\tilde{\sigma}$ is the dual cell of σ. All the vectors on $\partial(\tilde{\sigma} \cap \sigma^*)$ point outwards, so that gf' is the identity on it. All the vectors on $\partial(\sigma \cap \sigma^*)$ point inward so that gf' is the antipodal map on it. This map has degree $(-1)^k$, proving the claim. But now we have the following result:

$$\theta(f)(z) = \sum_{k=0}^{n} (-1)^k \operatorname{rank} C_k(K),$$

and the right hand side is simply the Euler characteristic $\chi(B)$. Thus, we have sketched a proof of the following famous result.

Theorem (Hopf index theorem). *Let M be a smooth compact oriented manifold and let V be a vector field on M with isolated zeros. Then*

$$\chi(M) = \sum_{V(x)=0} \operatorname{ind}_x(V).$$

Bibliographic notes
As noted in the preface, much of what I know about homotopy theory I learned from the book of Griffiths & Morgan [3]. The presentation of this chapter has its origins there. The book of Bott & Tu [14] was also very useful, especially for the computation of $\pi_4(S^3)$. The reader is urged to consult it for further calculations of this type.

Bibliography

[1] Greenberg MJ, Harper JR. Algebraic topology. vol. 58 of Mathematics Lecture Note Series. Benjamin/Cummings Publishing Co., Inc., Advanced Book Program, Reading, MA; 1981. A first course.

[2] Hatcher A. Algebraic topology. Cambridge University Press, Cambridge; 2002.

[3] Griffiths P, Morgan J. Rational homotopy theory and differential forms. vol. 16 of Progress in Mathematics. 2nd ed. Springer, New York; 2013. Available from: https://doi.org/10.1007/978-1-4614-8468-4.

[4] Spanier EH. Algebraic topology. Springer-Verlag, New York; [1995?]. Corrected reprint of the 1966 original.

[5] Munkres JR. Elements of algebraic topology. Addison-Wesley Publishing Company, Menlo Park, CA; 1984.

[6] Steenrod N. The topology of fibre bundles. Princeton Landmarks in Mathematics. Princeton University Press, Princeton, NJ; 1999. Reprint of the 1957 edition, Princeton Paperbacks.

[7] Weibel CA. An introduction to homological algebra. vol. 38 of Cambridge Studies in Advanced Mathematics. Cambridge University Press, Cambridge; 1994. Available from: https://doi.org/10.1017/CBO9781139644136.

[8] Gutiérrez J. Lecture notes. [Online; accessed 22-March-2024].

[9] Conrad K. $SL_2(\mathbb{Z})$. [Online; accessed 22-March-2024].

[10] Kassel C, Turaev V. Presentations of $SL_2(\mathbb{Z})$ and $PSL_2(\mathbb{Z})$. In: Braid groups. New York, NY: Springer New York; 2008. p. 311–4.

[11] Serre JP. Trees. Springer Monographs in Mathematics. Springer-Verlag, Berlin; 2003. Translated from the French original by John Stillwell, Corrected 2nd printing of the 1980 English translation.

[12] Edelsbrunner H, Harer JL. Computational topology. American Mathematical Society, Providence, RI; 2010. An introduction. Available from: https://doi.org/10.1090/mbk/069.

[13] Nanda V. Computational algebraic topology. [Online; accessed 22-March-2024].

[14] Bott R, Tu LW. Differential forms in algebraic topology. vol. 82 of Graduate Texts in Mathematics. Springer-Verlag, New York-Berlin; 1982.

https://doi.org/10.1515/9783111014852-005

Index

https://doi.org/10.1515/9783111014852-006